CW01371764

IET TRANSPORTATION SERIES 38

Applications of Machine Learning and Data Analytics Models in Maritime Transportation

Other related titles:

Volume 1	**Clean Mobility and Intelligent Transport Systems** M. Fiorini and J-C. Lin (Editors)
Volume 2	**Energy Systems for Electric and Hybrid Vehicles** K.T. Chau (Editor)
Volume 5	**Sliding Mode Control of Vehicle Dynamics** A. Ferrara (Editor)
Volume 6	**Low Carbon Mobility for Future Cities: Principles and Applications** H. Dia (Editor)
Volume 7	**Evaluation of Intelligent Road Transportation Systems: Methods and Results** M. Lu (Editor)
Volume 8	**Road Pricing: Technologies, economics and acceptability** J. Walker (Editor)
Volume 9	**Autonomous Decentralized Systems and their Applications in Transport and Infrastructure** K. Mori (Editor)
Volume 11	**Navigation and Control of Autonomous Marine Vehicles** S. Sharma and B. Subudhi (Editors)
Volume 12	**EMC and Functional Safety of Automotive Electronics** K. Borgeest
Volume 15	**Cybersecurity in Transport Systems** M. Hawley
Volume 16	**ICT for Electric Vehicle Integration with the Smart Grid** N. Kishor and J. Fraile-Ardanuy (Editors)
Volume 17	**Smart Sensing for Traffic Monitoring** Nobuyuki Ozaki (Editor)
Volume 18	**Collection and Delivery of Traffic and Travel Information** P. Burton and A. Stevens (Editors)
Volume 20	**Shared Mobility and Automated Vehicles: Responding to socio-technical changes and pandemics** Ata Khan and Susan Shaheen
Volume 23	**Behavioural Modelling and Simulation of Bicycle Traffic** L. Huang
Volume 24	**Driving Simulators for the Evaluation of Human-Machine Interfaces in Assisted and Automated Vehicles** T. Ito and T. Hirose (Editors)
Volume 25	**Cooperative Intelligent Transport Systems: Towards high-level automated driving** M. Lu (Editor)
Volume 26	**Traffic Information and Control** Ruimin Li and Zhengbing He (Editors)
Volume 30	**ICT Solutions and Digitalisation in Ports and Shipping** M. Fiorini and N. Gupta
Volume 32	**Cable Based and Wireless Charging Systems for Electric Vehicles: Technology and control, management and grid integration** R. Singh, S. Padmanaban, S.Dwivedi, M. Molinas and F. Blaabjerg (Editors)
Volume 34	**ITS for Freight Logistics** H. Kawashima (Editor)
Volume 36	**Vehicular ad hoc Networks and Emerging Technologies for Road Vehicle Automation** A. K. Tyagi and S Malik
Volume 38	**The Electric Car** M.H. Westbrook
Volume 45	**Propulsion Systems for Hybrid Vehicles** J. Miller
Volume 79	**Vehicle-to-Grid: Linking Electric Vehicles to the Smart Grid** J. Lu and J. Hossain (Editors)

Applications of Machine Learning and Data Analytics Models in Maritime Transportation

Ran Yan and Shuaian Wang

The Institution of Engineering and Technology

Published by The Institution of Engineering and Technology, London, United Kingdom

The Institution of Engineering and Technology is registered as a Charity in England & Wales (no. 211014) and Scotland (no. SC038698).

© The Institution of Engineering and Technology 2022

First published 2022

This publication is copyright under the Berne Convention and the Universal Copyright Convention. All rights reserved. Apart from any fair dealing for the purposes of research or private study, or criticism or review, as permitted under the Copyright, Designs and Patents Act 1988, this publication may be reproduced, stored or transmitted, in any form or by any means, only with the prior permission in writing of the publishers, or in the case of reprographic reproduction in accordance with the terms of licences issued by the Copyright Licensing Agency. Enquiries concerning reproduction outside those terms should be sent to the publisher at the undermentioned address:

The Institution of Engineering and Technology
Futures Place
Kings Way, Stevenage
Herts, SG1 2UA, United Kingdom

www.theiet.org

While the authors and publisher believe that the information and guidance given in this work are correct, all parties must rely upon their own skill and judgement when making use of them. Neither the author nor publisher assumes any liability to anyone for any loss or damage caused by any error or omission in the work, whether such an error or omission is the result of negligence or any other cause. Any and all such liability is disclaimed.

The moral rights of the author to be identified as author of this work have been asserted by him in accordance with the Copyright, Designs and Patents Act 1988.

British Library Cataloguing in Publication Data
A catalogue record for this product is available from the British Library

ISBN 978-1-83953-559-8 (hardback)
ISBN 978-1-83953-560-4 (PDF)

Typeset in India by Exeter Premedia Services Private Limited
Printed in the UK by CPI Group (UK) Ltd, Croydon
Cover Image: Longhua Liao/ Moment via Getty Images

Contents

About the Authors ix

1 Introduction of maritime transportation 1
   1.1 Overview of maritime transport 1
   1.2 World fleet structure 1
       1.2.1 Bulk carrier 1
       1.2.2 Oil tanker 2
       1.2.3 Container ship 2
   1.3 Key roles in the shipping industry 3
       1.3.1 Ship owner 3
       1.3.2 Ship operator 3
       1.3.3 Ship management company 3
       1.3.4 Flag state 3
       1.3.5 Classification society 4
       1.3.6 Charterer 4
       1.3.7 Freight forwarder 4
       1.3.8 Ship broker 5
   1.4 Container liner shipping 5

2 Ship inspection by port state control 9
   2.1 Key issues in maritime transport 9
       2.1.1 Maritime safety management 9
       2.1.2 Marine pollution control 10
       2.1.3 Seafarers' management 10
   2.2 Port state control 10
       2.2.1 The background and development of PSC 11
       2.2.2 Ship selection in PSC 11
       2.2.3 Onboard inspection procedure 12
       2.2.4 Inspection results 14
   2.3 Data set used in this book 15

3 Introduction to data-driven models 21
   3.1 Predictive problem and its application in maritime transport 21
       3.1.1 Introduction of predictive problem 21
       3.1.2 Examples of predictive problem in maritime transport 22

			3.1.3	Comparison of theory-based modeling and data-driven modeling	23
			3.1.4	Popular data-driven models	23

4 Key elements of data-driven models 29
  4.1 Comparison of three popular data-driven models 29
  4.2 Procedure of developing ML models to address maritime transport problems 29
        4.2.1 Problem specification 32
        4.2.2 Feasibility assessment 32
        4.2.3 Data collection 32
        4.2.4 Feature engineering 34
        4.2.5 Model construction 43
        4.2.6 Model refinement 48
        4.2.7 Model assessment, interpretation/explanation, and conclusion 50

5 Linear regression models 51
  5.1 Simple linear regression and the least squares 51
  5.2 Multiple linear regression 53
  5.3 Extensions of multiple linear regression 55
        5.3.1 Polynomial regression 55
        5.3.2 Logistic regression 56
  5.4 Shrinkage linear regression models 59
        5.4.1 Ridge regression 60
        5.4.2 LASSO regression 61

6 Bayesian networks 63
  6.1 Naive Bayes classifier 63
  6.2 Semi-naive Bayes classifiers 68
  6.3 BN classifiers 73

7 Support vector machine 79
  7.1 Hard margin SVM 79
  7.2 Soft margin SVM 83
  7.3 Kernel trick 86
  7.4 Support vector regression 90

8 Artificial neural network 93
  8.1 The structure and basic concepts of an ANN 93
        8.1.1 Training of an ANN model 97
        8.1.2 Hyperparameters in an ANN model 100
  8.2 Brief introduction of deep learning models 103

Contents vii

9 Tree-based models — 105
9.1 Basic concepts of a decision tree — 105
9.2 Node splitting in classification trees — 106
    9.2.1 Iterative dichotomizer 3 (ID3) — 106
    9.2.2 C4.5 — 109
    9.2.3 Classification and regression tree (CART) — 110
    9.2.4 Node splitting in regression trees — 113
9.3 Ensemble learning on tree-based models — 115
    9.3.1 Bagging — 117
    9.3.2 Boosting — 120

10 Association rule learning — 129
10.1 Large item sets — 129
10.2 Apriori algorithm — 131
10.3 FP-growth algorithm — 139

11 Cluster analysis — 143
11.1 Distance measure in clustering — 143
    11.1.1 Distance measure of examples — 143
    11.1.2 Distance measure of clusters — 146
11.2 Metrics for clustering algorithm performance evaluation — 147
    11.2.1 Clustering algorithms — 149
    11.2.2 K-means (partition-based method) — 149
    11.2.3 DBSCAN (density-based method) — 151
    11.2.4 Agglomerative algorithm (hierarchy-based methods) — 154

12 Classic and emerging approaches to solving practical problems in maritime transport — 161
12.1 Topics in maritime transport research — 161
12.2 Research methods and their specific applications to maritime transport research — 163
12.3 Issues of adopting data-driven models to address problems in maritime transportation — 167
    12.3.1 Data — 168
    12.3.2 Model — 169
    12.3.3 User — 170
    12.3.4 Target — 172

13 Incorporating shipping domain knowledge into data-driven models — 179
13.1 Considering feature monotonicity in ship risk prediction — 180
    13.1.1 Introduction of monotonicity in the ship risk prediction problem — 180
    13.1.2 Integration of monotonic constraint into XGBoost — 182

viii *Machine learning and data analytics for maritime studies*

    13.2 Integration of convex and monotonic constraints into ANN (artifical neural network)   184

14 Explanation of black-box ML models in maritime transport   **191**
    14.1 Necessity of black-box ML model explanation in the maritime industry   191
        14.1.1 What is the explanation for ML models   191
        14.1.2 Propose and evaluate explanations for black-box ML models   194
    14.2 Popular methods for black-box ML model explanation   196
        14.2.1 Forms and types of explanations   196
        14.2.2 Introduction of intrinsic explanation model using DT as an example   198
        14.2.3 SHAP method   200

15 Linear optimization   **209**
    15.1 Basics   209
    15.2 Classification of linear optimization models according to solutions   213
    15.3 Equivalence between different formulations   215
    15.4 Graphical method for models with two variables   218
    15.5 Using software to solve linear optimization models   225
    15.6 An in-depth understanding of linear optimization   227
    15.7 Useful applications of linear optimization solvers   228

16 Advanced linear optimization   **231**
    16.1 Network flow optimization   231
    16.2 Dummy nodes and links   245
    16.3 Using linear formulations for nonlinear problems   248
    16.4 Practice   250

17 Integer optimization   **275**
    17.1 Formulation I: natural integer decision variables   275
    17.2 Formulation II: 0–1 decision variables   278
    17.3 Formulation III: complex logical constraints   281
    17.4 Solving mixed-integer optimization models   282
    17.5 Formulation IV: challenging problems   283
    17.6 Formulation V: linearizing binary variables multiplied by another variable   288
    17.7 Practice   289

18 Conclusion   **297**
    18.1 Summary of this book   297
    18.2 Future research agenda   298

Index   **301**

About the Authors

Ran Yan is a research assistant professor in the Department of Logistics and Maritime Studies at The Hong Kong Polytechnic University (PolyU), China. Dr. Yan received her Bachelor of Science degree from Hohai University in China in 2018 and her Master of Philosophy and Doctor of Philosophy degrees from The Hong Kong Polytechnic University in 2020 and 2022, respectively. Dr. Yan's research interests include applying data analytics methods and technologies to improve shipping efficiency and green shipping management. Dr. Yan has published more than 30 papers in international journals and conference proceedings, such as *Transportation Research Part B/C/E, Transport Policy, Journal of Computational Science, Maritime Policy & Management, Ocean Engineering, Engineering, Sustainability*, and *Electronic Research Archive*, and won several times of best paper/student paper award from international conferences. Dr. Yan is an editorial assistant of *Cleaner Logistics and Supply Chain*.

Shuaian Wang is currently Professor at The Hong Kong Polytechnic University (PolyU), China. Prior to joining PolyU, he worked as a faculty member at Old Dominion University, USA, and the University of Wollongong, Australia. Dr. Wang's research interests include big data in shipping, green shipping, shipping operations management, port planning and operations, urban transport network modeling, and logistics and supply chain management. Dr. Wang has published over 200 papers in journals such as *Transportation Research Part B, Transportation Science*, and *Operations Research*. Dr. Wang is an editor-in-chief of *Cleaner Logistics and Supply Chain* and *Communications in Transportation Research*, an associate editor of *Transportation Research Part E, Flexible Services and Manufacturing Journal, Transportmetrica A*, and *Transportation Letters*, a handle editor of *Transportation Research Record*, an editorial board editor of *Transportation Research Part B*, and an editorial board member of *Maritime Transport Research*. Dr. Wang dedicates to rethinking and proposing innovative solutions to improve the efficiency of maritime and urban transportation systems, to promote environmental friendly and sustainable practices, and to transform business and engineering education.

TO MY PARENTS
IN MEMORY OF MY BELOVED MATERNAL GRANDPARENTS
—Ran

TO MY DAUGHTER AMY
—Shuaian

Chapter 1
Introduction of maritime transportation

1.1 Overview of maritime transport

Maritime transportation is the transport of passengers and cargoes by sea which can be dated back to ancient Egypt times. It is the backbone of global trade, manufacturing supply chain, and world economy, as about 80% of world trade by volume is carried out by ocean-going vessels [1]. Even during the ever-hard time brought about by the COVID-19 pandemic, the maritime transport still plays a key role in moving various goods and products, especially the necessary living and medical supplies, around the world. The world economic development has a critical impact on maritime transport. According to the review of maritime transport produced by the United Nations Conference on Trade and Development (UNCTAD), the international maritime trade volumes from 1970 to 2020 are shown in Table 1.1 [1].

1.2 World fleet structure

The total number of ships of 100 gross tons and above reached 99 800 in 2020. Different types of ships are responsible for carrying different types of commodities. The top three ship types with the largest market shares in the world fleet by deadweight tonnage (DWT) are bulk carriers (913 032 000 tons, 42.77%), oil tankers (619 148 000 tons, 29.00%), and container ships (281 784 000 tons, 13.20%) in 2021, according to the UNCTAD based on data from Clarksons Research [1]. A brief introduction of the three types of ships is given in the following subsections.

1.2.1 Bulk carrier

Bulk carriers mainly transport dry cargoes in bulk quantities in a loose manner, that is, the cargoes are free from any specific packaging. Major category of bulk carriers as per size includes handysize carriers (usually between 25 000 and 40 000 DWT), handymax carriers (usually between 40 000 and 60 000 DWT), Panamax carriers (usually between 60 000 and 100 000 DWT), post-Panamax carrier (usually between 80 000 and 120 000 DWT), capesize carrier (usually between 100 000 and 200 000 DWT), and very large bulk carrier (usually over 200 000 DWT). Bulk carriers are mainly used to transport ores, coals, cement, etc.

Table 1.1 International maritime trade from 1970 to 2020 (millions of tons loaded)

Year	Tanker trade*	Main bulk†	Other dry cargo‡	Total (all cargoes)
1970	1 440	448	717	2 605
1980	1 871	608	1 225	3 704
1990	1 755	988	1 265	4 008
2000	2 163	1 186	2 635	5 984
2005	2 422	1 579	3 108	7 109
2010	2 752	2 232	3 423	8 408
2011	2 785	2 364	3 626	8 775
2012	2 840	2 564	3 791	9 195
2013	2 828	2 734	3 951	9 513
2014	2 825	2 964	4 054	9 842
2015	2 932	2 930	4 161	10 023
2016	3 058	3 009	4 228	10 295
2017	3 146	3 151	4 419	10 716
2018	3 201	3 215	4 603	11 019
2019	3 163	3 218	4 690	11 071
2020	2 918	3 181	4 549	10 648

Source: UNCTAD secretariat, based on reports from countries, governments, port industries websites, and specialist sources.
*Includes crude oil, refined petroleum products, gas, and chemicals.
†Includes iron ore, grain, and coal.
‡Includes minor bulk commodities, containerized trade, and residual general cargo.

1.2.2 Oil tanker

Oil tanker is a type of tanker ship which is designed for the particular purpose of transporting liquefied goods, mainly including crude oil and its by-products. According to their sizes, oil tankers can be divided into Panamax (usually between 50 000 and 75 000 DWT), Aframax (coming from the Average Freight Rate Assessment system, usually between 75 000 and 120 000 DWT), Suezmax (usually between 120 000 and 180 000 DWT), very large crude carrier (usually between 200 000 and 320 000 DWT), and ultra large crude carrier (usually more than 320 000 DWT).

1.2.3 Container ship

Container ships have specially designed structures to hold a large quantity of cargoes compacted in different types of containers. Container ship capacity is measured in twenty-foot equivalent units (TEUs). They can be further divided into feeder (1 001–2 000 TEU), feedermax (2 001–3 000 TEU), Panamax (3 001–5 100 TEU), post-Panamax (5 101–10 000 TEU), new Panamax (10 000–14 500 TEU), and ultra large container vessel (over 14 500 TEU). Container shipping is revolutionary in the form of cargoes that are ferried and transported around the world, as it is the foundation of international multimodal transport that connects ship with truck and rail. Globalization would not have been possible without containerization. Common cargoes transported by container ships include manufactured goods, raw materials, food goods, agricultural products, and seafood in refrigerated containers. More information on container liner shipping is provided in section 1.4.

1.3 Key roles in the shipping industry

1.3.1 Ship owner

The owner of a merchant ship is called a ship owner, which can be companies, people, and investment funds. The ownership is achieved by building new ships or purchasing second-hand ships. Technical or commercial operations are provided by some ship owners, while others prefer to outsource these services to specialized companies. The party that offers commercial management service is called a ship operator, and the party that offers technical management service is called a ship management company.

1.3.2 Ship operator

Ship operator is the company in charge of daily matters on ships. It needs to tackle with several tactical problems such as determining the cargo orders as well as the freight rate, fleet deployment, ship routing and scheduling, and port service negotiation. In addition, ship operator also needs to deal with operational problems such as deciding ship loading, selecting sailing speeds, routes, and bunkering ports.

1.3.3 Ship management company

Ship owners can choose to delegate various managerial functions to a third party, which is the so-called ship management company or simply the manager. The delegated services of a management company include technical management, commercial management, and crew management. To be more specific, technical management services include but are not limited to insuring a ship that complies with international codes and conventions, arranging repairs and maintenance, and providing technical consultants. Crew management services include crew recruitment, training, insurance, payroll, etc. Ship managers also offer commercial management services, such as securing vessel employment, obtaining marine insurance, and providing risk management services. One ship management company can manage a large number of fleets on behalf of numerous ship owners.

1.3.4 Flag state

The ship flag state is the nationality of a ship under whose laws the ship is plying in the open sea. A flag state is responsible for conducting regular inspections of its ships to ensure the safety of its cargo and crew (which is called flag state control), collect taxes, and regulate the pollution levels by imposing maritime policies and laws. In turn, ships receive protections and preferential treatments such as tax, certification, trade rights, and security from their flags of registration. The flag state of a ship plays an important role in many aspects of ship management and operation, such as vessel leasing, sale and purchase, newbuilding deliveries, financing, and different priorities of owners and mortgagees. The practice of registering a ship to a state different from that of the ship owner is known as the flag of convenience.

1.3.5 Classification society

Classification societies are mainly responsible for setting technical standards for seagoing cargo vessels based on their experience, expertise, and research in the shipping industry while considering international maritime conventions and regulations. A classification society should also inspect the ships under its classing ever since the ship construction process. The periodic operation of their ships should also be provided to ensure that they continuously meet the related regulations. Classification society plays an important role in enhancing maritime safety and protecting the marine environment, and their work interacts with ship flag states, port state control, surveyors, and among other groups in addition to ship owners.

1.3.6 Charterer

Ship charterer is a company that hires ships from ship owners to transport their cargoes, and the contract between them is known as a charterparty. Time charter and voyage charter are the most popular charter party agreements, where the ship owner rents a ship to the charterer for a jointly determined period with restrictions on trading and cargoes imposed in the former, while a ship is hired to carry a particular cargo between specific places in the latter. Usually, under the time charterparty, the ship owner takes care of their crew, maintains ship stores and provisions, and conducts ship maintenance. Meanwhile, the time charterers are responsible for fuel arrangement, port and canal dues, wharfage, and many other types of operational costs. Whereas in a typical voyage charterparty, most of the operational costs are covered by the ship owner, and the voyage charterer is responsible for the costs in ports and terminals of the specific voyage. Another popular charterparty is a bareboat charter, where a ship charterer hires a ship without administration or technical maintenance provided, and it bears the full operating expenses. In all cases, both ship owners and charterers are free to negotiate their liabilities.

1.3.7 Freight forwarder

A freight forwarder is a person or an entity that manages shipments to transport the goods from origin to destination, and it takes several duties which could include consultancy on export and/or import costs and regulations, arrangement of goods transportation, cargo insurance, document translation, and communication with clients.

1.3.8 Ship broker

Ships are chartered every day through ship brokers to carry cargoes worldwide. A ship owner's broker takes the responsibility to arrange the most profitable employment of the vessels under the owner's control, and the ship charterer's broker helps the charter to find the most effective way to transport the cargoes with the right ship at the right price. Therefore, it is necessary that the ship broker to be professional and is aware of the current freight market and is also able to forecast its future status as well as acts as the bridge between the ship owner and the charterer.

1.4 Container liner shipping

A majority of cargoes in supermarkets, such as fruits and vegetables, kitchen appliances, furniture, garments, meats, fish, dairy products, and toys, are transported in containers by ship. Containers are usually expressed in terms of TEUs, a box that is 20 feet long (6.1 m). Throughout this book, unless otherwise specified, we use "TEU" to express "the number of containers" or "the volume of containers."

Containers are transported by ship on liner services, which are similar to bus services. Figure 1.1 is the Central China 2 (CC2) service operated by Orient Overseas Container Line (OOCL), a Hong Kong-based shipping company. We call it a service, a route, or a service route. A route is a loop, and the port rotation of a route is the sequence of ports of call on the route. Any port of call can be defined as the first port of call. For example, if we define Ningbo as the first port of call, then Shanghai is the second port of call, and Los Angeles is the third port of call. We can therefore represent the port rotation of the route as follows:

Ningbo (1) → Shanghai (2) → Los Angeles (3) → Ningbo (1)

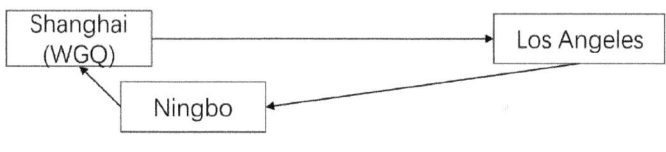

From/To	Los Angeles (Thu)
Shanghai (Fri)	13
Ningbo (Tue)	16

From/To	Ningbo (Sun)	Shanghai (Wed)
Los Angeles (Sun)	14	17

Figure 1.1   CC2 service by OOCL

6 *Machine learning and data analytics for maritime studies*

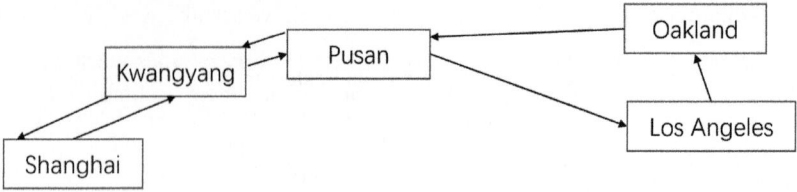

From/To	Los Angeles (Thu)	Oakland (Mon)
Pusan (Sun)	11	15
Kwangyang (Fri)	13	17
Shanghai (Thu)	14	18

From/To	Pusan (Sat)	Kwangyang (Mon)	Shanghai (Wed)
Oakland (Tue)	18	20	22
Los Angeles (Sun)	20	22	24

Figure 1.2   CC1 service by OOCL

Note that on a route, different ports of call may be the same physical port. For example, the Central China 1 (CC1) service of OOCL shown in Figure 1.2 has the port rotation below:

Shanghai (1) → Kwangyang (2) → Pusan (3) → Los Angeles (4) → Oakland (5) → Pusan (6) → Kwangyang (7) → Shanghai (1)

Both the second and the seventh ports of call are Kwangyang, and both the third and the sixth ports of call are Pusan.

A leg is the voyage from one port of call to the next. Leg i is the voyage from the i th port of call to port of call $i + 1$. The last leg is the voyage from the last port of call to the first port of call. On CC1, the second leg is the voyage from Kwangyang (the second) to Pusan (the third), and the seventh leg is the voyage from Kwangyang (the seventh) to Shanghai (the first).

The rotation time of a route is the time required for a ship to start from the first port of call, visit all ports of call on the route, and return to the first port of call. As can be read from Figures 1.1 and 1.2, the rotation time of CC2 is 35 days[*], and the rotation time of CC1 is 42 days. Each route provides a weekly frequency, which means that each port of call is visited on the same day every week. Therefore, a string of five ships are deployed on CC2, and the headway between two adjacent ships is 7 days. These five ships usually have the same TEU capacity

---

[*] 13 days from the departure from Shanghai to the arrival at Los Angeles + 17 days from the departure from Los Angeles to the arrival at Shanghai + 2 days spent at the port of Shanghai (Wednesday to Friday) + 3 days spent at the port of Los Angeles (Thursday to Sunday).

and other characteristics. Unless otherwise specified, we assume weekly frequencies for all routes.

On CC2, it may be the case that we are interested in container transportation from Ningbo to Los Angeles, Shanghai to Los Angeles, Los Angeles to Ningbo, and Los Angeles to Shanghai, and there is no container to transport between Ningbo and Shanghai. We therefore define an origin-destination (OD) pair as an ordered pair of origin port and destination port. In this example, there are four OD pairs: (Ningbo and Los Angeles), (Shanghai and Los Angeles), (Los Angeles and Ningbo), and (Los Angeles and Shanghai). The demand for an OD pair is the number of TEUs per week to transport. For example, if the demand for the OD pair (Ningbo and Los Angeles) is 5 000, then there are 5 000 TEUs from potential customers to be transported from Ningbo to Los Angeles every week. Nevertheless, the shipping line may choose to transport, e.g., 4 675 TEUs and reject the remaining 325 TEUs.

Unless otherwise specified, throughout this book we examine tactical planning problems rather than operational problems. For example, we might predict that in the next 6 months, there will be 2 000 TEUs per week to transport from Ningbo to Los Angeles, and we aim to estimate the profit we can make if the CC2 service is operated. As the "2 000 TEUs per week to transport" may not be accurate, it is generally not a problem to say "1845.3 TEUs will be transported"(a more detailed analysis can be found in Reference 2). In other words, we assume that the number of TEUs is a real number rather than an integer, unless otherwise specified.

References

[1] 'United Nations conference on trade and development' in *Review of Maritime Transport*. New York, NY: United Nations Publications; 2022.
[2] Wang S. 'Essential elements in tactical planning models for container liner shipping'. *Transportation Research Part B*. 2013;54:84–99.

Chapter 2
Ship inspection by port state control

2.1 Key issues in maritime transport

2.1.1 Maritime safety management

Maritime transport is a highly globalized industry in terms of operation and management. For ship operation, ocean-going vessels sail on the high seas from the origin port in one country/region to the destination port in another country/region. For ship management, parties responsible for ship ownership, crewing, and operating may locate in different countries and regions. Even the country of registration, i.e., ship flag state, may not have a direct link and connection with a ship's activities as the ship may not frequently visit the ports belonging to its flag state. For inland countries such as Mongolia, the ships registered under it never visit its ports. Such complex and disintegrated nature of the shipping industry makes it hard to control and regulate international shipping activities, and thus pose danger to maritime safety, the marine environment, and the crew and cargoes carried by ocean-going vessels.

Shipping is one of the world's most dangerous industries due to the complex and ever-changing environment at sea, the dangerous goods carried, and the difficulties in search and rescue. Safety at sea is always put at the highest priority in ship operation and management. It is widely believed that the most effective and efficient way of improving safety at sea is to develop international regulations that should be followed by all shipping nations [1]. A unified and permanent international body was expected to be established for regulation and supervision by several nations from the mid-19th century onward, and the hopes came true after the International Maritime Organization (IMO, whose original name was Inter-Governmental Maritime Consultative Organization) was established at an international conference in Geneva held in 1948. Through hard efforts of all parties, the members of IMO met for the first time in 1959, one year after the IMO convention came into force. The IMO's task was to adopt a new version of the most important conventions on maritime safety, i.e., the International Convention for the Safety of Life at Sea, which specifies minimum safety standards for ship construction, equipment, and operation. It covers comprehensive aspects of shipping safety, including vessel construction, fire safety, life-saving arrangements, radio communications, navigation safety, cargo carriage, dangerous goods transporting, the mandatory of the International Safety Management (ISM) code, verification of compliance, and measures for specific ships, and is constantly amended [2]. The Maritime Safety Committee is responsible for every aspect of maritime safety and security, and it is the highest technical body of the IMO.

2.1.2 Marine pollution control

Although safety has always been the first priority and the most important responsibility, pollution caused by maritime accidents and shipping activities began to merge and is receiving an increasing attention from the parties. The most important international convention is the International Convention for the Prevention of Pollution from Ships proposed in 1973 and modified in 1978 (MARPOL 73/78), which aims at preventing pollution of the marine environment caused by vessel operations and accidents. As of January 2022, six technical annexes are included in the MARPOL convention, including prevention of pollution by oil, noxious liquid substances in bulk, harmful substances in packaged form, sewage from ships, garbage from ships, and air pollution from ships [3]. The Marine Environment Protection Committee of the IMO takes the full responsibility to address matters concerned with prevention and control of pollution from ships by adopting and amending the related conventions and regulations and ensure their enforcement.

2.1.3 Seafarers' management

Seafarers are regarded as key workers by the IMO, who are essential to shipping and the world, in particular, during the COVID-19 pandemic. They are at the core of shipping's future. Basic requirements on the training, certification, and watchkeeping of seafarers on an international level are established in the International Convention on Standards of Training, Certification, and Watchkeeping for Seafarers by the IMO, which entered into force in 1984 [4]. Besides, the milestone to ensure a decent living and working environment for seafarers is the Maritime Labor Convention, 2006 (MLC2006), which entered into force in 2013 and is developed by the International Labor Organization [5]. Minimum requirements for seafarers regarding conditions of employment, work, accommodation, food and catering, health care and medical care onboard, as well as welfare and social security protection are specified in MLC2006. A sub-committee on Human Element, Training, and Watchkeeping deals with the human sides of shipping in the IMO.

2.2 Port state control

A ship whose hull, machinery, equipment, or operational safety is substantially below the standards required by the relevant convention or whose crew is not in conformity with the safe manning document is deemed as a substandard ship [6]. The ship flag state is the first line of defense against substandard shipping. It is obliged to ensure its ships are periodically surveyed and re-certified by carrying out inspections by its own surveyors or by an authorized party called a recognized organization (RO). However, due to the internationalism of shipping activities, many ships do not regularly call at ports in their flag states, and thus restrict the role that flag state plays in identifying substandard ships.

2.2.1 The background and development of PSC

PSC is then developed, aiming to inspect foreign visiting ships in national ports to verify that "the condition of the ship and its equipment comply with the requirements of international regulations and that the ship is manned and operated in compliance with these rules" as mentioned by the IMO [7]. During an inspection, a condition onboard that does not comply with the requirements of the relevant convention is called a deficiency. The number and nature of the deficiencies found onboard determine the corresponding action taken by the PSC officer(s) (PSCO[s]). Common actions include rectifying a deficiency at the next port within 14 days or before departure and ship detention. Especially, ship detention is an intervention action taken by the port state that prevents a severely substandard ship from proceeding to sea until it would not present danger to the ship or persons onboard as well as to the marine environment.

PSC inspection is carried out on a regional level. The Memorandum of Understanding (MoU) on PSC was first signed in 1982 by 14 European countries, which is called the Paris MoU and marks the establishment of PSC. Since then, the number of member states of the Paris MoU has constantly increased, and it contains 27 participating maritime administrations covering the waters of the European coastal States and the North Atlantic basin as of January 2022. Another large regional MoU is in the Far East responsible for the Asia Pacific region, which is called Tokyo MoU and was signed in 1993. It now contains 22 member states. In addition, there are another seven MoUs on PSC, namely Acuerdo de Viña del Mar (Latin America), Caribbean MoU (Caribbean), Abuja MoU (West and Central Africa), Black Sea MoU (the Black Sea region), Mediterranean MoU (the Mediterranean), Indian Ocean MoU (the Indian Ocean), and the Riyadh MoU. The main objectives of constructing MoUs are constructing an improved and harmonized PSC system, strengthening cooperation and information exchange among member states, and avoiding multiple inspections within a short period. Apart from the nine regional MoUs, the United States Coast Guard maintains the tenth PSC regime.

2.2.2 Ship selection in PSC

One critical step in ship inspection by PSC is the identification and selection of ships with a higher risk of inspection. This is because on the one hand, not all foreign visiting ships are able to be inspected due to the limited inspection resources, while on the other hand, not all ships are substandard. A PSC inspection is usually conducted by one or two inspectors, i.e., PSCOs, and are authorized by the port state to carry out PSC inspections. A typical PSC inspection usually lasts 2–3 hours. According to the estimation of Knapp [8], the average cost of an inspection free from deficiency is US$506 for the port. If any deficiency is found, the average cost per inspection increases to US$759. Therefore, carrying out an inspection can be both time-consuming and costly. Meanwhile, among all the foreign visiting ships, only part of them are substandard, or in other words, with a deficiency or deficiencies detected. For example, in the Tokyo MoU, among all the inspections between 2009 and 2019, about 40% of them are without deficiency detected, while the average detention rate is 4.18% [9]. It needs to be mentioned that both statistics are calculated based only on the ships selected for inspection. If all the foreign visiting ships are considered, both rates will be lower (as only ships with relatively high risk are selected for inspection). Therefore, accurately identifying substandard ships with a higher risk, i.e., with a larger number of deficiencies or a higher probability of detention, is of vital importance to enhance the effectiveness of PSC.

To achieve efficient PSC inspection, uniform ship selection methods are adopted by individual MoUs, while the methods adopted by different MoUs might be different. One fundamental ship selection method is based on expert experience. For example, in the Mediterranean MoU, ships can be exempted from further inspection if they have been inspected within the last 6 months and found to comply with the regulations [10]. A more advanced ship selection method is called the new inspection regime (NIR), which is adopted by the Abuja MoU, the Black Sea MoU, the Paris MoU, and the Tokyo MoU. The NIR takes ship age and type, performance of ship flag/RO/company*, and the number of deficiencies and detentions in the last 3 years to calculate ship risk profile (SRP). An example of the information sheet for ship risk calculation used in the Tokyo MoU is presented in Table 2.1. Under the NIR, ships are divided into three risk profiles in accordance with their risk levels calculated, namely high-risk ships (HRS), standard-risk ships (SRS), and low-risk ships (LRS). Inspection time window is then attached to each SRP as shown in Table 2.2. It is also noted that the NIR for ship risk calculation can be more or less different in different MoUs, including the information sheet used and the time windows attached.

2.2.3 Onboard inspection procedure

According to the guidance of the IMO, a general PSC inspection can include an initial inspection and a more detailed inspection. An initial inspection may start before a PSCO gets on board a ship by observing its appearance to gain an impression of the ship's overall maintenance standards, like the paintwork, corrosion or pitting, and unrepaired damage. After getting onboard, the PSCO checks and examines the ship's relevant documents and certificates in accordance with the ship's type, size, and year of build, as well as the certificates and documents of its crew. Then, the PSCO walks around the ship in an unsupervised manner to inspect its overall condition, including its equipment, navigation bridge, forecastle, cargo holds/areas, engine room, and pilot transfer arrangements. If the PSCO is satisfied with the general impression, the inspection can be confined, and the observed deficiency(ies) should be recorded.

Alternatively, if clear grounds regarding technical and managerial aspects of the ship's equipment and its crew or pollution prevention issues are found, the PSCO can decide to conduct a more detailed inspection, which covers the ship's construction, equipment, manning, living and working conditions, and compliance with onboard operational procedures. Similar to the initial inspection, any deficiency detected is recorded. If a detainable deficiency (i.e., serious deficiency ground for detention) is found, the ship can be detained. A ship can apply for a follow-up inspection, which is for verifying the rectification of deficiency(ies) found at the previous initial

*Here company refers to the ISM company

Table 2.1 Information sheet of NIR adopted by the Tokyo MoU

Parameters	Values	Weighting points	Criteria for LRS
Ship type	Chemical tanker, gas carrier, oil tanker, bulk carrier, passenger ship, and container ship	2	
Ship age (calculated based on the keel laid date)	All types with age >12 years	1	
Flag performance in Black-Gray-White list of Tokyo MoU	Black	1	White and should be IMO audit
RO performance evaluated by Tokyo MoU	Low/very low	1	High and should be an RO recognized by the Tokyo MoU
Company performance evaluated by Tokyo MoU	Low/very low/no inspection within previous 36 months	2	High
Deficiencies within previous 36 months	Inspections which recorded over five deficiencies	The number of inspections which recorded over five deficiencies	All inspections have five or less deficiencies, and the ship has at least one inspection within previous 36 months
Detentions within previous 36 months	Three or more detentions	1	No detention

Source: Tokyo MoU [11].

14 *Machine learning and data analytics for maritime studies*

Table 2.2   Inspection time window of each SRP

SRP	Criteria	Inspection time window
HRS	When the sum of weighting points ≥ 4	2–4 months
SRS	Neither HRS nor LRS	5–8 months
LRS	All the criteria for LRS are met	9–18 months

Source: Tokyo MoU [11].

inspection(s). It can be carried out onsite a ship or by remote mode (which is also called remote follow-up inspection). A general PSC inspection is not charged, while port state charges can be expected in follow-up inspections.

2.2.4   Inspection results

After a PSC inspection, inspection results, i.e., deficiency(ies) identified and the detention decision are recorded in the relevant forms and reported in the central database of the corresponding MoU. To be more specific, deficiency codes are derived from major maritime conventions and regulations and are specified and constantly updated by individual MoUs and might be slightly different among the MoUs. Deficiency codes adopted by the Paris MoU (effective from 1 July 2021) are shown in Table 2.3, and those adopted by the Tokyo MoU (effective from 2 December 2019) are shown in Table 2.4.

A ship is detained if the PSCO decides it is unseaworthy due to the identification of serious deficiency(ies), while the captain has the right to appeal against a detention order. Although the general purpose of PSC is to prevent a substandard ship from proceeding to sea, the detention of a ship is a serious matter involving many issues, and thus PSCOs are cautious of the detention decision to avoid undue delays.

Table 2.3   List of deficiency codes in Paris MoU

Code	Content	Code	Content
01	Certificates and documentation	02	Structural condition
03	Water/Weathertight condition	04	Emergency systems
05	Radio communication	06	Cargo operations including equipment
07	Fire safety	08	Alarms
09	Working and living conditions	10	Safety of navigation
11	Life-saving appliances	12	Dangerous goods
13	Propulsion and auxiliary machinery	14	Pollution prevention
15	ISM	16	ISPS (the International Ship and Port Facility Security Code)
18	MLC2006	99	Other

Source: Paris MoU [12].

Table 2.4 List of deficiency codes in Tokyo MoU

Code	Item	Code	Item
01	Certificates and documentation	02	Structural conditions
03	Water/Weathertight condition	04	Emergency systems
05	Radio communications	06	Cargo operations including equipment
07	Fire safety	08	Alarms
09	Working and living conditions	10	Safety of navigation
11	Life-saving appliances	12	Dangerous goods
13	Propulsion and auxiliary machinery	14	Pollution prevention
15	ISM	18	Labor conditions
99	Other		

Source: Paris MoU [12].

Regarding the inspection information reported in central databases, some MoUs offer rough inspection results, such as Caribbean MoU[†] and Riyadh MoU[‡]. More detailed ship information together with the inspection results are provided by some other MoUs, such as Paris MoU[§], Tokyo MoU[¶], Abuja MoU[**], Black Sea MoU[††], Indian Ocean MoU[‡‡], and Mediterranean MoU[§§]. Especially, statistics on inspections and their results are offered by Paris MoU in graph user interface, which is more convenient and intuitive for the users[¶¶].

2.3 Data set used in this book

Data of the examples of algorithm illustration and code implementation are mainly from PSC inspection records at the Hong Kong port. Unless otherwise stated, the case data set used in this book is 1 991 initial inspection records from 2015 to 2017 at the Hong Kong port. The following fields are collected from the website operated by the Tokyo MoU: inspection date, IMO number (IMO no.), ship type, keel laid date, deadweight tonnage (DWT), gross tonnage (GT), SRP, classification society, flag state, flag performance/RO/company performance given by the Tokyo MoU (flag performance, RO performance, and company performance), last date of initial inspection (last inspection date), number of deficiencies in last initial inspection (last deficiency number), detention in last initial inspection (last detention), deficiency

---

[†]http://www.caribbeanmou.org/content/inspection-detention-data
[‡]https://www.riyadhmou.org/basicsearch.html?lang=en
[§]https://www.parismou.org/inspection-search/inspection-search
[¶]https://www.tokyo-mou.org/inspections_detentions/psc_database.php
[**]http://www.abujamou.org/index.php?pid=125disclaimer
[††]http://www.bsmou.org/database/inspections/
[‡‡]https://www.iomou.org/HOMEPAGE/search_insp.php?l1=4&l2=26
[§§]http://www.medmouic.org/Advanced
[¶¶]https://www.parismou.org/inspection-search/inspections-results-kpis

Table 2.5 Five samples in the PSC case data set (part 1)

No.	Inspection date	IMO no.	Ship type	Keel laid date	DWT	GT	SRP	Classification society	Flag state
1	December 29, 2017	9332626	Oil tanker	February 2, 2008	74997	42893	SRP	DNV GL AS	Norway
2	December 27, 2017	9254484	Bulk carrier	May 24, 2002	52364	30011	HRS	Indian Register of Shipping	India
3	December 27, 2017	9760500	Oil tanker	October 2, 2015	106359	57164	SRS	Lloyd's Register	Liberia

Table 2.6 Five samples in the PSC case data set (part 2)

No.	Flag performance	RO performance	Company performance	Last inspection date	Last deficiency number	Last detention	Deficiency number	Detention	Deficiency codes	Detainable deficiency codes
1	White	High	High	September 30, 2013	0	No	4	No	01314/ 05118/ 10106/ 07110*	None
2	White	High	Low	November 19, 2015	9	No	14	Yes	03102/ 10118/ 03109/ 03107/ 11101/ 11101/ 11101/ 04102/ 04116/ 14402/ 03105/ 03108/ 03105/ 07113 †	11101/ 11101
3	White	High	Medium	None	None	None	0	No	None	None

*The nature of these codes is certificate and documentation - documents (SOPEP (The Shipboard Oil Pollution Emergency Plan), radio communications (operation of GMDSS (Global Maritime Distress and Safety System) equipment), safety of navigation (compass correction log), and fire safety (fire fighting equipment and appliances), respectively.

†The nature of these codes is water/weathertight conditions (freeboard marks), safety of navigation (speed and distance indicator), water/weathertight conditions (machinery space openings), water/weathertight conditions (doors), life-saving appliances (lifeboats), emergency systems (emergency fire pump and its pipes), emergency systems (means of communication between safety center and other control stations), pollution prevention - MARPOL annex IV (sewage treatment plant), water/weathertight conditions (Covers [hatchway-, portable-, tarpaulins, etc.]), water/weathertight conditions (ventilators, air pipes, and casings), and fire safety (fire pumps and its pipes), respectively.

number of the current inspection (deficiency number), and detention of the current inspection (detention). In addition, the specific deficiency codes and detainable deficiency codes are also collected. Among them, "deficiency number" and "detention" are potential target variables, and the others are potential input features. Three typical samples in the initial data set directly retrieved from the database are given in Table 2.5 and Table 2.6 Table 2.6.

The three records above have their own characteristics. Record no. 1 is a common PSC inspection in the data set, which is an un-detained inspection with last inspection available. Record no. 2 is a detained inspection record with a large number of deficiencies detected (at 14) and detainable deficiencies detected. Record no. 3 is an un-inspected ship by any port in the Tokyo MoU, and thus it has no historical inspection information in the database.

References

[1] *Brief history of IMO* [online]. London: International organization (IMO). Available from https://www.imo.org/en/About/HistoryOfIMO/Pages/Default.aspx [Accessed 9 Jan 2022].

[2] *International convention for the safety of life at sea (SOLAS), 1974* [online]. London: International maritime organization (IMO). 2019. Available from https://www.imo.org/en/About/Conventions/Pages/International-Convention-for-the-Safety-of-Life-at-Sea-(SOLAS)-1974.aspx [Accessed 12 Dec 2021].

[3] *International convention for the prevention of pollution from ships (MARPOL)* [online]. London: International maritime organization. 2019. Available from https://www.imo.org/en/About/Conventions/Pages/International-Convention-for-the-Prevention-of-Pollution-from-Ships-(MARPOL).aspx [Accessed 12 Dec 2021].

[4] *International convention on standards of training, certification and watchkeeping for seafarers (STCW)* [online]. London: International Maritime Organization (IMO). 2019. Available from https://www.imo.org/en/About/Conventions/Pages/International-Convention-on-Standards-of-Training-Certification-and-Watchkeeping-for-Seafarers-(STCW).aspx [Accessed 19 Jan 2022].

[5] *International labour organization* [online]. Geneva: Maritime labour convention. 2006. Available from https://www.ilo.org/global/standards/maritime-labour-convention/lang--en/index.htm [Accessed 12 Feb 2022].

[6] *Resolution A. 1155 (32) adopted on 15 december 2021 (agenda items 12 and 14) procedures for port state control, 2021* [online]. Geneva: International Maritime Organization. 2022. Available from https://www.register-iri.com/wp-content/uploads/A.115532.pdf [Accessed 2 Feb 2022].

[7] *Port state control* [online]. London: International Maritime Organization (IMO). Available from https://www.imo.org/en/OurWork/MSAS/Pages/PortStateControl.aspx [Accessed 25 Jul 2021].

[8] Knapp S. 'The econometrics of maritime safety'. *Recommendations to Enhance Safety at Sea*. 2007;96.

[9] *Annual report* [online]. Tokyo: Tokyo MOU. Available from http://www.tokyo-mou.org/doc/ANN19-f.pdf [Accessed 25 Aug 2022].

[10] *Selection of ships for inpsection* [online]. London: Mediterranean MoU. 2014. Available from http://www.medmou.org/Basic_Principlse.aspx#3 [Accessed 10 Dec 2020].

[11] *Information sheet on the new inspection refime (NIR)* [online]. Tokyo: Tokyo MoU. 2014. Available from http://www.tokyo-mou.org/doc/NIR-information%20sheet-r.pdf [Accessed 3 Mar 2022].

[12] *List of paris mou deficiency codes* [online]. Paris: Paris MOU. 2021. Available from https://www.parismou.org/list-paris-mou-deficiency-codes [Accessed 15 Jan 2022].

[13] *Tokyo MOU deficiency codes* [online]. Tokyo: Tokyo MoU. 2019. Available from http://www.tokyo-mou.org/publications/tokyo_mou_deficiency_codes.php [Accessed 15 Jan 2022].

Chapter 3
Introduction to data-driven models

This chapter aims to clarify the basic issues of data-driven modeling and its application in maritime transport. A predictive problem is first introduced, and the predictive analysis in the field of maritime transport is then discussed with typical examples given. A comparison between the classic method and data-driven modeling to deal with prediction tasks is also provided. Finally, typical data-driven modeling approaches are briefly discussed.

3.1 Predictive problem and its application in maritime transport

3.1.1 Introduction of predictive problem

Predictive problem is common in engineering and management. It aims to get the correct output (also called target, label, and dependent variable) to an example (also called instance and sample) based on the input (also called feature and independent variable). Prediction can be realized in both qualitative and quantitative manners. Qualitative prediction is mainly achieved by the judgment of practitioners based on their experience and expertise, while quantitative prediction is largely dependent on historical data and mathematical tools. In the era of big data, practical prediction problems are more often addressed by quantitative methods, with two classes of approaches widely used: theory-based approach and data-driven approach.

Generally speaking, theory-based modeling, as a classic approach, is derived by certain system processes following physical laws and rules and is represented by mathematical equations describing the whole process so as to estimate the target based on system parameters. Alternatively, data-driven modeling, as an emerging approach which has received increasing attention, is based on data analytics while emphasizing less on the underlying system processes. The target is estimated using historical data based on a fitted predictive model, which can be a statistical model or a machine learning (ML) model. The key part of data-driven modeling is to find

[1]The Hong Kong Polytechnic University, Hung Hom, Hong Kong, China

connections between the system state variables, i.e., the input, internal, and output variables.

3.1.2 Examples of predictive problem in maritime transport

In maritime transportation, prediction analysis is involved in many practical scenarios in both shipping and port operation and management, such as ship trajectory prediction, ship navigation risk prediction, and safety management risk prediction and safety management as a whole, which means to predict and reduce risks, ship inspection planning, ship energy efficiency prediction, ocean freight market condition prediction, ship destination and arrival time prediction, and port condition prediction [1], to name just a few. We use the following examples to illustrate the main parts of using theory-based modeling and data-driven modeling approaches to address predictive problems.

Example 3.1: Ship fuel consumption prediction problem:

Ship fuel consumption prediction is a hot research topic in green shipping management. In theory-based modeling, models that strictly follow hydrodynamics laws and physical principles of the vessel power system are developed. Generally, a basic white-box model starts from vessel calm water resistance and additional resistance calculation, which is the basis of total resistance calculation. Then, the engine power required to drive the vessel to a certain speed under such resistance, as well as the fuel consumption rate, is estimated. Typical methods include Holtrop-Mennen method and Kristensen-Lützen method [2]. As the model structure and parameters are known because they are calibrated based on physical and hydrodynamics laws, they can be applied in the initial stage of ship design and sea trials without using historical sailing data. Alternatively, data-driven models largely rely on historical vessel sailing records, which can be collected from ship noon reports and onboard sensors. A suitable regression model for fuel consumption rate prediction is constructed using proper data and features extracted from available historical sailing records, and its performance is then validated. The representative of data-driven model for ship fuel consumption prediction is the so-called "cubic law," where a ship's fuel consumption rate at sea is regarded to be proportional to its sailing speed to a power of $\alpha = 3$. Other popular data-driven models in existing literature include statistical modeling such as multiple linear regression analysis and polynomial regression, and ML models such as artificial neural networks (ANNs), tree-based models, and support vector regressor (SVR).

Example 3.2: Ship trajectory prediction problem:

Accurate ship trajectory prediction is the prerequisite for collision prevention and sailing risk analysis. In theory-based modeling, ship trajectory prediction is mainly achieved by simulation, considering the vessel's real-time sailing behavior, resistances, and hydrodynamic equations. Ship trajectory prediction using data-driven modeling approach is mainly powered by historical temporal trajectory points from vessel automatic identification systems data. These time-series data are then mined by classic ML models such as extreme learning machine, support vector machine (SVM), and *k*-nearest neighbors (KNNs), as well as more advanced deep-learning models like recurrent neural networks, so as to figure out the sailing patterns that can be further used for vessel trajectory prediction.

3.1.3 Comparison of theory-based modeling and data-driven modeling

We summarize the advantages and disadvantages of theory-based modeling and data-driven modeling based on the above two examples in Table 3.1.

3.1.4 Popular data-driven models

Two common approaches for data-driven modeling are statistical modeling and ML modeling. We introduce the definition, modeling process, examples, and when to use the two approaches in this subsection.

3.1.4.1 Statistical modeling

A statistical model is a mathematical model representing the data generation process based on a set of statistical assumptions. In statistical modeling, regression analysis is a set of statistical processes with the aim to estimate the relationship between a dependent variable (i.e., output) and a set of independent variables (i.e., input) using a training set that is constituted by historical data, so as to predict the output of an unseen this means an example that is not used for constructing and tuning the model example. After data acquisition, the statistical modeling process begins from data preprocessing and feature extraction. As statistical modeling relies on statistical assumptions, reasonable assumptions are made on the data generation process, and then suitable regression models are selected. Next, the parameters of the regression models are estimated using the historical data. Finally, the fit and generalization abilities of the models are tested. Popular statistical models for prediction tasks are linear regression, logistic regression, polynomial regression, stepwise regression, ridge regression, lasso regression, and elastic net regression. Given a dataset with *n* samples and *m* features denoted by $D = \{(x_i, y_i), i = 1, ..., n\}$, $x_i \in R^m$, $y_i \in R$, the basic forms and common parameter estimation methods of these popular statistical models are shown in Table 3.2.

3.1.4.2 ML modeling

ML modeling belongs to the area of artificial intelligence, which aims to enable a computer to think and act like humans and to solve complex tasks without or with only little human intervention. A commonly used definition is given by Tom Mitchell:

Table 3.1 Comparison between theory-based modeling and data-driven modeling

Quantitative prediction approach	Advantages	Disadvantages
Theory-based modeling	• System behavior and the prediction results are fully tractable and interpretable. • No or few historical data of the system are needed, and thus it can be applied in initial application stage with low-data acquisition costs. • It strictly complies with domain knowledge and thus can be more acceptable by practitioners.	• The prediction accuracy can be low, which highly depends on the assumptions made. • Much *a priori* knowledge is needed. • As the model structure and parameters are fixed, its performance cannot be improved as data accumulate.
Data-driven modeling	• With proper data, feature, and model, its prediction accuracy can be very high. • Little *a priori* knowledge is needed. • It can be constantly improved with the accumulation of data and refinement of model.	• It is usually in a black-box nature, and thus the model-working mechanism and the predictions generated are highly likely to be unexplainable. • Much historical data are needed for model construction, and the data acquisition costs can be very high. • It can be very sensitive to data quality and quantity. • The training data can be over-fitted, leading to poor generalization ability. • The predictions might be contradictory to domain knowledge.

Table 3.2 Popular statistical models

Statistical model	Basic form	Common parameter estimation method
Linear regression	$y = \beta_1 x_1 + ... + \beta_m x_m + \epsilon$. If $m = 1$, it is simple linear regression. Otherwise, it is multiple linear regression.	Least squares
Logistic regression (for binary classification problems)	Given input $\mathbf{x}$ and model parameters θ, the probability of $y = 1$ is calculated by $$P(y = 1|\mathbf{x}; \theta) = g(\theta^T \mathbf{x}) = \frac{1}{1 + e^{-\theta^T \mathbf{x}}}.$$	Weighted least squares and maximum likelihood estimation
Polynomial regression	$y = \beta_0 + \beta_1 X + \beta_2 X^2 + \beta_3 X^3 + ... + \beta_h X^h + \epsilon$, where $\mathbf{X}$ is the feature matrix of the training set.	Least squares
Stepwise regression	Stepwise regression is used to address the multicollinearity of variables in linear regression models by automatically choosing predictive variables. There are three main approaches for stepwise regression. • Forward selection: Starting from no variable in the model and adding the variables from the largest to the least contribution evaluated by F-test until no more variable(s) can be added. • Backward elimination: Starting from all candidate variables in the model and deleting each variable whose loss gives the least statistically significant deterioration of the model fit evaluated by F-test until no more variable(s) can be deleted. • Bidirectional elimination: A combination of forward selection and backward elimination to ensure adding the variable with the largest contribution in each step, while all the added variables in all runs are statistically significant (otherwise, statistically insignificant variable(s) will be deleted).	Least squares combined with F-test and t-test
Ridge regression (regularized by an L2 term)	The regression function format is similar to linear regression models. An L2 regularization term is included in the cost function to address the over-fitting problem as $$l = \sum_{i=1}^{n}(y_i - \sum_{j=1}^{m} \beta_j x_{ij})^2 + \lambda \sum_{j=1}^{m} \|\beta_j\|_2^2,$$ $\lambda > 0$.	Least squares

(Continues)

Table 3.2   Continued

Statistical model	Basic form	Common parameter estimation method
Lasso regression (regularized by an L1 term)	The regression function format is similar to linear regression models. An L1 regularization term is included in the cost function to address the over-fitting problem as $l = \sum_{i=1}^{n}(y_i - \sum_{j=1}^{m}\beta_j x_{ij})^2 + \lambda \sum_{j=1}^{m}\|\beta_j\|_1$, $\lambda > 0$.	Least squares

ML is a computer program that is said to learn from experience E with respect to some task T and some performance measure P, if its performance on T, as measured by P, improves with experience E.

ML models can be further divided into three categories, namely supervised learning, unsupervised learning, and reinforcement learning. In supervised learning, a training set consisting of examples with features as well as targets is used to train ML models, while in unsupervised learning, a training set containing examples with features while without targets is used for ML model training. Meanwhile, reinforcement learning aims to find suitable actions to be taken in a given situation to maximize a reward. Among them, supervised learning is the most widely used method to address predictive problems. Popular supervised ML models include ANN, decision tree (DT), tree ensemble models, support vector classifier (SVC)/SVR, and KNN. The basic idea and common forms of these popular supervised ML models are shown in Table 3.3.

Although statistical modeling and ML modeling have different forms and features, both are widely used to solve predictive models in multiple disciplines. In summary, statistical modeling takes a form of a mathematical function, and ML modeling uses computer algorithms to extract patterns from data without relying on rule-based programming not mentioned before; but this is how it works. It is hard to compare their advantages and disadvantages. Nevertheless, they have their own suitable application scenarios. Generally speaking, statistical models are more suitable to be applied when the input and output variables take a linear or near-linear form. In addition, as they are dependent on statistical theory, they are more suitable to be applied when high-model interpretability, as well as the model and variable significant tests, are required. Meanwhile, ML models should be adopted when the requirement of model accuracy is high or when the data size is large and with many uncertainties. However, it should also be noted that there is no rule of thumb in deciding which approach is more suitable to address a specific prediction problem. Instead, the one that can best meet the actual needs should be found in a trial-and-error manner.

Table 3.3   Popular supervised ML models

Model	Basic idea	Common forms
ANN	An ANN model aims to extract linear combinations of the input features so as to derive more effective features, and then calculate the target using a nonlinear function of these derived features. An ANN model usually consists of an input layer, a hidden layer, and an output layer.	Back-propagation neural network, feed forward neural network, radial basis network, and deep neural networks.
DT	The construction process of a DT mainly involves consecutive node splitting from the root node to the leave nodes, aiming to minimize the loss function. Each split requires selecting a feature together with one of its values.	Popular DT algorithms include ID3 (short for iterative dichotomizer 3) and C4.5 (an extension of ID3) for classification tasks, and CART (standing for classification and regression tree) for both classification and regression tasks.
Tree ensemble models	To reduce the tendency of over-fit and local optimization of a single DT and thus to improve model generalization ability, tree ensemble models are developed by applying bagging (bootstrap aggregating) or boosting on a set of DTs.	Random forest, extremely randomized trees, adaptive booster, gradient boost, and extreme gradient boosting.
SVC/SVR	As the names suggest, SVC is for classification task, and SVM is for regression task, while both methods belong to the class of SVM. An SVC aims to produce a nonlinear boundary (hyperplane) that maximizes the distance margin between two classes, and an SVR aims to minimize the generalization bound by developing a linear regression function in a high-dimensional feature space.	Nonlinearity in SVM is achieved by kernel function. Popular kernel functions include Gaussian kernel, Gaussian kernel radial basis function, sigmoid kernel, and polynomial kernel.
KNN	KNN is memory-based with no model to be fit. For a given point to be predicted, the k training points nearest to the point of concern are first identified. Then, for classification problem, the output is predicted by majority vote among the k neighbors; for regression problem, the predicted target of the point is the average output of the k neighbors.	Distance metric is an important issue affecting the performance of KNN algorithm. Popular distance metrics are Euclidean distance, Minkowski distance, Manhattan distance, Cosine distance, Jaccard distance, and Hamming distance.

References

[1] Yan R., Wang S., Zhen L., Laporte G. 'Emerging approaches applied to maritime transport research: past and future'. *Communications in Transportation Research*. 2021;**1**:100011.

[2] Yan R., Wang S., Psaraftis H.N. 'Data analytics for fuel consumption management in maritime transportation: status and perspectives'. *Transportation Research Part E*. 2021;**155**:102489.

Chapter 4
Key elements of data-driven models

This chapter aims to thoroughly introduce and discuss the major issues of data-driven models, with a focus on machine learning (ML) models and their construction procedure. We first compare three popular data-driven models, namely, statistical model, ML model, and deep learning (DL) model. Then, the whole procedure of developing a data-driven model will be provided. Key elements from various aspects of the whole process will be covered in detail.

4.1 Comparison of three popular data-driven models

We first compare the differences among the three popular data-driven models, namely, statistical model, ML model, and DL model from various perspectives as presented in Table 4.1. It is noted that although DL is a type of ML as DL models are based on deep neural networks (DNNs) consisting of multiple hidden layers, we treat them separately as they have several major differences from other ML models. In the remainder of this book, our main focus will be on using ML models to address practical problems in maritime transport.

4.2 Procedure of developing ML models to address maritime transport problems

The whole procedure of developing an ML model to address a specific practical problem in maritime transport is shown in Figure 4.1. It starts by specifying the problem to be solved, which needs extensive communications and discussions with the parties of interest. The main goal is to understand the business scenarios where the ML model will be applied and the expected outcomes of the stakeholders. Then, the feasibility of addressing the problem should be evaluated considering the available data, technology, and human resources. If needed, the requirements and outcomes can be adjusted after reaching an agreement among the parties. Then, data needed to fulfill the requirements and outcome are to be collected from available (and also may be various) resources. After that, feature engineering is conducted to enable the available data to better describe the essence of the problem studied, so as to improve the prediction accuracy. Next, ML models are developed and are then evaluated, assessed, and refined. It should be noted that although the basic sequence

Table 4.1 Comparison of statistical, ML, and DL models

Model name	Basic idea	Data and computing power requirement	Prediction accuracy level	Explainability	Typical applications
Statistical models	A statistical model is based on statistical learning theory, with the goal of finding a predictive function in an explicit form by leveraging historical data	Structured data, low computing power required	Medium, highly depending on whether reasonable assumptions are made and proper structures are adopted	Explainable like a "white box"	Time-series analysis, attribution analysis, survival analysis
ML models	An ML model is constructed to make a computer to be able to perform tasks with the process of "learning" while without explicit "programming"	Mainly structured data, medium computing power required	Medium to high, depending on data characteristics. Generally, ensemble models have higher predictive accuracy	Most are un-explainable like a "black box", but some are explainable, such as linear regression and decision tree	Medical and healthcare management, fraud detection, product recommendation, dynamic pricing, and social network mining
DL models	A DL model aims to mimic the behavior of human brain (regarding structures and functions) and is mainly achieved by constructing multi-layer DNNs	Structured and unstructured data, high computing power required [where GPU (graphics processing unit) is usually needed]	Medium to high, depending on data quantity and the model structure	Un-explainable like a "black box"	Automated driving, sentimental analysis, natural language processing, digital image processing, video and audio processing, and automatic game playing

Key elements of data-driven models 31

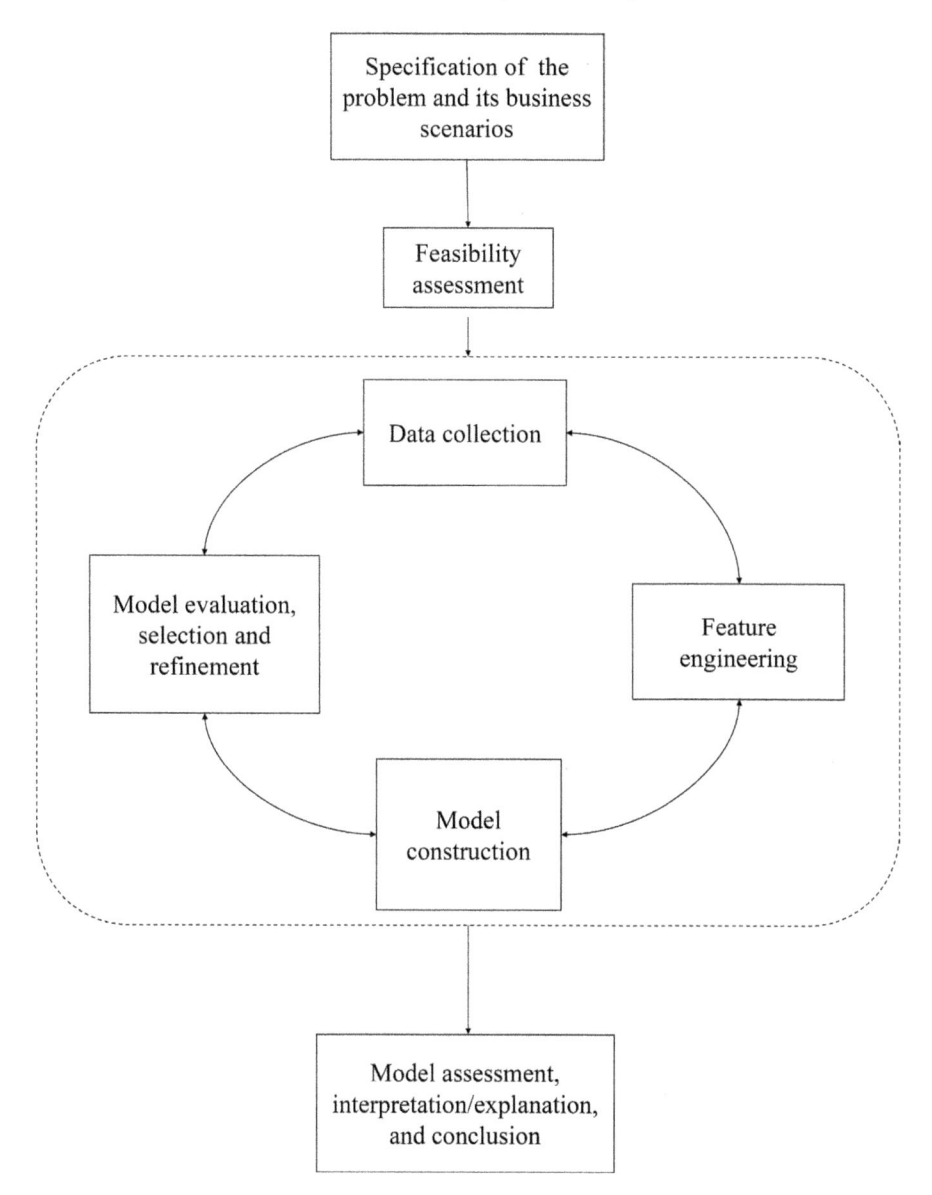

Figure 4.1 Procedure of developing ML models for maritime transport problems*

*Note that the basic sequence of the four steps in the rounded rectangle is data collection → feature engineering → model construction → model evaluation, selection and refinement. In practice, some of them may be repeatedly operated to satisfy the requirements.

of these four steps (from data collection to model evaluation, assessment, and refinement) is generally like this, they might also be alternating and repetitive considering the gap between the requirements as well as the expected outcomes and the developed ML models on hand. Finally, prediction results and model explanations are obtained, and managerial insights are derived for improving the operation efficiency. In the next subsections, we will discuss the key elements of the procedure in detail.

4.2.1 Problem specification

The problem specification step aims to understand the business problem to be solved and then turn the business problem into a data science problem, which is mainly achieved by communicating with and interviewing the parties of interest and understanding the problem backgrounds. In this process, the following three aspects should be explored, namely, objectives, resources, and requirements. To be more specific, objectives mainly refer to the prediction targets of the problem and the problem's specific form. For example, to predict ship risk in PSC inspection, the prediction target can be ship deficiency number or detention rate, among other prediction targets. Resources include the available data sources and computational power, especially when big data, real-time or near-real-time prediction, and complex prediction models are needed. Requirements regarding prediction accuracy, presentation form, and model interpretability and explainability can be varied considering different users and application scenarios. For example, academic and technical staff may expect more details of the fundamental theories and the detailed implementation procedure of the prediction model, while concise and user-friendly graphical user interface might be more likely to be expected by onboard crew. Based on the information gathered, the problem is then broken down into several smaller problems, so as to shed more light on the data to be collected and the models/methods to be adopted.

4.2.2 Feasibility assessment

Feasibility assessment is usually conducted after the requirements are specified, and the data to be collected and methodology to be adopted are planned. It aims to determine whether the requirements of the project can be met given the budget, the schedule, the anticipated technology, and the human resources. For any infeasibility, requirements might be updated or the technologies adopted should be modified based on mutual agreements between the stakeholders and the data scientists.

4.2.3 Data collection

Data can be regarded as the most important component in data-driven modeling, as the quality and quantity of data used could have a significant impact on model accuracy and its generalization ability. In this step, required data are collected and fused from multiple data sources. The data collected can be structured and unstructured data. Especially, structured data are usually in a quantitative and formatted form, which are mainly represented by numbers and strings. In contrast, unstructured data are usually in a qualitative form and can be text files, audio and video files, and figures. We briefly introduce common data sources that are popular in maritime transport research as follows.

4.2.3.1 Automatic identification system

Automatic identification system (AIS) is required to be equipped on vessels involving international voyages with a gross tonnage above 300 and on all passenger ships. It was first developed in the 1990s as a short-range identification and tracking system for the purpose of collision avoidance. It then has a wider application in, e.g., fishing fleet monitoring and control, vessel traffic management, maritime security, aids to navigation, search and rescue, accident investigation, and fleet and cargo tracking. Vessel dynamic information (geographical location, sailing speed, navigation status, course, heading degree, and rate of turn), static information [ship name, type, size, draught, maritime mobile service identity (MMSI), IMO number, and call sign], and voyage information (destination and the estimated time of arrival) are given in AIS reports. The update interval of dynamic AIS reports, ranging from 2 to 10 seconds, highly depends on the vessel sailing speed, and the static and voyage AIS reports are updated every 6 minutes or when any field contained by the static and voyage AIS reports is changed. For a detailed introduction to AIS, readers are referred to References 1 and 2.

4.2.3.2 Ship specifications

Ship specification information, including but not limited to basic ship characteristics [current and former vessel names, vessel identification numbers (IMO number, MMSI, and call sign), vessel type, year of built and builder, tonnages and dimensions, and engine information], ship operation information (current and former flag states, management companies, operators, and owners), ship survey and maintenance records, and PSC detention records, can be found in commercial vessel databases such as Lloyd's List Intelligence[*], Shipping Intelligence Network[†], and MarineTraffic[‡]. Features of ships with certain characteristics, e.g., in certain classification societies (e.g., American Bureau of Shipping[§]) and the sunken ships (e.g., International Registry of Sunken Ships[¶]) can also be found in specific databases.

4.2.3.3 Ship sailing records

Ship sailing records can come from automated onboard sensors based on the Internet of Things technology and the manually filled ship noon report. Common sensors on ships to monitor their performance are GPS receiver, speed log, revolutions per minute and torque meter, shaft power tester, fuel flow meter, wind anemometer, and

---

[*]https://ihsmarkit.com/products/maritime-ships-register.html
[†]https://www.clarksons.net/portal/
[‡]https://www.marinetraffic.com/en/ais/home/centerx:-12.0/centery:25.0/zoom:4
[§]https://ww2.eagle.org/en.html
[¶]http://www.shipwreckregistry.com/

water depth sensor. Data are automatically collected and then stored in the vessel's central database and transmitted to port authorities.

4.2.3.4 Ship accident data

Public ship accident and incident data around the world can be found in the Global Integrated Shipping Information System (GISIS) provided by the IMO[**]. Factual data collected from various sources and more elaborated information extracted from the reports of investigation of casualties received by the IMO are provided by the GISIS. Ship casualties are classified into "very serious casualties," "serious casualties," "less serious casualties," and "marine incidents." Regional marine accident reports and statistics are also provided by local governments and departments. For example, the Marine Department of Hong Kong established the Marine Accident Investigation Section, which is responsible to investigate all marine accidents occurring in Hong Kong and onboard Hong Kong registered ships and to conduct official inquiries under certain ordinances. Reports and summaries of accident investigation of individual ships carried out by the Marine Accident Investigation Section are published on the Marine Department's website[††].

4.2.3.5 Port statistics

Key performance indicators are published by some ports that we refer to as port statistics. Common fields are visiting vessel statistics, port container throughput, port cargo throughput, marine accidents, and bunker sales. Fields published by different ports might differ from each other. For example, the Hong Kong Port[‡‡] publishes monthly, quarterly, and yearly vessel statistics, port container and cargo throughput, and cross-boundary ferry terminal statistics.

In addition to the common databases discussed so far, structured and unstructured data can be collected via other means, such as surveys, interviews, questionnaires, and observations.

4.2.4 Feature engineering

Feature engineering is the most time-consuming and important step in the procedure of ML model development. Feature engineering aims to prepare proper input data that are compatible with an ML model, so as to improve its performance. Feature engineering is a huge project, and we only cover its main parts in this subsection concisely from the following aspects: data cleaning, feature extraction, feature pre-processing, feature encoding and scaling, and feature selection.

[**]https://gisis.imo.org/Public/MCI/Default.aspx
[††]https://www.mardep.gov.hk/en/publication/publications/reports/ereport.html
[‡‡]https://www.mardep.gov.hk/en/fact/portstat.html

4.2.4.1 Data cleaning

After the raw data are collected, data quality should be (initially) checked (and will then be carefully checked and improved after feature extraction) considering their integrity, uniqueness, authoritativeness, legitimacy, and consistency. The explanation of the above factors using PSC data as an example and the corresponding solutions are given in Table 4.2. In addition to the above factors, some details also need attention. For example, numbers stored as strings should be converted to numeric types; extra spaces in data should be deleted; and the format of timestamp should be converted from string format to time format.

4.2.4.2 Feature extraction

Feature extraction aims to create new features from the existing ones and delete the original ones to make the features more informative and concise. Feature extraction is one approach for dimension reduction (and the other is feature selection which will be covered later in subsection 4.2.4.5), which refers to a process of reducing the number of features in the data set while keeping as much variation in the original data set as possible. Feature extraction methods can be linear and nonlinear. We introduce principal components analysis (PCA) and linear discriminant analysis (LDA) as the representatives of linear methods, and kernel PCA and multidimensional scaling (MDS) as the representatives of nonlinear methods.

PCA: PCA is an unsupervised linear transformation to project the original data to lower-dimension orthogonal directions of high variance, aiming at using less features in lower dimensions (i.e., principal components) to represent the original features in higher dimensions. The orthogonality property results in very low or almost close to zero correlation in the projected data.

LDA: LDA is a supervised learning-based feature extraction method and also an ML classifier. It aims to maximize the distance between the mean of each class and minimize the spreading within each class. The idea is in line with that of classification, and thus it can be expected that LDA can lead to better performance in classification ML models.

Kernel PCA: The assumption of PCA is that there is a linear hyperplane for data projection. However, when a data set is not linearly separable, linear PCA might not be a good choice. To address this issue, kernel PCA using kernel function to project data into a higher dimensional feature space is proposed, where the nonlinear data are linearly separable in the kernel space.

MDS: MDS has two major types: metric MDS and non-metric MDS, where the former attempts to calculate the similarity/dissimilarity of data measured by distances based on geometric coordinates, and the latter is designed to deal with ordinal data. In general, dissimilarities of each pair of points in the original high-dimensional space are first calculated, and then the points are mapped to lower-dimensional space while the dissimilarities of the point pairs are preserved as much as possible.

Table 4.2 Explanation and solution of data cleaning

Factor	Explanation	Example in PSC data	Solution
Integrity	Data integrity is the overall accuracy, completeness, and consistency of data.	Ship age is missing in the data set.	The missing features can be added by searching for more databases or calculating from other features.
Uniqueness	Data uniqueness means that a single record only appears once in the data set.	One ship inspection record is recorded more than once in the data set.	Only one of the duplicate records is kept.
Authoritativeness	Data authoritativeness requires that the data should be legislated and regulated, and should come from authoritative source or reference that is widely recognized by experts in the field or industry.	The number of deficiencies of a ship in one inspection recalled by the captain is different from that searched on the MoU's official website.	Data from authoritative sources should be used.
Legitimacy	Data legitimacy means that data values should comply with ground truth and expert knowledge.	The age of an inspected ship is 1,000 years.	Rules considering ground truth or based on domain knowledge are set to specify the data range (ways to process outliers are covered in the following steps).
Consistency	Data consistency requires that the same type of data appears in different places throughout the data set to be matched.	The format of inspection date is varied, e.g., yyyy-mm-dd, dd-mm-yyyy, and mm-dd-yyyy.	Data format should be unified.

Table 4.3   Methods to handle missing values in data sets

Method	Description	Note
Search the missing values from various sources and databases	Query different sources and databases to search for specific missing values.	Data format of external sources should be in compliance with the original data format.
Fill with a global constant	Fill missing values of all features using a global constant, such as "−1," "0," and "N.A."	The assumption of this method is that null value is informative in itself, that is, it is indeed different from valid values. Although this method is intuitive and easy to operate, it cannot be used or will cause confusion sometimes.
Fill with statistics of the feature	Fill missing values of a certain feature using its mean, median, or mode over all or similar record(s) in the data set.	Record similarity can be evaluated by k-means clustering and distance measures. An extra binary feature indicating whether each record has a missing value of a certain feature can be added.
Fill using the forward and backward method	Use the feature value of the next or previous record (or the mean feature value of the next or previous few records) in time-series data or a quantity that is stable over a period of time (e.g., ship draft value over a voyage).	N.A.
Fill with regression-based approach	Predict the missing value of a certain feature using other features based on regression models.	One drawback is that the correlation between features can increase.

4.2.4.3 Feature pre-processing

Feature pre-processing aims to deal with missing values and outliers in features. Missing values cannot be simply ignored or deleted in the original data set; instead, they must be carefully handled. One exception is dealing with a record (i.e., row data) or a feature (i.e., column data) where most data fields are missing. Consequently, the record or feature is meaningless. In other cases, there are several ways to handle missing data in features, which we list in Table 4.3.

An outlier is an observation that is far away in a random sample from the data set. Outliers can have a significant impact on the assumptions of data distribution and the prediction models constructed, and thus they need to be detected and properly treated. Outliers can be univariate and multivariate, where the former is an extreme value of a specific feature, whereas the latter is a combination of values that are not likely to occur together with an observation. For example, if there is a PSC

38 *Machine learning and data analytics for maritime studies*

inspection record saying that ship type is "handymax carriers" and the dead weight tonnage is 80,000, it is an outlier, as the dead weight tonnage of ship type "handymax carriers" is between 40,000 and 60,000. We only discuss univariate outliers in this subsection. Common approaches of univariate outlier detection are listed in Table 4.4.

After an outlier is detected, whether to drop the outlier needs careful consideration. Some outliers are detected due to inherent variations in feature values, and they are normal values instead of noise. For example, ship age 42 is detected as an outlier in the PSC data set used in Reference 3. Nevertheless, the oldest ship still afloat in the world is more than 200 years old, and thus the 42-year-old ship should not be regarded as an outlier. Actually, given all other conditions equal, older ships are usually associated with a larger number of deficiencies and a higher detention probability and is thus the focus on PSC. Therefore, this outlier detected should not be deleted or modified. Actually, the above example is just one case where the detected outlier should be kept. When there are lots of outliers (e.g., more than 30%) or the outliers can be rectified by capping, assigning new values, and data transforming, they should not be dropped. In some other cases, the outliers should be dropped, e.g., when it is sure that they are wrong, there are many data while outliers are few, and they come from exceptional samples that should not be included.

4.2.4.4 Feature encoding and scaling

Feature encoding is needed for categorical features, which are non-numeric and are divided into groups and categories, with the aim to turn categorical data into numerical data that are compatible with most ML models. Meanwhile, feature scaling is applied to numerical features with varying scales when certain ML models are adopted. Their details will be introduced in the following paragraphs.

There are basically three types of categorical features: binary feature (e.g., yes and no; true and false), ordinal feature (e.g., low, medium, and high; bad, average, and good), and nominal feature (e.g., container ship, bulk carrier, and passenger ship). Encoding a binary feature is the easiest: just encode positive samples to 1 and negative samples to 0. For ordinal feature and nominal feature, we summarize the ideas, examples, and application scenarios of popular encoding methods in Table 4.5 that are widely applicable to PSC-related research.

In addition to the common feature encoding methods shown in Table 4.5, there are some other popular feature encoding methods that can be adopted to address other practical problems in maritime transportation. One is hash encoding, which aims to convert categorical values into numerical values using hashing functions and is applicable to nominal feature. Another is bin counting, which is based on statistics that represent the relationship between the categorical feature and the target, where the relationship is measured by a conditional probability of the target under a given value of the categorical variable and is applicable to both nominal and ordinal features.

Feature scaling should be applied in developing ML models based on geometric distance or with weights assigned to features, such as linear regression (when it is

Table 4.4 Methods to handle outliers in data sets

No.	Method	Description
1	Method based on standard deviation	If a feature value has a distance to the average higher than a given threshold (usually between 2 and 4) of the standard deviation, it is regarded as an outlier.
2	Method based on percentiles	Assume a certain percentage of feature values from the top or from the bottom as outliers considering data distribution and domain knowledge.
3	Interquartile range (IQR)	Sort the values of a feature in ascending order and split them into four equal parts by $Q1$, $Q2$, and $Q3$ (i.e., the 25th, 50th, and 75th percentile of data). $IQR = Q3 - Q1$. Values below $Q1 - 1.5 \times IQR$ and above $Q3 + 1.5 \times IQR$ are regarded as outliers.
4	Box plot	Box plot can demonstrate the data visually, which is a data distribution figure with median, quartiles of $Q3$ and $Q1$, lower bounds calculated by $Q1 - 1.5 \times IQR$ and $Q1 - 3 \times IQR$ and upper bounds calculated by $Q3 + 1.5 \times IQR$ and $Q3 + 3 \times IQR$ shown. It should be noted that which records are regarded to be outliers is highly dependent on subjective judgment.

Table 4.5   Methods of feature encoding in data sets

Encoding method	Idea	Example	Application scenarios
One-hot encoding	Convert each category value into a new column and assign "1" or "0" to the new columns, where "1" is assigned to the column of the category the sample belongs to, and "0," otherwise.	Feature "ship type" has three values: "container," "bulk carrier," and "passenger ship." We first convert this feature into three features: "is_container," "is_bulk_carrier," and "is_passenger_ship" and encode "container" as "(1, 0, 0)," "bulk carrier" as "(0, 1, 0)," and "passenger ship" as "(0, 0, 1)" where each digit is for one new feature.	Nominal feature. Note that although it can also be applied to ordinal features, the order cannot be retained.
Target encoding	Convert each value in a column to a number considering their inherent orders.	Feature "ship age" has three values: "young," "middle-aged," and "old." We encode "young," "middle-aged," and "old" to "1," "2," and "3."	Ordinal features

solved by gradient descent algorithm), k-means, k nearest neighbors, and PCA. In contrast, feature scaling is not needed in ML models that are not distance-based and no weights are assigned, such as Naive Bayes and tree-based models. There are mainly three types of feature scaling methods, which are presented as follows. Suppose that we have a set of values of a feature denoted by $\{x_1, ..., x_n\}$:

Min–max scaling: $x'_i = \dfrac{x_i - \min(x_1, ..., x_n)}{\max(x_1, ..., x_n) - \min(x_1, ..., x_n)}$, where x'_i is the feature value after min–max scaling, $\min(x_1, x_2, ..., x_n)$ and $\max(x_1, x_2, ..., x_n)$ are the minimum and the maximum values of all the feature's values, respectively. The range after min–max scaling is [0, 1].

Mean normalization: $x'_i = \dfrac{x - \text{mean}(x_1, ..., x_n)}{\max(x_1, ..., x_n) - \min(x_1, ..., x_n)}$, where $\text{mean}(x_1, x_2, ..., x_n)$ is the mean of the feature's values.

Standardization: $x'_i = \dfrac{x - \text{mean}(x_1, ..., x_n)}{\text{std}(x_1, ..., x_n)}$, where $\text{std}(x_1, x_2, ..., x_n)$ is the standard deviation of the feature. It makes the feature values have zero-mean and unit variance.

4.2.4.5   Feature selection

Feature selection is a critical problem in feature engineering. It aims to delete irrelevant and redundant features and find the optimal feature subset (combination of features), so as to improve model performance and reduce computation time. Both supervised and unsupervised methods can be used for feature selection. Supervised feature selection techniques use the prediction target, and it can be further divided into three categories: filter, wrapper, and embedded, while unsupervised feature selection techniques ignore the target. We present common approaches for supervised feature selection as follows. Readers are referred to [4] for more information on unsupervised feature selection techniques.

4.2.4.6 Pearson's correlation coefficient (Pearson's r)

Pearson's r is a supervised method based on the filter that measures the linear correlation between two numerical variables. Pearson's r between feature x and target y is calculated by $\rho_{x,y} = \dfrac{\text{cov}(x, y)}{\sigma_x \sigma_y}$, where cov(x, y) is the covariance of feature x and target y, and σ_x and σ_y are the standard deviations of feature x and target y, respectively. $\rho_{x,y} \in [-1, 1]$, with $-1(+1)$ indicates complete negative(positive) correlation and 0 indicates no linear correlation. The basic idea of using Pearson's r for feature selection is first to calculate Pearson's r between each of the numerical features and the numerical target. Then, a certain number of features with the highest absolute Pearson's rs or features with absolute Pearson's r above a given threshold are selected as the optimal feature subset.

4.2.4.7 Spearman's rank coefficient (Spearman's ρ)

Spearman's ρ is a supervised method based on the filter that measures the nonlinear correlation between two numerical variables. Spearman's ρ between feature x and target y is calculated by $\rho_{R(x),R(y)} = \dfrac{\text{cov}(R(x), R(y))}{\sigma_{R(x)} \sigma_{R(y)}}$, where cov(R(x), R(y)) is the covariance of the rank variables, and $\sigma_{R(x)}$ and $\sigma_{R(y)}$ are the standard deviations of the rank variable of feature x and target y. The basic idea of using Spearman's ρ for feature selection is first to calculate Spearman's ρ between each of the numerical features and the numerical target. Then, certain number of features with the highest absolute Spearman's ρ or features with absolute Spearman's ρ above a given threshold are selected as the optimal feature subset.

4.2.4.8 Chi-squared test (χ^2 test)

χ^2 test is a supervised method based on the filter that measures the correlation between two categorical variables (one is a categorical feature and the other is the categorical target). Suppose that there are n observations belonging to K categories and category k has m_k samples. A null hypothesis gives the probability $p_k = \frac{1}{K}$, meaning that a sample falls into category k and the k categories are independent. Then, $\chi^2 = \sum_{k=1}^{K} \dfrac{(m_k - np_k)^2}{np_k}$. The basic idea of using χ^2 test for feature selection is first to calculate χ^2 test between each of the categorical features and the categorical target. Then, a certain number of features with the highest χ^2 statistic or features with χ^2 statistic above a given threshold are selected as the final feature subset.

4.2.4.9 Mutual information
Mutual information (MI) is a supervised method based on filter that measures the correlation between two categorical variables. $MI(x,y) = \sum_{x \in X} \sum_{y \in Y} p(x,y) \log \frac{p(x,y)}{p(x)p(y)}$, where X and Y are the domains of x and y, $p(x,y)$ is the joint probability of $X = x$ and $Y = y$, and $p(x)(p(y))$ is the probability of $X = x(Y = y)$. The basic idea of using MI for feature selection is first to calculate MI between each of the categorical features and the categorical target. Then, a certain number of features with the highest MI or features with MI above a given threshold are selected as the optimal feature subset.

4.2.4.10 Recursive feature elimination
Recursive feature elimination (RFE) is a supervised method based on the wrapper that removes unimportant features iteratively based on the evaluation metric. An ML model that assigns weights to features is trained on the entire initial set with all features. Then, the least important feature is removed from the current feature set, and a new ML model is trained using the remaining features. The above step is repeated until the desired number of features is selected as the optimal feature subset.

4.2.4.11 Feature selection based on regularization
This feature selection method belongs to the embedded method, which is usually integrated as a part of an ML algorithm. Regularization can shrink the coefficients (i.e., weights) of the unimportant/low impact features toward zero, and thus achieve the goal of feature selection. Common approaches include least absolute shrinkage and selection operator (LASSO), ridge, and ElasticNet where L1 regularization, L2 regularization, and the combination of L1 and L2 regularization are used, respectively. These three approaches are introduced in Table 4.2 of section 4.2.4.1.

4.2.4.12 Feature selection based on feature importance
This feature selection method belongs to the embedded method and is usually achieved by tree-based models and permutation. In tree-based models, such as decision tree, random forest, gradient boosting decision tree, XGBoost, and LightGBM, feature importance is usually estimated by the impurity reduction in node split process using the selected features, and features leading to higher impurity reduction should be more important. In permutation-based feature selection, an initial ML model with original features is first trained. Then, the features are permuted one by one to form new data set for model training, and the increase in the prediction error is observed. Feature leading to a higher increase in error should be more important.

4.2.5 Model construction

The development of ML models is one of the most crucial modeling steps. One critical issue is that as there are many algorithms to develop different ML models to choose from, how to choose a proper one? Another critical issue is how to tune the hyperparameter values so as to improve the ML model's performance? These two issues are discussed in this subsection.

4.2.5.1 Model evaluation metrics

Different metrics should be adopted to evaluate regression and classification ML models. For regression models where the output is a numeric value, common metrics are mean-squared error (MSE), root-mean-squared error (RMSE), mean absolute error (MAE), mean absolute percentage error (MAPE, where none of the output values can be zero), and r-squared (R^2). Suppose that there is a test set with n samples where the real output and predicted output of each sample i are denoted by y_i and $\hat{y}_i$, $i = 1, ..., n$. The metrics are calculated as follows:

$$\text{MSE} = \frac{1}{n}\sum_{i=1}^{n}(y_i - \hat{y}_i)^2,$$

$$\text{RMSE} = \sqrt{\text{MSE}} = \sqrt{\frac{1}{n}\sum_{i=1}^{n}(y_i - \hat{y}_i)^2},$$

$$\text{MAE} = \frac{1}{n}\sum_{i=1}^{n}|y_i - \hat{y}_i|,$$

$$\text{MAPE} = \frac{100\%}{n}\sum_{i=1}^{n}|\frac{y_i - \hat{y}_i}{y_i}|,$$

$$R^2 = 1 - \frac{\sum_{i=1}^{n}(y_i - \hat{y}_i)^2}{\sum_{i=1}^{n}(y_i - \bar{y})^2}, \text{ where } \bar{y} = \frac{1}{n}\sum_{i=1}^{n}y_i.$$

For the metrics mentioned above, MSE is the most widely used metric for regression problems as it is older and well-established. It is also differentiable, and thus there are many good algorithms that can be applied to optimize loss functions based on MSE. However, the unit of MSE is the square of the original data unit, and hence this makes it hard to be interpreted. To remedy this issue, RMSE is proposed where the unit is in compliance with the original data unit. Both metrics can be very sensitive to the outliers in the test set, especially for the MSE where the square operation is applied. This problem can be largely addressed by the MAE, whose unit also complies with the original data and the metric is not that sensitive to outliers compared to MSE. However, as it contains an absolute function that is not differentiable, the optimization process is difficult. MAPE is a relative error rate that also considers the original output, but the original output values cannot be zero. R^2 is also widely used in statistics analytics, which represents the proportion of the variance in the output that can be explained by the features. It can be seen that the meanings as well as the pros and cons of the metrics for regression problems are different from each other.

Table 4.6 Confusion matrix for a binary classification problem

Predicted value/Actual value	Positive	Negative
Positive	True positive (TP)	False positive (FP)
Negative	False negative (FN)	True negative (TN)

To evaluate ML models developed to address practical problems, it is recommended that more than one metric should be used, and the metrics should comply with the requirements.

For classification models where the output is a categorical variable, the situation can be much more complex, which is mainly because the problem can be binary, i.e., the target can take two values (which are also called positive and negative classes; e.g., detained or not detained), or multiclass, i.e., the target can take more than two values (e.g., ship type of container, bulk carrier, and passenger ship). As multiclass prediction problems can be turned into binary class prediction problems, we only introduce the metrics for binary classification problem evaluation.

The metrics used for binary classification problems are based on the confusion matrix as shown in Table 4.6.

4.2.5.1.1 Accuracy

Accuracy is a classic and intuitive classification metric, which measures the ratio of the samples that are correctly classified to all the samples in the test set calculated as follows:

$$Accuracy = \frac{TP + TN}{TP + FP + FN + TN}.$$

Accuracy is suitable to be applied in balanced classification problems where the numbers of samples in the positive class and the negative class are similar. In case that the problem is imbalanced, where the positive class is very sparse, always predicting the target to be negative will get a very high accuracy.

Example 1. When the ratio of ships that are detained (positive class) to ships that are not detained (negative class) is 1:99, predicting all the ships to be not detained could achieve a 99% accuracy rate, but the model is meaningless as the detained ships are indeed our focus.

Therefore, metrics suitable to be applied to address imbalanced problems should be proposed.

4.2.5.1.2 Precision

Precision aims to evaluate the proportion of predicted positive samples that are indeed positive. It takes the following form:

$$Precision = \frac{TP}{TP + FP}.$$

For the above Example 4.2.1, *Precision* is undefined as $TP + FP = 0$ is a cautious metric that focuses on truly positive samples.

4.2.5.1.3 Recall
Precision is from the perspective of samples predicted to be positive. From the perspective of samples that are actually positive, the metric *Recall* aims to evaluate the proportion of actual positive samples that are correctly predicted. It takes the following form:

In Example 4.2.1, Recall = 0 as TP = 0 aims to capture as many positive samples as possible. It is also noted that if we predict all samples to be positive, we have Recall = 1.

Precision and *Recall* are somewhat contradictory as they view the same problem from opposite angles: *Precision* only looks at the samples predicted to be positive, while *Recall* takes all the actual positive samples into account.

4.2.5.1.4 F_1 score
To reach a trade-off between *Precision* and *Recall*, the metric of F_1 score, which is the harmonic mean of *Precision* and *Recall*, takes the following form:

When $TP + FP = 0$ where *Precision* is undefined and $TP + FN = 0$ where *Recall* is undefined, $F_1 = 0$. The range of F_1 score is between 0 and 1, and a higher value indicates better performance of a classification model. F_1 score treats *Precision* and *Recall* equally. In case that domain knowledge of different weights for *Precision* and *Recall* should be included, a more generic F_β score can be applied, which takes the following form:

β is chosen such that *Recall* is considered β times as important as *Precision*. When $\beta = 1$, F_β is degenerated to F_1. When $\beta > 1$, *Recall* has a larger impact on F_β; when $\beta < 1$, *Precision* has a larger impact.

4.2.5.1.5 Receiver operating characteristic (ROC) curve
In many classification ML models, a value between 0 and 1 or the probability of being positive is predicted for one sample first, and the value/probability is then compared with a threshold (which is usually 0.5): if the value/probability is larger than the threshold, it is predicted to be positive; otherwise, it is predicted to be negative. For samples predicted to be positive, they can be either positive (TP) or negative (FP) actually, and the rate of positive samples predicted to be positive [TP rate (TRP)] and the rate of negative samples predicted to be positive [FP rate (FPR) are calculated as follows:

$$TPR = \frac{TP}{TP + FN}, FPR = \frac{FP}{TN + FP}.$$

Given a data set (with limited samples) and a threshold, a pair of (*FPR*, *TPR*) can be obtained. By changing the threshold from the smallest to the largest predicted values/probabilities for all the samples, different (*FPR*, *TPR*) pairs can be obtained, and linking all the consecutive pairs together with (0, 0) and (1, 1) can produce the

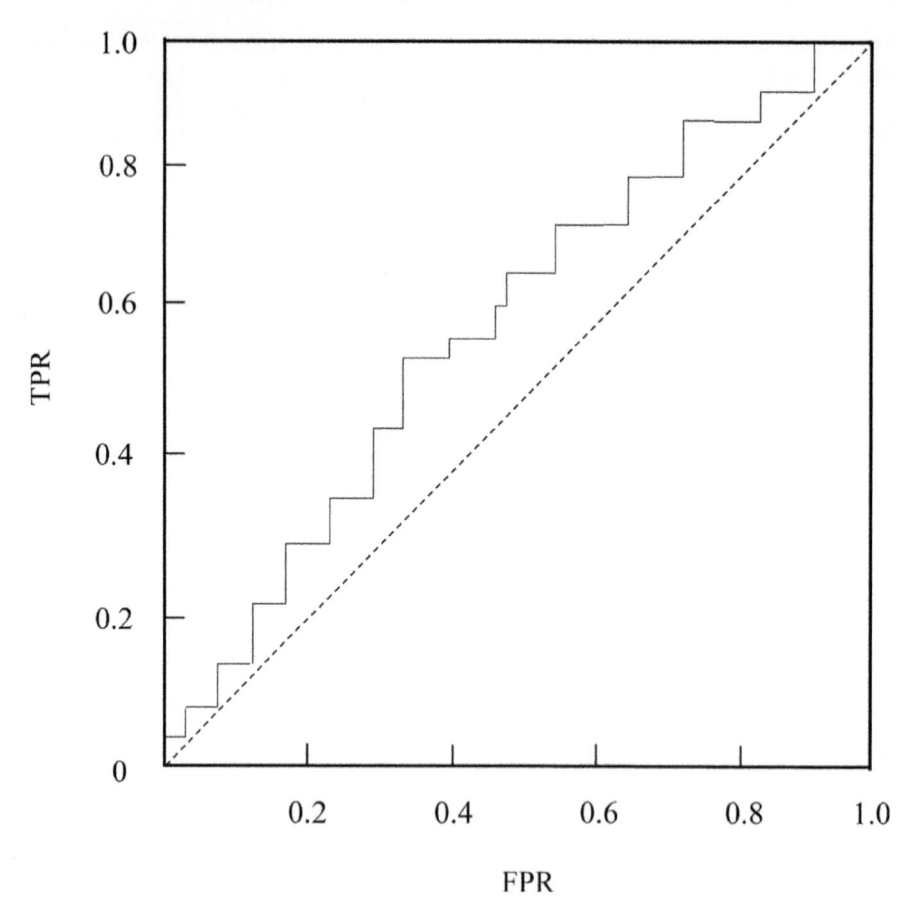

Figure 4.2 An illustration of ROC curves

ROC curve, where an illustration is given by the solid line in Figure 4.2. As limited samples are contained in the test set, the solid line is ragged. The dotted diagonal line corresponds to the ROC curve of random guessing.

Figure 4.2 shows that as the threshold decreases, more samples are predicted to be positive, and *FPR* and *TPR* increase simultaneously from $(0,0)$, where the threshold is 1 with no sample predicted to be positive, to $(1,1)$, where the threshold is 0 with no sample predicted to be negative. A perfect classifier has $(FPR, TPR) = (0, 1)$. Therefore, the closer an ROC curve to $(0,1)$, the better a classifier is.

4.2.5.1.6 Area under ROC curve (AUC)

When comparing two classifiers, if the ROC curve of one classifier is fully above the other, it can be safely concluded that the performance of the former is better than that of the latter. However, if their ROC curves intersect, it can be hard to tell which is better. To remedy this issue, the concept of area under ROC curve, or AUC, is proposed. It is a single value within range $[0.5, 1]$ to evaluate the performance of a classifier, and a larger value indicates a better performance. If AUC is smaller than 0.5, one can simply swap the predicted negative and positive classes. Suppose that the ROC curve is formulated by several pairs (FPR_1, TPR_1), (FPR_2, TPR_2), ..., (FPR_n, TPR_n) where FPR_i increases from $(FPR_1 = 0)$ to $(FPR_n = 1)$, AUC can be calculated by

4.2.5.2 Model selection

An ideal ML model should have good generalization ability, or low generalization error, that performs well on unseen data. Therefore, there is a trade-off between over-fitting and under-fitting when an ML model is constructed: if the model is too complex, training error can be very low or even zero. Meanwhile, the possibility of over-fitting the training data can be high, leading to a poor generalization ability of the ML model constructed. In contrast, a too simple ML model might lead to under-fitting as well as a poor generalization ability. An ideal ML model can be selected by the metrics introduced in section 4.2.5.1 considering different scenarios and requirements. Model selection aims to estimate the performance of different ML models using proper metrics in order to choose the best one. To achieve effective selection, a validation set/several validation sets that is/are mutually exclusive with the training set and a test set should be formulated and used for model evaluation and selection. The ML model with the lowest prediction error on the validation set(s) (where we expect that it also has the lowest prediction error on the unseen test set) should be selected. Common data set division methods used for model selection are given below.

A common approach is to randomly divide the whole data set into a training set, a validation set, and a test set, where the training set is used for model construction, the validation set is used for model selection, and the test set should not be used until the final ML model is constructed and the generalization error of the final chosen model is assessed. There is no general rule to decide the number or ratio of samples in these three sets, as it depends on the size and features of the whole data set. Common ratios of these sets are 50:25:25, 60:20:20, and 70:15:15% To reduce the uncertainties brought by the division of data set, multiple times of random partition should be conducted, and the results generated should be averaged.

Another popular approach is called cross validation. After splitting a test set from the whole data set D, the remaining samples in D, denoted by D', are divided into k ($k > 2$, and the common values are 3, 5, and 10) mutually exclusive subsets, i.e., $D' = D_1 \cup D_2 \cup ... \cup D_k$ and $D_i \cap D_j = \emptyset, i = 1, ..., k, j = 1, ..., k, i \neq j$. Cross validation in this form is also called k-fold cross validation. For each of the k times' training, one subset is selected as the validation set and the remaining subsets form the training set. Then, k model performance results can be obtained, and the average performance can be calculated. An illustration of the 10-fold cross validation is given in Figure 4.3. When $k = n'$, where n' is the size of D', each sample is given the opportunity to be used as the hold out validation set. This approach is also called leave-one-out cross validation. In each time of training, all the samples except the one in the validation set are used to train the ML model, and the ML models'

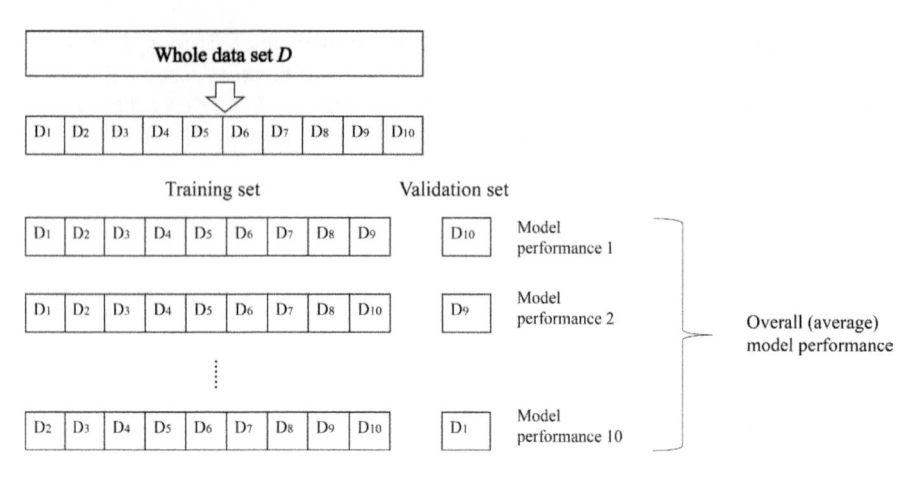

Figure 4.3 Procedure of 10-fold cross validation

performance is validated by all samples. Therefore, the results can be regarded to be very similar to the actual performance on the whole data set. However, when the sample size is too large, the computational burden can be very large and even intolerable.

The size of training set in the above approaches is more or less smaller than the original data set. Consequently, inconsistency in data distribution due to different sample sizes might be brought. To remedy this issue, bootstrapping that is based on bootstrap sampling is applied for model selection. Given a data set D containing n samples, we use bootstrap sampling to formulate a training set $\bar{D}$ by randomly selecting a sample from D and copying it to $\bar{D}$, and then the sample is put back to D, and thus this sample is also likely to be picked in the next round of sampling. The above steps are repeated n times, and finally, the training set $\bar{D}$ also has n samples. It is highly likely that $\bar{D}$ has duplicate samples, and some samples in D are not in $\bar{D}$. It can be estimated that after n rounds of sampling, the possibility that one sample is not picked is $(1 - \frac{1}{n})^n$, and we have

$$\lim_{n \to +\infty} (1 - \frac{1}{n})^n \to \frac{1}{e} \approx 0.368,$$

which means that about 36.8% of the samples do not appear in $\bar{D}$. Therefore, we can use $\bar{D}$ as the training set and $D \setminus \bar{D}$ as the validation set. The validation error obtained in this way is also called out-of-bag error.

4.2.6 Model refinement

We mainly introduce hyperparameter tuning in this subsection for refining the performance of ML models. A hyperparameter is a parameter used to control the overall learning process and is manually set before model learning. Hyperparameters are important to ML models as model performance is highly dependent on them. Hyperparameters from an ML model itself are responsible to control the capacity of the model, that is, how flexible the model is in fitting the data. Therefore, they can be used to control over-fitting and to avoid under-fitting. Although default values of the hyperparameters are given in ML models implemented by open source packages, finding their best values given a specific data set is not a trivial task, especially in ML models that have multiple hyperparameters that interact in nonlinear ways.

Hyperparameter tuning aims to search for the best values for a set of hyperparameters that result in the best model performance on a given data set. Denote the number of hyperparameters that need to be optimized by p. The tuning procedure starts by defining a search space for the p hyperparameters to be tuned, and then different hyperparameter settings constructed from the search space are evaluated by a tuning algorithm. Finally, the hyperparameter setting leading to the best model performance on the validation sets is output, which is used to train the final model.

The most commonly used hyperparameter tuning algorithm is grid search, which picks out a grid of values of the hyperparameters concerned and evaluates them one by one. It is also noted that many hyperparameters take values in the range of real numbers. However, in grid search, a range of their values and the searching interval are first given, and thus finite candidate values are contained in the search space. For example, for an ML model with p hyperparameters h_1 to h_p, the search space is

$$h_1 \in \left\{h_1^1, h_1^2, ..., h_1^{v_1}\right\},$$
$$\vdots$$
$$h_p \in \left\{h_p^1, h_p^2, ..., h_p^{v_p}\right\}.$$

Then, a grid of $v_1 \times ... \times v_p$ hyperparameter settings can be formulated with each constituted by one possible value of h_1 to h_p. All hyperparameter settings are then tested, and the hyperparameter setting with the best performance on the validation set or in cross validation is returned.

Grid search is quite simple and intuitive. However, it is very expensive in terms of computation power. When the ML model is complex and the number of hyperparameters is large, the computation time might become unaffordable. Instead of searching for the entire search space, random search is proposed which only evaluates a certain number of random hyperparameter settings (the number is much smaller than $v_1 \times ... \times v_p$) in the grid. Theoretically, the performance of the best hyperparameter setting found by grid search is highly likely to be worse than that found by grid search. Nevertheless, in many cases of public datasets, Bergstra and Bengio (2012) [5] found that random search can perform about as well as grid search with much less computation power needed.

In recent years, smart hyperparameter tuning algorithms are developed. Unlike the grid search and random search algorithms that form the candidate hyperparameter settings in advance and then evaluate all or part of them one by one, smart hyperparameter tuning algorithms only pick a few hyperparameter settings and evaluate their quality, and then decide where to sample next. The drawback is that the former two tuning algorithms can be parallelizable, while the latter is sequential. Popular algorithms are Bayesian optimization [6], derivative-free optimization [7], and random forest smart tuning [8].

4.2.7 Model assessment, interpretation/explanation, and conclusion

After selecting the most suitable model whose performance is refined by hyperparameter tuning, the final chosen model's performance should be assessed by a hold-out test set that should never be seen and used in the above processes (e.g., in model construction, selection, and refinement). The model performance, including its working mechanism and prediction results, might need to be further interpreted and explained. ML model interpretation and explanation is another large and increasingly important topic, and we will not cover it in this chapter. Readers are referred to Reference 9. Finally, based on the prediction model constructed and the prediction results, conclusions are generated and managerial insights to improve the efficiency of operation and business are derived.

References

[1] Tu E., Zhang G., Rachmawati L., Rajabally E., Huang G.B. 'Exploiting AIS data for intelligent maritime navigation: a comprehensive survey from data to methodology'. *IEEE Transactions on Intelligent Transportation Systems*. 2017;*19*(5):1559–82.

[2] Yang D., Wu L., Wang S., Jia H., Li K.X. 'How big data enriches maritime research—a critical review of automatic identification system (AIS) data applications'. *Transport Reviews*. 2019;39(6):755–73.

[3] Yan R., Wang S., Peng C. 'An artificial intelligence model considering data imbalance for SHIP selection in port state control based on detention probabilities'. *Journal of Computational Science*. 2021;*48*:101257.

[4] Solorio-Fernández S., Carrasco-Ochoa J.A., Martínez-Trinidad J.F. 'A review of unsupervised feature selection methods'. *Artificial Intelligence Review*. 2020;53(2):907–48.

[5] Bergstra J., Bengio Y. 'Random search for hyper-parameter optimization'. *Journal of Machine Learning Research*. 2012;13(2):281–305.

[6] Snoek J., Larochelle H., Adams R.P. 'Practical bayesian optimization of machine learning algorithms'. *Advances in Neural Information Processing Systems*. 2012;25.

[7] Conn A.R., Scheinberg K., Vicente L.N. *Introduction to derivative-free optimization [online]*. SIAM; 2009 Jan. Available from http://epubs.siam.org/doi/book/10.1137/1.9780898718768

[8] Hutter F., Hoos H.H., Leyton-Brown K. 'Sequential model-based optimization for general algorithm configuration'. *International Conference on Learning and Intelligent Optimization*; 2011. pp. 507–23.

[9] *Interpretable machine mearning: a guide for making black box models explainable* [online]. 2022 Mar 29. Available from https://christophm.github.io/interpretable-ml-book [Accessed 31 Mar 2022].

Chapter 5
Linear regression models

Linear regression aims to learn a linear model that can predict the target using the features as accurately as possible. The assumption of linear regression models is that the target is linearly correlated with the features, i.e., the regression function $E(y|\mathbf{x})$ is linear in $\mathbf{x}$, where $E(\cdot)$ denotes expectation. If the assumption can (almost) be satisfied, linear regression can be comparable or can even outperform fancier non-linear models. The linear regression model is one of the most classic models for prediction tasks, and it is still widely used in the computer and big data era, thanks to its intuitiveness and interpretability in particular. In the following sections, we first introduce simple linear regression models (with a single feature) and the least squares method, which aims to find the optimal parameters of a linear regression model by minimizing the sum of squares of the residuals. Then, we discuss multiple linear regression (with more than one feature) and its extension. Finally, we introduce shrinkage linear regression models.

5.1 Simple linear regression and the least squares

Simple linear regression uses only one feature to predict the target. For example, we use ship age to predict the number of deficiencies of a PSC inspection. Denote the training set with n samples by $D = \{(x_1, y_1), (x_2, y_2), ..., (x_n, y_n)\}$ and the feature vector by x. Simple linear regression aims to develop a model taking the following form:

$$\hat{y}_i = wx_i + b ,$$

where $\hat{y}_i$ is the predicted target for sample i, w is the parameter weight and b is the bias. w and b need to be learned from D. Then, a natural question is: what are good w and b? Or in other words, how to find the values of w and b such that the predicted target is as accurate as possible? The key point of developing a simple linear regression model is to evaluate the difference between $\hat{y}_i$ and y_i, $i = 1, ..., n$ using the loss function and to adopt the values of w and b that minimize the loss function. In a regression problem, the most commonly used loss function is the mean squared error (MSE), where $MSE = \frac{1}{n}\sum_{i=1}^{n}(y_i - \hat{y}_i)^2$. Therefore, the learning objective of simple linear regression is to find the optimal (w^*, b^*) such that the MSE is minimized. The above idea can be presented by the following mathematical functions:

$$(w^*, x^*) = \arg\min_{(w,b)} \sum_{i=1}^{n} (y_i - \hat{y}_i)^2$$
$$= \arg\min_{(w,b)} \sum_{i=1}^{n} (y_i - wx_i - b)^2 \quad (5.1)$$

This idea is called the least squares method. The intuition behind it is to minimize the sum of lengths of the vertical lines between all the samples and the regression line determined by w and b. It can easily be shown that MSE is convex in w and b, and thus (w^*, b^*) can be found by

$$\frac{\partial MSE}{\partial w} = 2\left(\sum_{i=1}^{n} x_i \left[wx_i - (y_i - b)\right]\right) = 0$$
$$\Rightarrow w^* = \frac{\sum_{i=1}^{n} y_i(x_i - \frac{1}{n}\sum_{i=1}^{n} x_i)}{\sum_{i=1}^{n} x_i^2 - \frac{1}{n}(\sum_{i=1}^{n} x_i)^2} \quad (5.2)$$

The optimal w^* is first found by Equation (5.2), and then it can be used to calculate the optimal value of b, denoted by b^*, as follows:

$$\frac{\partial MSE}{\partial b} = 2\left(\sum_{i=1}^{n} w^* x_i + b - y_i\right) = 0$$
$$\Rightarrow b^* = \frac{1}{n}\sum_{i=1}^{n}(y_i - w^* x_i) \quad (5.3)$$

Simple linear regression can easily be realized by *scikit-learn* API [1] in Python. Here is an example of using ship age to predict ship deficiency number using simple linear regression.

Example 5.1: In this problem, the only feature is ship age and the target is ship deficiency number. (w^*, b^*) is estimated on the training set using the LinearRegression method in scikit-learn API. The core code is as follows:

```
from sklearn import linear_model
reg_def = linear_model.LinearRegression()
reg_def.fit(x_train_age, y_train_def)
print(reg_def.coef_, reg_def.intercept_)
# output: [[0.26484284]] [1.5306191]
```

Therefore, the relationship between ship age and deficiency number learned by simple linear regression can be presented in:

$$\text{deficiency number} = 0.2648 \times \text{age} + 1. \quad (5.4)$$

Model performance is then validated on the hold-out test set, which can also easily be achieved by the scikit-learn API. Finally, we have $MSE = 20.3471$, $RMSE = 4.5108$, $MAE = 2.9650$, and $R^2 = 0.0713$. Recall that the definitions of MSE, RMSE, and MAE are given in section 4.2.4 of chapter 4. Note that in this example, R^2 is very small, indicating that the model is highly likely to be ineffective, and simple linear regression might not be suitable to address this problem.

5.2 Multiple linear regression

In more general cases, the feature dimension denoted by m is more than one, i.e., two or more features are used to predict the target. For example, we use a linear regression model with five features: ship age, type, GT, last inspection time, and last deficiency number to predict the ship deficiency number of the current PSC inspection, and this model is called multiple linear regression. Mathematically, multiple linear regression can be written as follows:

$$\begin{aligned} \hat{y}_i &= w_1 x_{i1} + w_2 x_{i2} + \ldots + w_m x_{im} + b \\ &= \mathbf{x}_i \mathbf{w} + b, \\ \mathbf{x}_i &= (x_{i1}, x_{i2}, \ldots, x_{im}), \\ \mathbf{w} &= (w_1; w_2; \ldots; w_m). \end{aligned} \quad (5.5)$$

To unify the parameters, Equation (5.5) can be written as

$$\begin{aligned} \hat{y}_i &= \tilde{\mathbf{x}}_i \tilde{\mathbf{w}}, \\ \tilde{\mathbf{x}}_i &= (x_{i1}, x_{i2}, \ldots, x_{im}, 1), \\ \tilde{\mathbf{w}} &= (w_1; w_2; \ldots; w_m; b). \end{aligned} \quad (5.6)$$

In a training set D with n samples, the matrix form of Equation (5.6) is

$$\begin{pmatrix} \hat{y}_1 \\ \hat{y}_2 \\ \vdots \\ \hat{y}_n \end{pmatrix} = \begin{pmatrix} x_{11} & x_{12} & \cdots & x_{1m} & 1 \\ x_{21} & x_{22} & \cdots & x_{2m} & 1 \\ \vdots & \vdots & \ddots & \vdots & \vdots \\ x_{n1} & x_{n2} & \cdots & x_{nm} & 1 \end{pmatrix} \begin{pmatrix} w_1 \\ w_2 \\ \vdots \\ w_m \\ b \end{pmatrix}. \quad (5.7)$$

or $\hat{\mathbf{y}} = \tilde{\mathbf{x}} \tilde{\mathbf{w}}$. The dimension of $\hat{\mathbf{y}}$ is $n \times 1$, the dimension of $\tilde{\mathbf{x}}$ is $n \times (m+1)$, and the dimension of $\tilde{\mathbf{w}}$ is $(m+1) \times 1$. A closed-form solution of the optimal $\tilde{\mathbf{w}}^*$ can be found by the least squares method but the case is more complex than that in simple linear regression. The MSE is calculated by $MSE = (\mathbf{y} - \tilde{\mathbf{x}}\tilde{\mathbf{w}})^{\mathrm{T}}(\mathbf{y} - \tilde{\mathbf{x}}\tilde{\mathbf{w}})$, and its derivative to $\tilde{\mathbf{w}}$ is

$$\frac{\partial MSE}{\partial \tilde{\mathbf{w}}} = 2\tilde{\mathbf{x}}^{\mathrm{T}}(\tilde{\mathbf{x}}\tilde{\mathbf{w}} - \mathbf{y}). \quad (5.8)$$

If the matrix $\tilde{\mathbf{x}}^{\mathrm{T}}\tilde{\mathbf{x}}$ is a full-rank matrix or a positive definite matrix, $\tilde{\mathbf{w}}^*$ can be calculated by setting Equation (5.8) to zero, and we can have

$$\tilde{\mathbf{w}}^* = (\tilde{\mathbf{x}}^{\mathrm{T}}\tilde{\mathbf{x}})^{-1}\tilde{\mathbf{x}}^{\mathrm{T}}\mathbf{y}. \quad (5.9)$$

here $(\tilde{\mathbf{x}}^{\mathrm{T}}\tilde{\mathbf{x}})^{-1}$ is the inverse matrix of $\tilde{\mathbf{x}}^{\mathrm{T}}\tilde{\mathbf{x}}$. The optimal multiple linear regression model solved by least squares is

$$\hat{y}_i = \tilde{\mathbf{x}}_i(\tilde{\mathbf{x}}^{\mathrm{T}}\tilde{\mathbf{x}})^{-1}\tilde{\mathbf{x}}^{\mathrm{T}}\mathbf{y}. \quad (5.10)$$

If $\tilde{x}^T\tilde{x}$ is not a full-rank matrix, which indicates that the features outnumber the samples, then there will be several $\tilde{w}^*$, and the optimal one can be selected by introducing regularization in section 5.4.

Multiple linear regression can also be realized by *scikit-learn* API [1] in Python. Here is an example of using ship age, type, GT, last inspection time, and last deficiency number, whose details are provided in section 2.3 of chapter 2, to predict ship deficiency number using multiple linear regression.

Example 5.2: Ship type is a nominal categorical feature. We first use one-hot encoding to encode this feature to six features: type_bulk_carrier, type_container_ship, type_general_cargo/multipurpose, type_passenger_ship, type_tanker, and type_other. For ships without the last inspection record, we set −1 to their features "last_inspection_time" and "last_deficiency_no." Then, we have a total of ten features, which are of varied ranges. For example, in the whole dataset, age ranges from 0 to 46, while GT ranges from 0 to 210,678. Therefore, min-max scaling (readers are referred to section 4.2.4 of chapter 4) is used for feature scaling. It should be noted the min-max scaler should be first fit on the features in the training set, and then applied to the features in the training and test sets. Then, the training set and test set after feature scaling using a min-max scaler are used to train and test the multiple linear regression model. Similar to Example 5.1, $\tilde{w}^*$ can be estimated on the training set using the LinearRegression in scikit-learn API. In this example, the ship deficiency number is predicted by Equation (5.11), where the features have been scaled by the min-max scaler:

$$\begin{aligned}
\textit{deficiency number} =\ & 8.1786 \times age + 0.1671 \times \textit{type\_bulk\_carrier} \\
& - 0.4938 \times \textit{type\_container\_ship} + 3.9357 \times \textit{type\_general\_cargo/multipurpose} \\
& + 0.9053 \times \textit{type\_other} - 2.9534 \times \textit{type\_passenger\_ship} \\
& - 1.5609 \times \textit{type\_tanker} - 2.9414 \times GT \\
& - 2.8232 \times \textit{last\_inspection\_time} + 13.6901 \times \textit{lastdeficiency\_no} + 2.3880.
\end{aligned}$$
(5.11)

The weights of the features show their contributions to the number of ship deficiencies detected. For example, ship deficiency number increases rapidly as ship age or the last deficiency number increases. In contrast, given all other conditions equal, a larger ship has less deficiencies. The longer the last inspection time, the fewer deficiencies detected. This is because ships in worse conditions are more likely to be selected for inspection in the current SRP ship selection scheme, and thus they have shorter inspection intervals. For different types of ships, being types general_cargo/multipurpose, other, and bulk carrier increases the deficiency number, while being passenger ship, tanker, and container ship decreases the deficiency number. The above analysis shows that linear regression is totally explainable and its compliance with domain knowledge can easily be validated.

Model performance is then validated on the hold-out test set, and we have $MSE = 15.9987$, $RMSE = 3.9998$, $MAE = 2.5948$, and $R^2 = 0.2698$. It is obvious that the performance of the multiple linear regression model for ship deficiency number prediction is much better than that of the simple linear regression model in Example 5.1, as many more features that can influence the target are considered.

5.3 Extensions of multiple linear regression

This section introduces two popular extensions of multiple linear regression: polynomial regression and logistic regression. As a relatively simple and basic ML model considering a linear relationship between the features and the target, multiple linear regression is highly likely to lead to the problem of underfitting. To address this problem, one viable approach is to incorporate non-linearity into the model. To retain the linear function form (and thus model interpretability), higher order terms of the original features (e.g. quadratic, cubed, and product terms) are formed and included in the multiple linear regression model. The model formulated is called polynomial regression. Besides, multiple linear regression can only address regression problems. To extend multiple linear regression to address classification problems, the continuous output given by multiple linear regression is mapped to classes using a surrogate function. For binary classification problem, logistic function is a commonly used mapping function, and the classification model is called logistic regression. Details of the two extensions of multiple linear regressions are given in the following subsections.

5.3.1 Polynomial regression

As linear regression assumes a linear relationship between the original features and the target by constructing a straight line, underfitting occurs if the features and the target take a non-linear relationship, where the complexity of the model needs to be improved to enhance model performance. One approach is first to add powers to the original features to form new (and more complex) features, and then to form a linear function between the new features and the target. This approach is called polynomial regression, and the features constructed are called polynomial features. For example, given two features (X_1, X_2) and power degree 2, each of them will be transformed to 3 features with orders from 0 to 2 (note that both features have the same value 1 when the order is 0), respectively, in addition to their product term. Therefore, the new features will be $(1, X_1, X_1^2, X_2, X_2^2, X_1 X_2)$. Applying polynomial regression to Example 5.2 yields Example 5.3.

Example 5.3: The 6 binary features encoded from the nominal feature ship type are not processed and are directly used in polynomial regression. Min-max scaling is applied to numerical features age, GT, last inspection time, and last deficiency number, and their polynomial features of degree 2 are constructed by scikit-learn API using the following codes:

56 *Machine learning and data analytics for maritime studies*

```
from sklearn.preprocessing import PolynomialFeatures
# Specify the power degree of the polynomial features
polynomial_features= PolynomialFeatures(degree=2)
# Apply to training data
x_train_numetical_scaled_poly =
    pd.DataFrame(polynomial_features.
    fit_transform(x_train_numetical_scaled))
# Apply to test data
x_test_numetical_scaled_poly =
    pd.DataFrame(polynomial_features.
    fit_transform(x_test_numetical_scaled))
```

Then, a total of 15 new features can be constructed from the 4 original features, including 1 constant term, 4 linear terms, 4 quadratic terms, and 6 product terms. Totally, there are 21 features (15 numerical features and 6 categorical features). Then, multiple linear regression is applied to all the 21 features. We omit the specific form as it is similar to that of Example 5.2. Model performance is then validated on the hold-out test set, and we have *MSE* = 15.4055, *RMSE* = 3.9250, *MAE* = 2.5140, and R^2 = 0.2969. The performance of polynomial regression is better than that of the multiple linear regression model in Example 5.2 and much better than that of the simple linear regression model in Example 5.1, as a more complex feature form is used.

5.3.2 Logistic regression

If the prediction target is whether a ship is detained in an inspection, which is a binary variable with "1" indicating ship detention and "0," otherwise. Neither simple nor multiple linear regression model can be directly applied to this classification problem, as the output is continuous and unbounded, while we expect the output to be categorical and bounded. An intuitive method is to set a threshold to predict the probability of $y = 1$ given the input features **x**, i.e., $P(y = 1|\mathbf{x})$. The unit-step function is a popular method to map a continuous output (denoted by z) to a probability (denoted by $\tilde{y}$), which takes the following form as shown in Figure 5.1.

However, Figure 5.1 shows that the final output given by the unit-step function is discontinuous, making it hard to be optimized. Therefore, a continuous, monotonic, and differentiable surrogate function of the unit-step function called logistic function taking the following form is used:

$$\tilde{y} = \frac{1}{1 + e^{-z}}, \tag{5.12}$$

here $z = \tilde{\mathbf{x}}\tilde{\mathbf{w}}$ is the continuous output given by a multiple linear regression model. An illustration of the logistic function is shown in Figure 5.2.

Equation (5.12) can also be transformed as follows:

Linear regression models 57

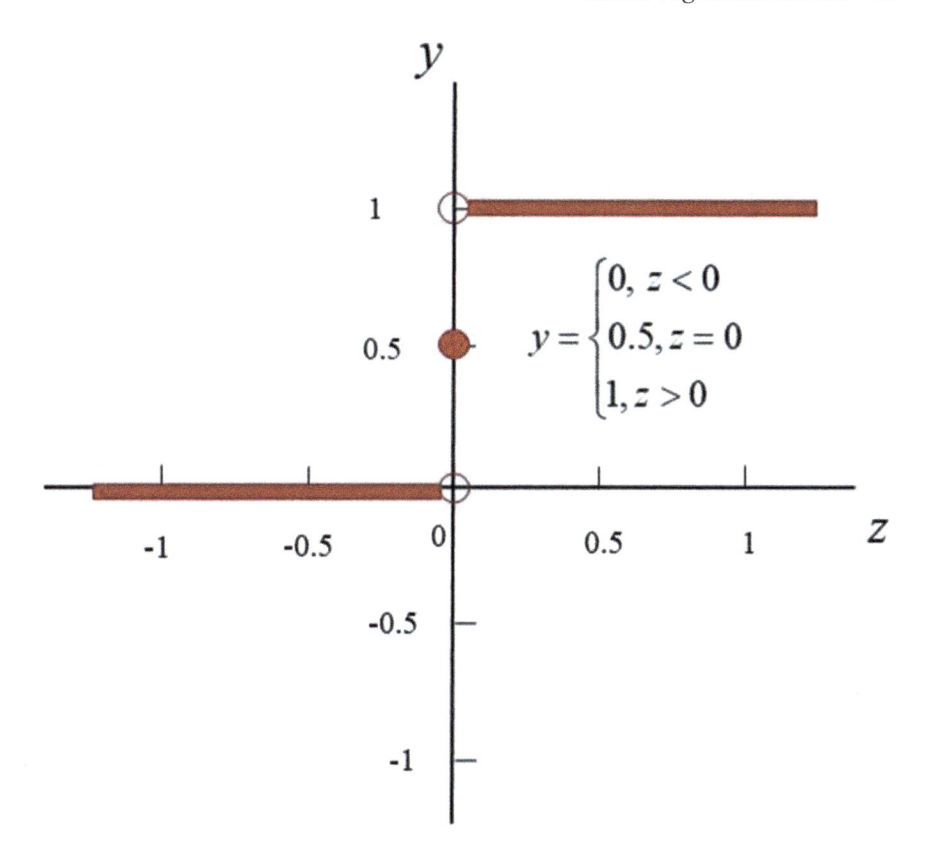

Figure 5.1 An illustration of unit-step function

$$\tilde{y} = \frac{1}{1+e^{-z}}$$
$$\Rightarrow z = \tilde{x}\tilde{w} = \ln\frac{\tilde{y}}{1-\tilde{y}}. \quad (5.13)$$

In Equation (5.13), $\tilde{y}$ is the probability of a sample with features **x** to be of class "1" and $1-\tilde{y}$ is the probability to be of class "0." Therefore, $\frac{\tilde{y}}{1-\tilde{y}}$ is the relative probability of sample **x** to be of class "1," which is called odds. In $\frac{\tilde{y}}{1-\tilde{y}}$ is the natural log of odds, and is called log odds, or logit. Therefore, Equation (5.13) can be interpreted as using the output of a multiple linear regression model to approximate the log odds, so as to map a continuous target to a probability.

Similar to simple linear regression and multiple linear regression, the optimal values of $\tilde{w}$ need to be estimated by minimizing the estimation error. Equation (5.13) can be further written as

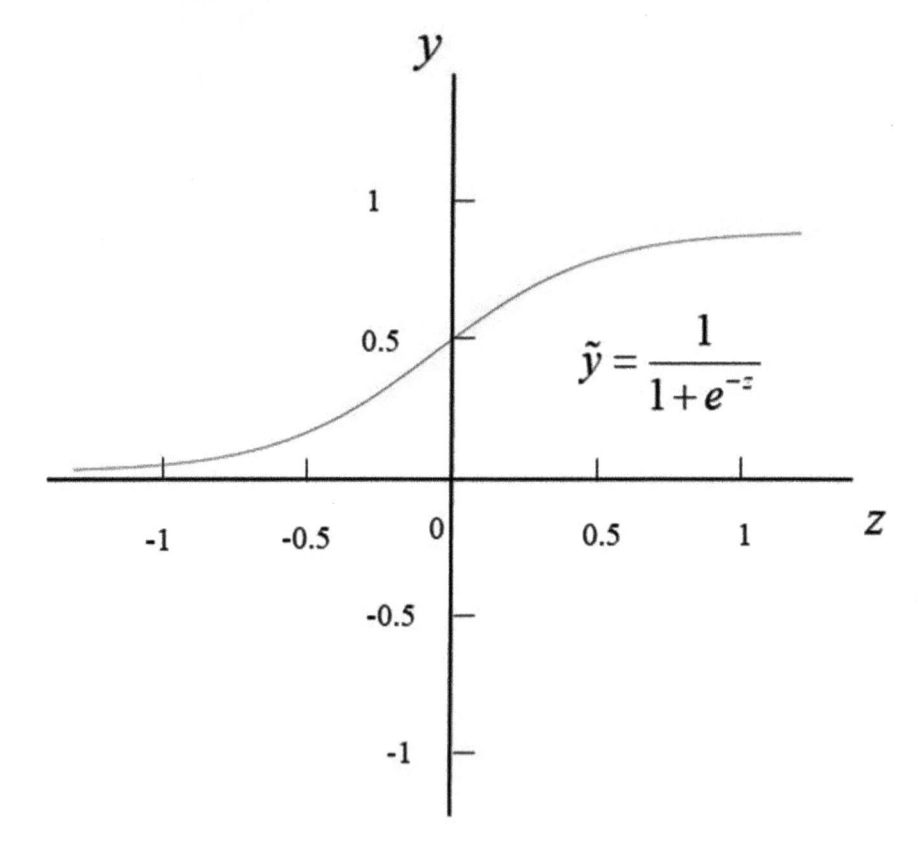

58 *Machine learning and data analytics for maritime studies*

Figure 5.2   An illustration of the logistic function

$$z = \tilde{x}\tilde{w} = \ln \frac{\tilde{y}}{1-\tilde{y}}$$
$$\Rightarrow \tilde{x}\tilde{w} = \ln \frac{P(y=1|x)}{P(y=0|x)}. \tag{5.14}$$

and we can have

$$P(y=1|x) = \frac{e^{\tilde{x}\tilde{w}}}{1+e^{\tilde{x}\tilde{w}}},$$
$$P(y=0|x) = \frac{1}{1+e^{\tilde{x}\tilde{w}}}. \tag{5.15}$$

Then, the maximum likelihood method [2] can be used to estimate $\tilde{w}$ to maximize the probability that the predicted target of each example is equal to its actual target. Algorithms in numerical optimization, such as the gradient descent method and Newton method can be used to obtain the optimal $\tilde{w}$, i.e., $\tilde{w}^*$. We use the same features of Example 5.2 to predict ship detention. The procedure and results are shown in Example 5.4.

Example 5.4: Min-max scaling is also first applied to numerical features age, GT, last inspection time, and last deficiency number. However, one critical issue in this example is that ship detention is a rare event. For example, among all the 1,592 samples in the training set, the target of only 85 of them is "1," indicating that the detention rate is about 5.34%. To overcome the problem of data imbalance, we first use the synthetic minority oversampling technique (SMOTE) to oversample the minority class, with the aim to construct a balanced dataset. Then, logistic regression is applied to predict ship detention. We use a pipeline combining SMOTE and logistics regression to first construct a balanced data set, and then predict ship detention based on scikit-learn API with main code as follows:

```
from imblearn.over_sampling import SMOTE
from imblearn.pipeline import make_pipeline
from sklearn.linear_model import LogisticRegression
pipe = make_pipeline(
SMOTE(),
LogisticRegression()
) # Construct a pipeline
pipe.fit(X_train, y_train_det) # Model fitting
y_pred = pipe.predict(X_test) # Model testing
```

Then, the constructed model is validated on the hold-out test set containing 399 examples (15 examples with detention), and the following confusion matrix given in Table 5.1 is obtained. The average *Precision*, *Recall*, and F_1 scores are 0.55, 0.79, and 0.54, respectively.

5.4 Shrinkage linear regression models

Shrinkage in linear regression models is often achieved by imposing a penalty on coefficients to reduce the absolute values of coefficients or the number of valid features in the regression model, so as to reduce the problem of overfitting, enhance the prediction accuracy, and increase model interpretability (as less features are finally contained in the model, meaning that less important features can be excluded from the model and the more important ones are left). In this section, we introduce two popular shrinkage linear regression models: ridge regression and LASSO regression.

Table 5.1   Confusion matrix for ship detention prediction using logistic regression

Predicted value/Actual value	**Detained**	**Not detained**
Detained	12	88
Not detained	3	296

5.4.1 Ridge regression

Ridge regression imposes a penalty on the size of the regression coefficients using L2 regularization, where the loss function takes the following form:

$$l = \sum_{i=1}^{n}(y_i - b - \sum_{j=1}^{m} x_{ij}w_j)^2 + \lambda \sum_{j=1}^{m} w_j^2, \text{ where } \lambda > 0. \quad (5.16)$$

λ is a complexity parameter to control the degree of shrinkage: a larger λ means a greater amount of shrinkage. The objective of ridge regression is to find the optimal $\mathbf{w}^*$ such that

$$\mathbf{w}^* = \arg\min_{\mathbf{w}} \left\{ \sum_{i=1}^{n}(y_i - b - \sum_{j=1}^{m} x_{ij}w_j)^2 + \lambda \sum_{j=1}^{m} w_j^2 \right\}. \quad (5.17)$$

This is equivalent to solving the following optimization problem:

$$\mathbf{w}^* = \arg\min_{\mathbf{w}} \left\{ \sum_{i=1}^{n}(y_i - b - \sum_{j=1}^{m} x_{ij}w_j)^2 \right\},$$
$$s.t. \sum_{j=1}^{m} w_j^2 \leq t, \quad (5.18)$$

here there is a one-to-one relationship between λ and t, and the size constraint on the parameters (i.e. constraint on parameter values) is imposed explicitly in Equation (5.18). Ridge regression is effective to alleviate the problem of high variance brought about by correlated variables in multiple linear regression by shrinking coefficients close to (but not exactly) zero. It is also noted that bias b, which is not directly related to the parameters, is excluded from the penalty terms, as they aim to regularize the coefficients of parameters. Its value should also be determined in Equation (5.18). An example of using ridge regression to predict ship deficiency number using the features of Example 5.2 based on *scikit-learn* API is as follows.

Example 5.5: Min-max scaling is also first applied to numerical features age, GT, last inspection time, and last deficiency number. Ridge regression with hyper-parameter tuning for λ based on 5-fold cross-validation can easily be implemented by the RidgeCV method provided by scikit-learn API. The main code is as follows:

```
from sklearn.linear_model import RidgeCV
# Specify the search space of alpha (i.e. lambda)
reg_def = RidgeCV(alphas=[0.01, 0.05, 0.1, 0.2, 0.5, 1, 3,
    5])
# Automatically find the best alpha value within the search
    space, and use the value to fit the ridge regression
    model using the training set
reg_def.fit(x_train_ridge_scaled,y_train_ridge)
print(reg_def.alpha_) # Output: 0.5 (which is the best
    value of lambda within the search space determined by
    leave-one-out cross validation)
print(reg_def.coef_,reg_def.intercept_)
```

The constructed ridge regression model for ship deficiency number prediction takes the following form:

deficiencynumber = 8.1786 × *age* + 0.1671 × *type_bulk_carrier*
− 0.4938 × *type_container_ship* + 3.9357 × *type_general_cargo/multipurpose*
+ 0.9053 × *type_other* − 2.9534 × *type_passenger_ship*
− 1.5609 × *type_tanker* − 2.9414 × *GT*
− 2.8232 × *last_inspection_time* + 13.6901 × *lastdeficiency_no* + 2.3880.
(5.19)

Compared to Equation(5.11) for ship deficiency number prediction using multiple linear regression, it can be seen that the absolute weight values associated with the parameters are smaller in Equation(5.19) in ridge regression. Meanwhile, the relative importance of the parameters is similar in both models. Model performance is then validated on the hold-out test set, and we have *MSE* = 15.9981, *RMSE* = 3.9998, *MAE* = 2.5948, and R^2 = 0.2698, which is slightly better than the performance of the multiple linear regression model as shown in Example 5.2.

5.4.2 LASSO regression

LASSO regression imposes L1 regularization on the size of the regression coefficients, whose loss function takes a similar form of ridge regression except for the regularization term:

$$l = \sum_{i=1}^{n}(y_i - b - \sum_{j=1}^{m} x_{ij}w_j)^2 + \lambda \sum_{j=1}^{m} |w_j|, \text{ where } \lambda > 0. \quad (5.20)$$

And it is equivalent to solving the following optimization model to find the optimal **w***:

$$\mathbf{w}^* = \arg\min_{\mathbf{w}} \left\{ \sum_{i=1}^{n}(y_i - b - \sum_{j=1}^{m} x_{ij}w_j)^2 \right\},$$
$$s.t. \sum_{j=1}^{m} |w_j| \leq t. \quad (5.21)$$

The optimal value of *b* can also be determined by Equation (5.21). As the L1 regularization term with absolute operation involved is used in LASSO regression, it can lead to zero coefficient, which means that some features are totally ignored by the regression model. The larger value of λ, the fewer features are retained. Therefore, LASSO regression can also be used for feature selection in addition to overfitting control. An example of using LASSO regression to predict ship deficiency number using the features of Example 5.2 based on *scikit-learn* API is as follows.

Example 5.6: Min-max scaling is also first applied to numerical features age, GT, last inspection time, and last deficiency number. Like ridge regression, LASSO regression with hyperparameter tuning for λ based on 5-fold cross validation can easily be implemented by the LassoCV method provided by scikit-learn API. The main code is as follows:

```
from sklearn.linear_model import LassoCV
reg_def = LassoCV(alphas=[0.005, 0.01, 0.02, 0.05, 0.1,
    0.2, 0.5, 1, 3, 5])
reg_def.fit(x_train_LASSO_scaled, y_train_LASSO)
print(reg_def.alpha_)  # Output: 0.05
print(reg_def.coef_, reg_def.intercept_)
```

The constructed LASSO regression model for ship deficiency number prediction takes the following form:

$$\text{deficiency number} = 8.0205 \times age + 0 \times type\_bulk\_carrier$$
$$- 0.6216 \times type\_container\_ship + 3.7846 \times type\_general\_cargo/multipurpose$$
$$+ 0.7083 \times type\_other - 2.7820 \times type\_passenger\_ship$$
$$- 1.6691 \times type\_tanker - 2.9266 \times GT$$
$$- 1.9954 \times last\_inspection\_time + 13.6331 \times last\_deficiency\_no + 2.5125. \quad (5.22)$$

Equation (5.22) shows that the coefficient of feature *type_bulk_carrier* is shrunk to exactly 0, indicating that the LASSO algorithm can be used for feature selection. However, it is noted that the model performance of LASSO is slightly worse than multiple linear regression and ridge regression, with $MSE = 16.0133$, $RMSE = 4.0017$, $MAE = 2.5908$, and $R^2 = 0.2691$. This shows that LASSO may not be a suitable method for ship deficiency number prediction, as in this problem, the number of examples is much larger than the number of features and these features are not highly correlated with each other.

From the results of Examples 5.5 and 5.6, it can be seen that ridge regression tends to produce small but non-zero weights, while LASSO tends to produce sparse weights (to exactly zero). Therefore, when there are too many features, especially when the number of features is larger than the number of examples, LASSO regression might be more suitable to be applied for feature selection. Otherwise, ridge regression might be more suitable to be applied for feature coefficient regulation.

References

[1] Pedregosa F., Varoquaux G., Gramfort A., *et al.* 'Scikit-learn: machine learning in python'. *Journal of Machine Learning Research*. 2011;12:2825–30.
[2] Rossi R.J. *Mathematical statistics* [online]. Hoboken, NJ: John Wiley & Sons; 2018 Jul 18. Available from http://doi.wiley.com/10.1002/9781118771075

Chapter 6

Bayesian networks

This chapter introduces the basics of Bayesian network (BN) classifiers that are used to address classification problems. Naive Bayes classifier is first presented, where a simplified (but unrealistic) assumption that the features are conditionally independent and are of equal importance is made. To weaken the assumption so as to improve the classification accuracy, semi-naive Bayes classifiers are then presented, where part of the dependencies between the features is considered. Finally, BN in more general form is introduced.

6.1 Naive Bayes classifier

Consider a classification problem where the target has a total of K possible values denoted by $\mathbf{C} = \{c_1, ..., c_K\}$. We have a set of samples (observations) $D = \{(\mathbf{x}_i, y_i), i = 1, ..., n\}$, where $\mathbf{x}_i \in R^m$ and $y_i \in \mathbf{C}$. For observations whose set of features is $\mathbf{x}_i$, due to the limited types of features considered and the errors in measurement, they may be in different classes in the dataset D. Therefore, for an unseen observation with a feature set $\mathbf{x}_i$, the probability of this observation to be of class c_k can be estimated by Bayes' rule as follows:

$$P(c_k|\mathbf{x}_i) = \frac{P(\mathbf{x}_i, c_k)}{P(\mathbf{x}_i)} = \frac{P(c_k)P(\mathbf{x}_i|c_k)}{P(\mathbf{x}_i)}, \tag{6.1}$$

where $P(c_k)$ is the prior probability of class c_k, which can be calculated directly from the targets of the given dataset. $P(c_k)$ is the weight given to the class conditional probability or the likelihood of sample $\mathbf{x}_i$ given class c_k, i.e., $P(\mathbf{x}_i|c_k)$ is the evidence, which is the occurrence probability of the observation $\mathbf{x}_i$ and acts as a normalizing constant for the joint probability $P(\mathbf{x}_i, c_k)$ calculated. The main reason of using is that the dataset is limited. Consequently, for a given example $\mathbf{x}_i$, there can be rare or even no example with the same feature values as $\mathbf{x}_i$ in the training set especially when there are a lot of features. In naive Bayes classifier, a strong assumption is made to simplify the prediction of $P(c_k|\mathbf{x}_i)$: for a known class, all the features are assumed to be independent and of equal importance. Therefore, (6.1) can be written as

$$P(c_k|\mathbf{x}_i) = \frac{P(\mathbf{x}_i, c_k)}{P(\mathbf{x}_i)}$$
$$= \frac{P(c_k)}{P(\mathbf{x}_i)} \prod_{j=1}^{m} P(x_{ij}|c_k). \tag{6.2}$$

The predicted target given the set of features $\mathbf{x}_i$ should be the class leading to the largest value of Equation (6.2). For a given $\mathbf{x}_i$, $P(\mathbf{x}_i)$ is fixed. The classification problem is turned to estimate the product of $P(c_k)$ and the class conditional probability $P(\mathbf{x}_i|c_k)$, and find the class c_k leading to the largest product as the predicted target $\hat{y}_i$. Mathematically, this process can be presented as

$$\hat{y}_i = \arg\max_{c_k \in C} P(c_k) \prod_{j=1}^{m} P(x_{ij}|c_k). \tag{6.3}$$

$P(c_k)$ can be calculated by $P(c_k) = \frac{|D_{c_k}|}{|D|}$, where $|D_{c_k}|$ is the number of samples in class c_k, $|D|$ is the total number of samples in the dataset, and $P(x_{ij}|c_k)$ is the conditional probability of each feature j, $j = 1, \ldots, m$ of sample $\mathbf{x}_i$. Especially, for categorical features, we denote the number of samples belonging to class c_k and taking value j' for feature j by $D_{c_k, j, j'}$. Then, $P(x_{ij}|c_k)$ can be calculated by

$$P(x_{ij}|c_k) = \frac{D_{c_k, j, j'}}{|D_{c_k}|}. \tag{6.4}$$

We use an example considering five features, namely ship age (age), type, last inspection time (l_ins_time), last deficiency number (l_def_no), and last inspection state (whether a ship is detained in the last inspection; l_ins_state) to predict whether a new coming ship has six or more deficiencies. Therefore, as the target variable, the deficiency number (denoted by def_no for short) is decoded to two states: 0to5 and 6+. The training set contains 200 random samples from the whole port state control (PSC) dataset.

Example 6.1 Age, l_ins_time, and l_def_no are numerical features. For the sake of simplicity, we treat them as categorical features by discretizing their values. The values for age after discretization are 0to5, 6to10, 11to15, and 16+. The values for l_ins_time after discretization are no_inspection, 0to5, 6to10, 11to15, and 16+. The values for l_def_no after discretization are no_inspection, 0to5, and 6+. For the two categorical features type and l_ins_state, they have five states (bulk_carrier, container_ship, general_cargo/multipurpose, tanker, and other) and three states (no_inspection, no_detention, and detention), respectively. For the three features related to last inspection, namely l_ins_time, l_def_no, and l_ins_state, if a ship has no last inspection record in the database, its state is set to "no inspection." According to Equation (6.3), we first calculate the prior probability $P(c_k)$ for each class $c_k \in C$. The frequency and distribution of the target are shown in Table 6.1.

Then, the conditional probability table, or the CPT, for each feature given the target is calculated and presented as follows in Tables 6.2–6.6.

Table 6.1   Frequency and distribution of def_no

State of def_no	Frequency	Proportion (%)
0to5	154	76.733
6+	46	23.267

Suppose here comes a 3-year-old (the value of age is 0to5) tanker ship (the value of type is tanker), the last inspection time is four months ago (the value of l_ins_time is 0to5) with eight deficiencies identified (the state of l_def_no is 6+) and is not detained (the state of l_ins_state is no detention). Then, the product of the probability that the ship has 6+ deficiencies and the conditional probabilities of the feature values given the condition that deficiency_no is 6+ can be calculated by

$$P(def\_no = 6+) \times P(type = tanker|def\_no = 6+) \times P(age = 0to5|def\_no = 6+)$$
$$\times P(l\_ins\_time = 0to5|def\_no = 6+) \times P(l\_def\_no = 6+|def\_no = 6+)$$
$$\times P(l\_ins\_state = nodetention|def\_no = 6+)$$
$$= 0.23267 \times 0.09804 \times 0.08 \times 0.4902 \times 0.26531 \times 0.87755$$
$$\approx 2.0827 \times 10^{-4}. \tag{6.5}$$

Similarly, the product of the probability that the ship has 0 to 5 deficiencies and the conditional probabilities of the feature values given the condition that the state of deficiency_no is 0to5 can be calculated by

$$P(def\_no = 0to5) \times P(type = tanker|def\_no = 0to5) \times P(age = 0to5|def\_no = 0to5)$$
$$\times P(l\_ins\_time = 0to5|def\_no = 0to5) \times P(l\_def\_no = 6+|def\_no = 0to5)$$
$$\times P(l\_ins\_state = nodetention|def\_no = 0to5)$$
$$= 0.76733 \times 0.11950 \times 0.25949 \times 0.42767 \times 0.05733 \times 0.94268$$
$$\approx 5.4995 \times 10^{-4}. \tag{6.6}$$

Table 6.2   Frequency and distribution of age

State of def_no	State of age 0to5	6to10	11to15	16+
0to5	40 (25.949%)*	57 (36.709%)	32 (20.886%)	25 (16.456%)
6+	3 (8%)	12 (26%)	9 (20%)	22 (46%)

*The format of this cell is "frequency (proportion)." This applies for the remaining cells of this table and the following tables.

66 *Machine learning and data analytics for maritime studies*

Table 6.3 Frequency and distribution of l_def_no

State of def_no	Value of l_def_no		
	0to5	6+	No inspection
0to5	141 (90.446%)	8 (5.733%)	5 (3.822%)
6+	32 (67.347%)	12 (26.531%)	2 (6.122%)

Table 6.4 Frequency and distribution of l_ins_state

State of def_no	Value of l_ins_state		
	Detention	No detention	No inspection
0to5	147 (94.268%)	2 (1.911%)	5 (3.822%)
6+	42 (87.755%)	2 (6.122%)	2 (6.122%)

Table 6.5 Frequency and distribution of l_ins_time

State of def_no	Value of l_ins_time				
	0to5	6to10	11to15	16+	No inspection
0to5	67 (42.767%)	55 (35.220%)	17 (11.321%)	10 (6.918%)	5 (3.774%)
6+	24 (49.020%)	12 (25.490%)	2 (5.882%)	6 (13.725%)	2 (5.882%)

As 5.4995×10^{-4} is larger than 2.0827×10^{-4}, it is predicted that the coming tanker ship has 0 to 5 deficiencies.

The above classification process can be graphically shown in the form of a BN, which is a directed acyclic graph consisting of a set of nodes and a set of directed arcs. Nodes in a BN can be features, latent variables (i.e. variables that are not directly observed), and the target to be predicted. The target is the

Table 6.6 Frequency and distribution of type

State of def_no	Value of type				
	Bulk carrier	Container ship	general_cargo/ multipurpose	Tanker	Other
0to5	12 (8.176%)	110 (69.811%)	8 (5.660%)	18 (11.950%)	6 (4.403%)
6+	12 (25.490%)	13 (27.451%)	11 (23.529%)	4 (9.804%)	6 (13.726%)

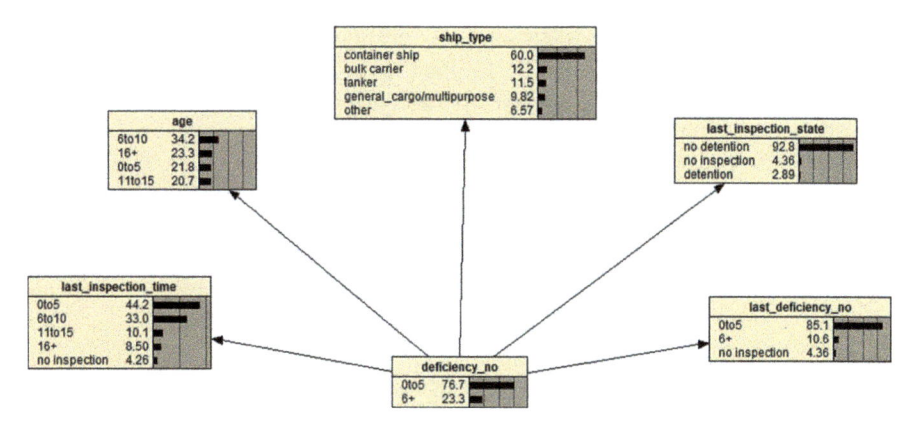

Figure 6.1 Construction of a naive Bayes classifier for ship deficiency condition prediction

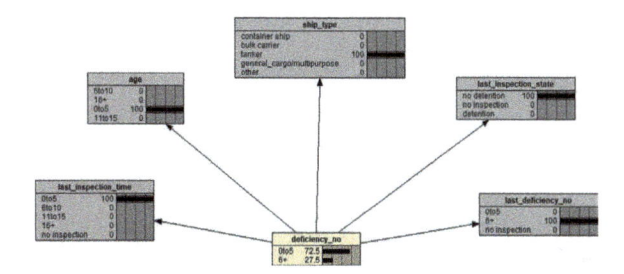

Figure 6.2 The constructed naive Bayes classifier for ship deficiency condition prediction

class variable, and the features are the attribute variables. The directed arcs connecting two nodes represent conditional dependencies, where the node at the tail of an arc is the parent node and the node at the head is the child node. A parent node represents the condition, and its child node is the consequence of that condition. The BN model to address the ship deficiency number prediction problem in Example 6.1 can be constructed by the software Netica* and is shown in Figure 6.1. A box represents a node, and the title in the first line of a box is the name of a feature or the target. The following lines are their values (or states). The marginal distributions of the features or the target are given by the belief bars on the right of the states.

For the coming tanker ship, clicking each of its feature values in Figure 6.1 yields the probability of the ship to have 0 to 5 deficiencies or 6 and more deficiencies as

*https://www.norsys.com/netica.html

shown by the states of node "deficiency_no" in Figure 6.2. The probability of this ship to have 0 to 5 deficiencies is 72.5%, and the probability of this ship to have 6 or more deficiencies is 27.5%. Therefore, it is predicted that the ship has no more than five deficiencies in the current inspection.

6.2 Semi-naive Bayes classifiers

As estimating $P(\vec{x}|c_k)$, $k = 1, ..., K$ is non-trivial, a strong assumption that the features are independent given the class variable is made in the naive Bayes classifier. Although it is an oversimplified assumption and can be unrealistic in many practical situations, the naive Bayes classifier exhibits competitive accuracy compared to other classifiers, such as logistic regression and decision tree, especially when the features are not highly correlated. Meanwhile, great efforts have been made to improve its performance. One common approach is to partially relax the feature independence assumption in naive Bayes, and the resulting model is called a semi-naive Bayes classifier. A review and test of a group of semi-naive Bayes classifiers are given in [1], where the authors subdivide the semi-naive Bayes classifiers into two groups: the first group uses a new feature set to construct naive Bayes classifiers, and the second group adds explicitly directed arcs between nodes to represent their interdependence. Especially, the new set of features in the first group is achieved by deleting and joining features, and the interdependence of features in the second group is achieved by introducing the concept of x-dependence BN classifier, where x is the maximum number of features allowed to be dependent on each feature in addition to the class. Obviously, naive Bayes is a 0-dependence BN classifier. We summarize the typical semi-naive Bayes classifiers in these two groups in Table 6.7.

In what follows, we use TAN, which is proposed by Friedman et al. [2], as an example to show the process of constructing a semi-naive Bayes classifier. Similar to the construction of a naive Bayes classifier, all the numerical features and the target if it is numerical are discretized. The procedure includes the construction of the qualitative part and quantitative part of a TAN classifier. In particular, the qualitative part refers to the network structure of a TAN classifier and the quantitative part refers to the CPTs of the nodes. Constructing the qualitative part of the TAN classifier has the following six steps.

Step 1 Specify the prediction target as the class variable and the features as the attribute variables.

Step 2 The conditional mutual information between each two of the features (denoted by A_i and A_j, and $A_i \neq A_j$) given the class variable (denoted by C) is calculated (denoted by $I(A_i; A_j|C)$) to identify the interdependence of the features.

Step 3 Build a complete undirected graph with attribute variables as the nodes and the conditional mutual information $I(A_i; A_j|C)$ as the weight of the arc between nodes A_i and A_j.

Table 6.7 Summary of typical semi-naive Bayes classifiers

Semi-naive Bayes classifier	Group	Basic idea
Backward sequential elimination (BSE)	Group 1	Starting from the full feature set, BSE eliminates the features one by one as long as the classification accuracy of the constructed naive Bayes classifier can be improved according to leave-one-out cross-validation.
Forward sequential elimination (FSE)	Group 1	Starting from an empty feature set, FSE adds the features that can improve the classification accuracy the most one by one to a naive Bayes classifier according to leave-one-out cross-validation.
BSE and joining (BSEJ)	Group 1	A new Cartesian product of features is formulated in BSEJ by using prediction accuracy as the merging criterion. Then, BSEJ repeatedly adds or deletes the original and newly constructed features so as to improve the prediction accuracy as much as possible according to leave-one-out cross-validation.
Tree augment naive Bayes (TAN)	Group 2	TAN uses an intermediate solution between naive Bayes and unrestricted BN. In TAN, one feature is allowed to depend on at most another feature in addition to the class. Therefore, TAN is a type of 1-dependence BN classifier. Conditional mutual information is used to evaluate feature interdependence.
Naïve-Bayes Decision Tree (NBTree)	Group 2	NBTree is a hybrid approach that combines a naive Bayes classifier and decision tree. The training data are partitioned using a tree structure and a naive Bayes classifier is constructed in each leaf node of the tree.

70 Machine learning and data analytics for maritime studies

Step 4 Build the maximum weighted spanning tree by sorting the weights of the arcs in descending order and then select the arcs from the largest weight to the smallest weight to link the nodes without forming a circle. That is, if adding the current largest weighted arc to the nodes forms a circle, it will not be selected anymore. Then, the availability of the arc with the second largest weight will be tested. Keep the chosen arcs and delete the others.

Step 5 Transform the undirected spanning tree into a directed tree by choosing an arbitrary feature as the root node and setting the direction of all arcs to other attribute variables to be outward from it.

Step 6 Add the class variable to the tree and arcs from the class variable to all attribute variables. The construction of the qualitative part of a TAN classifier is completed.

Especially, in Step 2, the conditional mutual information of attribute variables A_i and A_j can be calculated by

$$I(A_i; A_j | C) = \sum_{s'=1}^{N_i} \sum_{s''=1}^{N_j} \sum_{s=1}^{N_c} P(a_{i,s'}, a_{j,s''}, C_s) \log \frac{P(a_{i,s'}, a_{j,s''}|C_s)}{P(a_{i,s'}|C_s)P(a_{j,s''}|C_s)}, \qquad (6.7)$$

here "log" means the logarithmic operation with base 2. N_i, N_j, and N_c are the numbers of the possible values of A_i, A_j, and C, respectively, and $a_{i,s'}$, $a_{j,s''}$, and C_s are specific values of A_i, A_j, and C, respectively. $P(a_{i,s'}, a_{j,s''}, C_s)$ is the joint probability, and $P(a_{i,s'}, a_{j,s''}|C_s)$, $P(a_{i,s'}|C_s)$ and $P(a_{j,s''}|C_s)$ are the conditional probabilities. $P(a_{i,s'}, a_{j,s''}, C_s)$ should be understood as the proportion of samples in the whole training set whose states of attribute variables A_i and A_j and the class variable C are $a_{i,s'}$, $a_{j,s''}$, and C_s, respectively. Similarly, $P(a_{i,s'}, a_{j,s''}|C_s)$ means that among all the samples in the training set whose target is C_s, the proportion of samples whose states of attribute variables A_i and A_j are $a_{i,s'}$ and $a_{j,s''}$, respectively. The range of $I(A_i; A_j|C)$ is $[0, +\infty]$, where a large value of $I(A_i; A_j|C)$ shows that attribute variables A_i and A_j are more strongly correlated.

After finishing the construction of the qualitative part of the TAN classifier, its quantitative part is then addressed. The quantitative part has two components: the marginal distribution and the CPT of each variable, and both can be learned from the training set. The constructed TAN classifier can then be used to address classification tasks. An example of developing a TAN classifier to predict whether a new coming ship has six or more deficiencies is given in Example 6.2.

Example 6.2 The feature processing method is the same as that of Example 6.1. Netica software is used to construct the TAN classifier. The qualitative part of the TAN classifier is shown in Figure 6.3 and the complete TAN classifier is shown in Figure 6.4. Figure 6.3 shows that ship type is selected as the root variable, which only has the class variable as its parent. In addition, the last inspection time is highly related to age and the last deficiency number, while the last deficiency number is highly related to the last inspection state. It is noted that the marginal distributions shown in Figure 6.4 are a little different from those

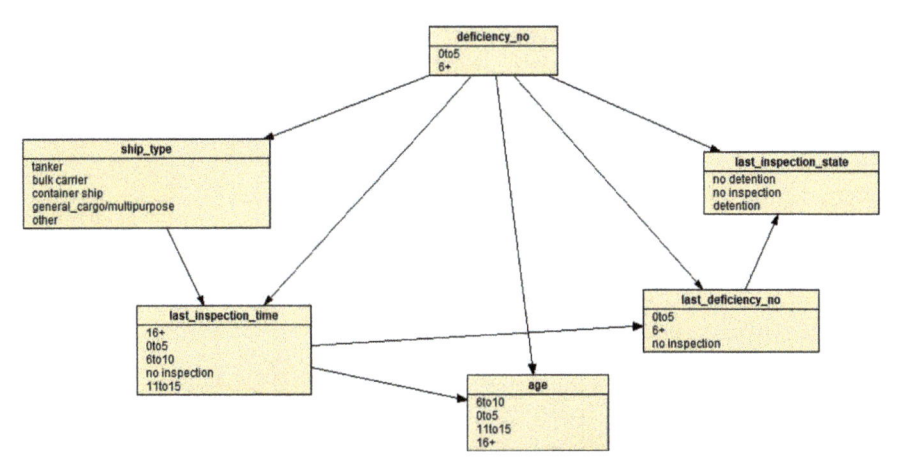

Figure 6.3   The qualitative part of the TAN classifier for ship deficiency condition prediction

shown in Figure 6.2 except for nodes deficiency_no and ship_type, where node deficiency_no has no parent node and ship_type has only deficiency_no as its parent node in both figures. This is because the marginal distribution of a node is calculated based on its CPT, i.e., based on its conditional distributions, in software Netica, which would be influenced by its parent nodes. Therefore, when the parent variable(s) of an attribute variable change(s), it is likely that its marginal

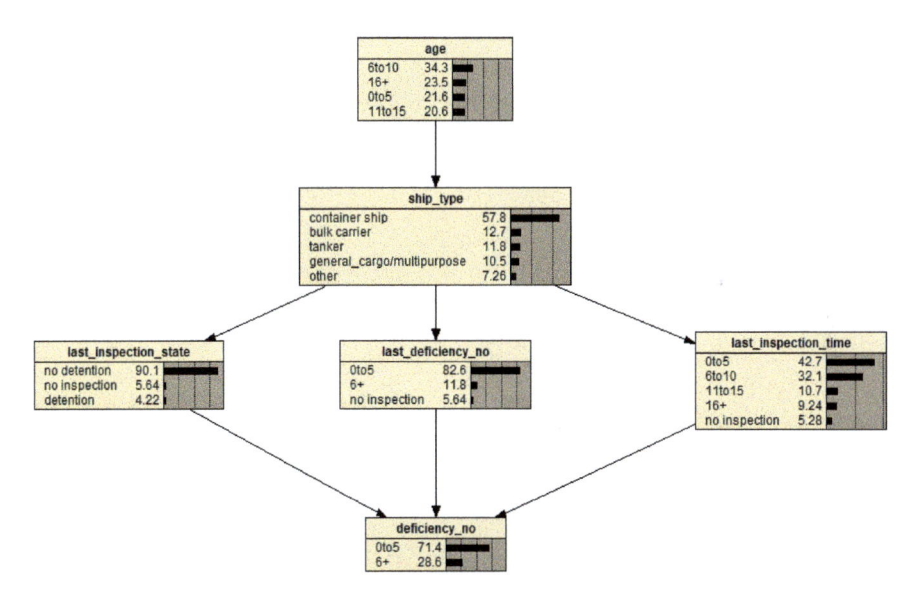

Figure 6.4   The constructed TAN classifier for ship deficiency condition prediction

Table 6.8 Distribution of last_inspection_time

deficiency_no	ship_type	0 to 5 (%)	6 to 10 (%)	11 to 15 (%)	16+ (%)	No inspection (%)
0to5	container ship	44.348	38.261	9.565	5.217	2.609
0to5	bulk carrier	35.294	35.294	17.647	5.882	5.882
0to5	tanker	26.087	21.739	21.739	17.391	13.043
0to5	general_cargo/multipurpose	61.539	7.692	15.385	7.692	7.692
0to5	other	9.091	36.364	9.091	27.273	18.182
6+	container ship	44.444	16.667	16.667	11.111	11.111
6+	bulk carrier	47.059	23.529	5.882	17.647	5.882
6+	tanker	22.222	33.333	11.111	11.111	22.222
6+	general_cargo/multipurpose	37.500	31.250	6.250	18.750	6.250
6+	other	45.455	18.182	9.091	18.182	9.091

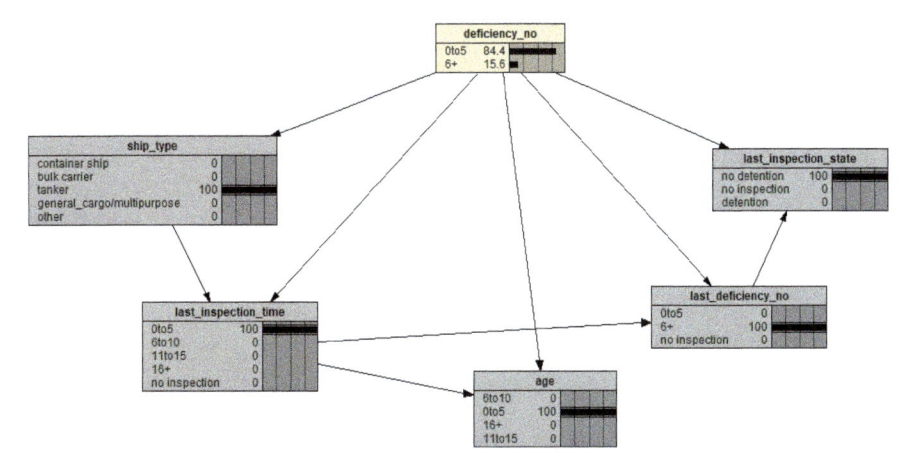

Figure 6.5   The TAN classifier for ship deficiency condition prediction

distribution as shown in the TAN classifier is slightly changed. However, this does not mean that those nodes' underlying marginal distributions are changed. Actually, their marginal distributions are fixed given a dataset. As an illustration, we present the CPT of node last_inspection_time in Table 6.8.

If the constructed TAN classifier is used to predict the deficiency condition of the new coming 3-year-old tanker ship whose last inspection time is four months ago with eight deficiencies identified and without detention, the result is shown in Figure 6.5.

It shows that the coming tanker ship has an 84.4% probability to have 0 to 5 deficiencies and a 15.6% probability to have no less than 6 deficiencies. Therefore, the tanker ship is predicted to have 0 to 5 deficiencies in the current PSC inspection.

6.3   BN classifiers

Naive Bayes classifier and semi-naive Bayes classifier are special cases of BN classifiers, where no interdependence and partial interdependence of the features are considered by them, respectively. In general BN classifiers, the interdependence of each pair of features are presented by directed arcs connecting the attribute variables. An example of the BN classifier to predict ship deficiency conditions using the same features as Examples 6.1 and 6.2 are shown in Figure 6.6.

The local Markov property [3] is satisfied in BN classifiers, where a node is conditionally independent of its non-descendants given its parents. Therefore, the calculation of joint probability is simplified and is shown in **Equation** (6.8), where only each node and the parent node(s) of the node need to be considered. This property can largely reduce the computation power required to calculate CPTs and to predict the target, especially in large BNs.

74 *Machine learning and data analytics for maritime studies*

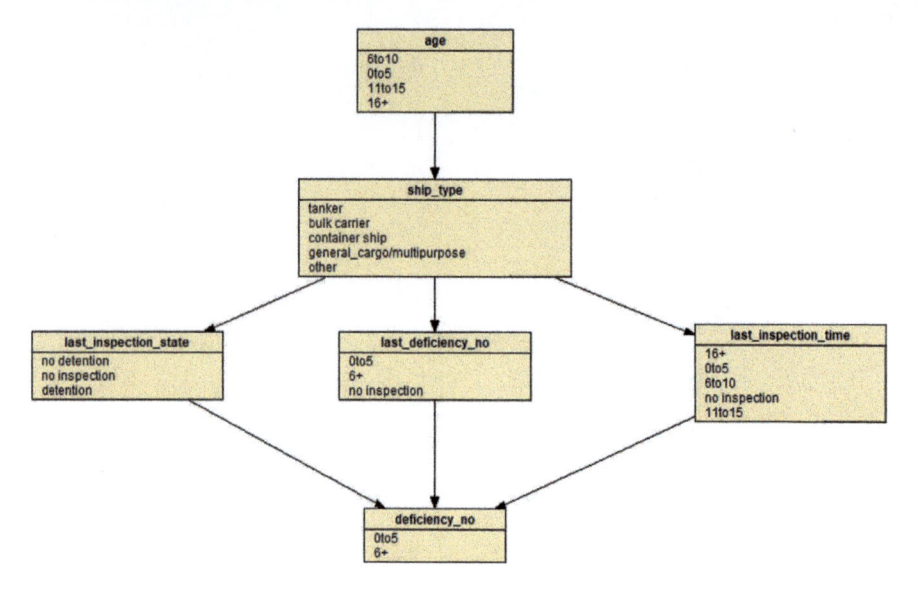

Figure 6.6 The structure of a BN classifier for ship deficiency condition prediction

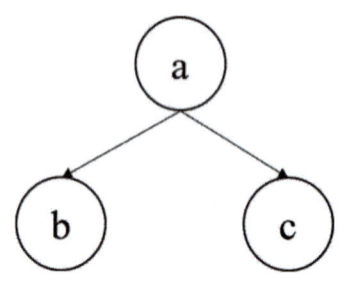

Figure 6.7 Common parent structure in BN classifier

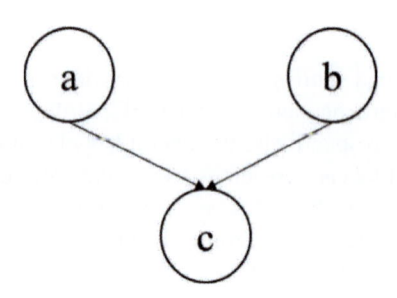

Figure 6.8 V-shape structure in BN classifier

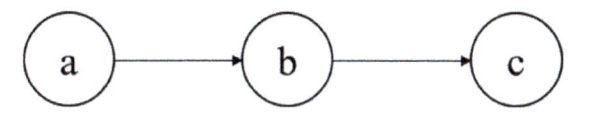

Figure 6.9   Sequential structure in BN classifier

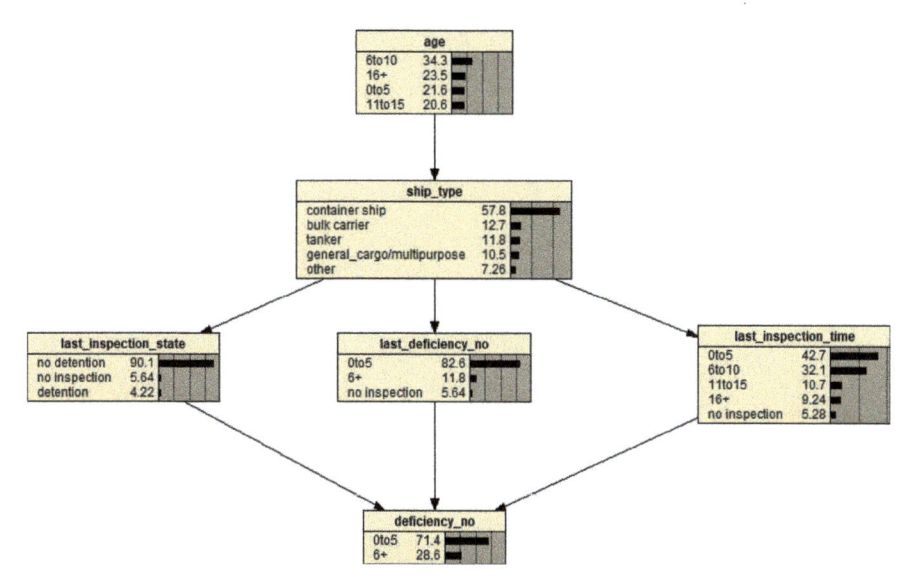

Figure 6.10   The structure of a BN classifier for ship deficiency condition prediction

$$P(c_k, x_{i1}, ..., x_{im}) = P(c_k|\text{parent}(c_k)) \times \prod_{j=1}^{m} P(x_{ij}|\text{parent}(x_{ij})), \quad (6.8)$$

Where parent(·) is the set of the parent node(s) of the node in the parentheses. If parent(x_{ij}) = Ø, then $P(x_{ij}|\text{parent}(x_{ij})) = P(x_{ij})$.

There are three typical interdependence relationships of three nodes in the BN classifier, which are also presented in Figure 6.6: common parent structure, where at least two child nodes have the same parent node as shown in nodes ship_type, last_inspection_state, and last_deficiency_no; V-shape structure, where one child node has at least two parent nodes as shown in nodes last_inspection_state, last_deficiency_no, and deficiency_no; and sequential structure, where three nodes are connected sequentially as shown in nodes age, ship_type, and last_deficiency_no. The calculation of the joint probability as well as the insights generated in each of the three typical structures is shown in Figures 6.7–6.9.

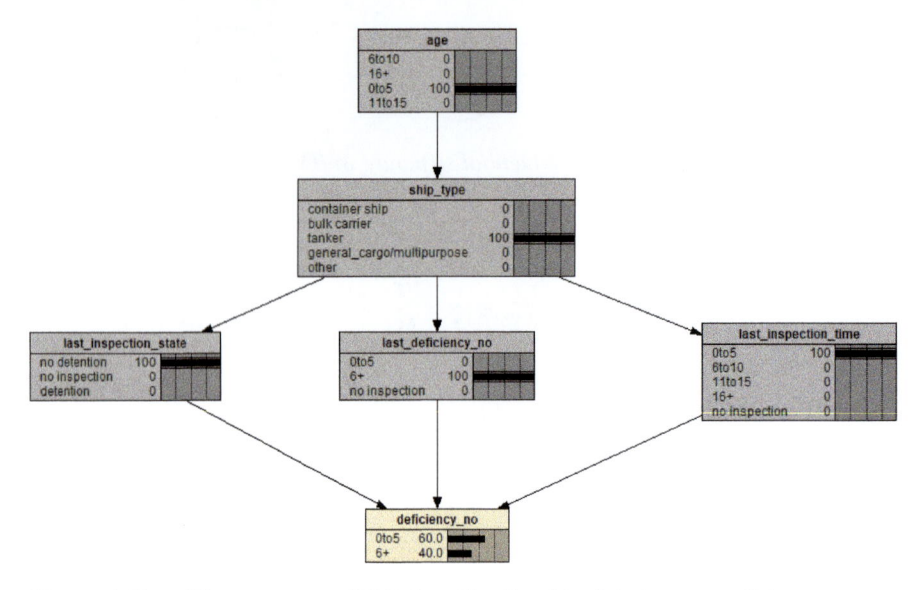

Figure 6.11 The constructed BN classifier for ship deficiency condition prediction

Common parent structure

$$P(a, b, c) = P(a)p(b|a)p(c|a),$$
$$P(b, c|a) = \frac{P(a, b, c)}{p(a)}, \tag{6.9}$$
$$\Rightarrow P(b, c|a) = P(b|a)P(c|a).$$

This shows that nodes b and c are conditionally independent given their parent node a.

V-shape structure

Given the value of node c, nodes a and b are interdependent as node c is influenced by both of them. However, when the value of node c is unknown, we have

$$\begin{aligned} P(a, b) &= \sum_c P(a, b, c), \\ &= \sum_c P(c|a, b)P(a)P(b), \\ \Rightarrow P(a, b) &= P(a)P(b), \end{aligned} \tag{6.10}$$

which shows that nodes a and b are independent with each other.

Sequential structure

$$\begin{aligned} P(a, c|b) &= \frac{P(a, b, c)}{P(b)}, \\ &= \frac{P(b)P(a|b)P(c|b)}{P(b)}, \\ &= P(a|b)P(c|b). \end{aligned} \tag{6.11}$$

which shows that nodes *a* and *c* are conditionally independent given the middle node *b*.

The BN presented in Figure 6.6 can also be used to predict ship deficiency condition. The BN classifier learned from the training set is shown in Figure 6.10. The prediction result of the new coming tanker ship is given in Figure 6.11, which also suggests that the new coming tanker ship is more likely to have 0 to 5 deficiencies.

References

[1] Zheng F., Webb G.I. 'A comparative study of semi-naive bayes methods in classification learning'. *AUSDM05*. 2005:141–55.
[2] Friedman N., Geiger D., Goldszmidt M. 'Bayesian network classifiers'. *Machine Learning*. 1997;**29**(2/3):131–63.
[3] Sebastiani P., Abad M.M., Ramoni M.F. 'Bayesian networks'. *Data Mining and Knowledge Discovery Handbook*. 2009:175–208.

Chapter 7

Support vector machine

This chapter first introduces one of the most popular machine learning models for classification tasks called support vector machine (SVM). Then, kernel trick to improve its prediction accuracy while reducing the computation burden is discussed. The extension of SVM to address regression tasks, which is called support vector regression, is then presented.

For a classification problem on data set $D = \{(\mathbf{x}_i, y_i), i = 1, ..., n\}$, where $\mathbf{x} \in R^m$ is the vector of features and $y \in C$ is categorical and to be predicted, the basic idea of the SVM algorithm is to find a proper hyperplane that can accurately separate examples in different classes. Consider a binary classification problem, i.e., $y \in \{c_1, c_2\}$, and $\mathbf{x}_i$ is two-dimensional, i.e., $\mathbf{x} = (x_1, x_2)$. A hyperplane that can distinguish the two classes of examples, where squares represent examples in positive class and circles represent examples in negative class, is shown by the lines in Figure 7.1.

Figure 7.1 shows that hyperplanes that can separate the two classes of examples are not unique; instead, the number of them can be infinite. Then, how to choose the most proper one? Intuitively, a hyperplane that has the maximum distance from the examples in both classes is the best, as it is more tolerant to examples close to the hyperplane, which are more likely to be misclassified, and thus is more robust to interference or noises in data. An SVM model aims to find such a hyperplane, as the hyperplane with the maximum margin can be expected to have the highest classification accuracy on unseen data, especially for the examples close to it. Therefore, in Figure 7.1, hyperplane h_0 should be the most suitable one. The dimension of the hyperplane depends on the number of features: if the number of features is m, then the hyperplane is of $(m-1)$ dimensions. In what follows, we consider the case of binary classification (where 1 indicates a positive example and -1 indicates a negative example) with two features, where the hyperplane is a line.

7.1 Hard margin SVM

We first consider the case where examples in the two classes are linearly separable, that is, they can be perfectly separated by a straight line. The hyperplane that can perfectly separate the two classes is a hard margin hyperplane, which is presented by

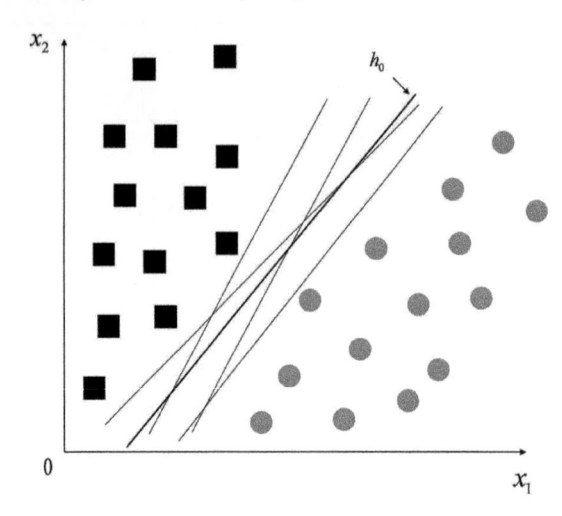

Figure 7.1 Hyperplanes to separate two classes of examples

$$\mathbf{w}^T\mathbf{x} + b = 0. \quad (7.1)$$

where $\mathbf{w} = (w_1, w_2)$ is the normal vector of the hyperplane determining its direction and b is the offset determining the distance between the hyperplane and the origin. A hyperplane is determined by the normal vector and the offset, and we denote it by $(\mathbf{w}, b)$ for simplicity. For example $\mathbf{x}$, its distance to $(\mathbf{w}, b)$ is

$$d(\mathbf{x}) = \frac{|\mathbf{w}^T\mathbf{x} + b|}{\|\mathbf{w}\|}. \quad (7.2)$$

For the positive and negative examples in data set D, the following constraints are satisfied to correctly classify them:

$$\begin{aligned} \mathbf{w}^T\mathbf{x}_i + b \geq +1, \; y_i = +1; \\ \mathbf{w}^T\mathbf{x}_i + b \leq -1, \; y_i = -1. \end{aligned} \quad (7.3)$$

Visually, these hyperplanes are shown in Figure 7.2. For all the positive examples, $\mathbf{w}^T\mathbf{x}_i + b \geq +1$ is satisfied, and the positive examples that are the closest to hyperplane h_o satisfy $\mathbf{w}^T\mathbf{x}_i + b = +1$. For negative examples, $\mathbf{w}^T\mathbf{x}_i + b \leq -1$ is satisfied, and the negative examples that are the closest to hyperplane h_o satisfy $\mathbf{w}^T\mathbf{x}_i + b = -1$. Examples in both classes that are the closest to the hyperplane are marked in black (the other examples are in gray) in Figure 7.2. These examples are called support vectors. The margin is twice the distance between a support vector to the hyperplane, which can be calculated by

$$\gamma = \frac{2}{\|\mathbf{w}\|}. \quad (7.4)$$

According to the basic idea of SVM where the maximum margin should be found, the problem is then turned to solve the following optimization model:

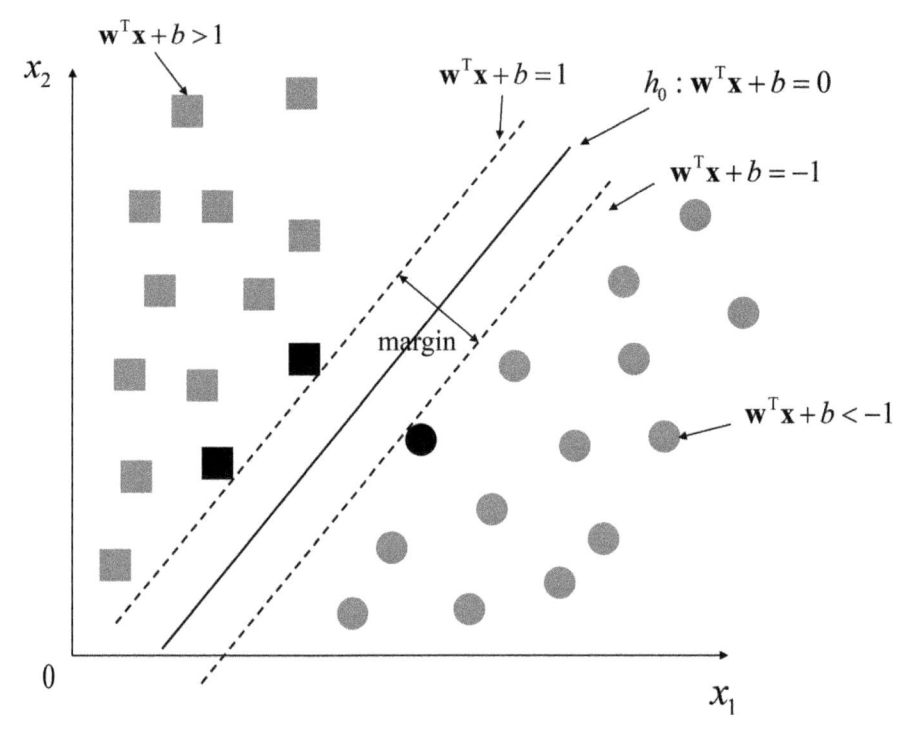

Figure 7.2 Hyperplane with the maximum margin and the support vectors

$$\max \gamma = \frac{2}{\|\mathbf{w}\|}. \tag{7.5}$$

subject to

$$\begin{aligned}\mathbf{w}^T\mathbf{x}_i + b \geq +1, \ y_i = +1, i = 1,...,n,\\ \mathbf{w}^T\mathbf{x}_i + b \leq -1, \ y_i = -1, i = 1,...,n.\end{aligned} \tag{7.6}$$

The optimization problem is equivalent to

$$[M_0] \quad \min \frac{1}{2}\|\mathbf{w}\|^2 \tag{7.7}$$

subject to

$$y_i(\mathbf{w}^T\mathbf{x}_i + b) \geq 1, \ i = 1,...,n. \tag{7.8}$$

The optimal hyperplane is denoted by $(\mathbf{w}^*, b^*)$. Optimization model $[M_0]$ is a convex quadratic programming problem that can be solved by off-the-shelf packages. Nevertheless, solving its dual problem by introducing the Lagrange multiplier can largely improve the computational efficiency. By adding the Lagrange multiplier α_i, $i = 1,...,n$, to each of the constraints in Equation (7.8), the Lagrangian function of optimization model $[M_0]$ can be written as

82 *Machine learning and data analytics for maritime studies*

$$[M_1] \quad L(\mathbf{w}, b, \alpha) = \frac{1}{2}\|\mathbf{w}\|^2 + \sum_{i=1}^{n} \alpha_i(1 - y_i(\mathbf{w}^T\mathbf{x}_i + b)) \quad (7.9)$$

subject to

$$\alpha_i \geq 0, i = 1, ..., n \quad (7.10)$$

where $\alpha = (\alpha_1, \alpha_2, ..., \alpha_n)$. The optimal $(\mathbf{w}^*, b^*)$ can be obtained by setting the partial derivatives of $L(\mathbf{w}, b, \alpha)$ (denoted by L for short) to $\mathbf{w}$ and b to zero. We first calculate the partial derivative of $L(\mathbf{w}, b, \alpha)$ to b as follows:

$$\begin{aligned} \frac{\partial L}{\partial b} &= -\sum_{i=1}^{n} \alpha_i y_i = 0, \\ \Rightarrow \sum_{i=1}^{n} \alpha_i y_i &= 0. \end{aligned} \quad (7.11)$$

By incorporating Equation (7.11) into objective function (7.9), we can have

$$\begin{aligned} L &= \frac{1}{2}\mathbf{w}^T\mathbf{w} + \sum_{i=1}^{n}\alpha_i - \sum_{i=1}^{n}\alpha_i y_i \mathbf{w}^T\mathbf{x}_i - \sum_{i=1}^{n}\alpha_i y_i b \\ &= \frac{1}{2}\mathbf{w}^T\mathbf{w} + \sum_{i=1}^{n}\alpha_i - \sum_{i=1}^{n}\alpha_i y_i \mathbf{w}^T\mathbf{x}_i. \end{aligned} \quad (7.12)$$

Then, the derivative of Equation (7.12) to $\mathbf{w}$ is

$$\frac{\partial L}{\partial \mathbf{w}} = \mathbf{w} - \sum_{i=1}^{n} \alpha_i \mathbf{x}_i y_i. \quad (7.13)$$

and by setting it to zero, we can have

$$\mathbf{w} = \sum_{i=1}^{n} \alpha_i \mathbf{x}_i y_i. \quad (7.14)$$

By introducing Equations (7.12) and (7.13) to optimization model $[M_1]$ so as to cancel out $\mathbf{w}$ and b, the following optimization problem can be obtained:

$$[M_2] \quad \max_{\alpha} \sum_{i=1}^{n} \alpha_i - \frac{1}{2}\sum_{i=1}^{n}\sum_{j=1}^{n} \alpha_i \alpha_j y_i y_j \mathbf{x}_i^T \mathbf{x}_j \quad (7.15)$$

subject to

$$\begin{aligned} \sum_{i=1}^{n} \alpha_i y_i &= 0, \\ \alpha_i \geq 0, i &= 1, ..., n. \end{aligned} \quad (7.16)$$

The Karush–Kuhn–Tucker condition should be satisfied in $[M_2]$, as all of the constraints are linear. Therefore, the following set of equations holds:

$$\begin{cases} \alpha_i \geq 0, \\ y_i(\mathbf{w}^T\mathbf{x}_i + b) - 1 \geq 0, \\ \alpha_i(y_i(\mathbf{w}^T\mathbf{x} + b) - 1) = 0. \end{cases}$$

This means that for any example $(\mathbf{x}_i, y_i)$, we have $\alpha_i = 0$ or $y_i(\mathbf{w}^T\mathbf{x}_i + b) = 1$. If $\alpha_i = 0$, it will not appear in Equation (7.12), and thus would not influence the hyperplane. Otherwise, $y_i(\mathbf{w}^T\mathbf{x}_i + b) = 1$ should be satisfied, which means that the example is on the maximum margin, i.e., it is a support vector. Therefore, it can be concluded that the hyperplane obtained by an SVM is only determined by the support vectors while having nothing to do with the other examples in the training set. After turning the objective function of optimization model [M_2] to

$$\min_\alpha \frac{1}{2} \sum_{i=1}^{n} \sum_{j=1}^{n} \alpha_i \alpha_j y_i y_j \mathbf{x}_i^T \mathbf{x}_j - \sum_{i=1}^{n} \alpha_i. \tag{7.17}$$

several efficient algorithms, such as sequential minimal optimization, can be used to solve [M_2] so as to obtain the optimal values of α (denoted by α^*). Then, the optimal $\mathbf{w}$ is

$$\mathbf{w}^* = \sum_{i=1}^{n} \alpha_i^* y_i \mathbf{x}_i. \tag{7.18}$$

To calculate the optimal value of b, a support vector $(\mathbf{x}_s, y_s)$ is first found and combined with the optimal $\mathbf{w}$, where we can have

$$\begin{aligned} & y_s(\mathbf{w}^*\mathbf{x}_s + b^*) = 1, \\ & y_s^2(\mathbf{w}^*\mathbf{x}_s + b^*) = y_s, \\ & y_s^2 = 1, \\ & \Rightarrow b^* = y_s - \mathbf{w}^*\mathbf{x}_s. \end{aligned} \tag{7.19}$$

To improve the robustness of the obtained value of b, the set of all support vectors, denoted by S, can be used to calculate b^*:

$$b^* = \frac{1}{|S|} \sum_{s \in S} (y_s - \mathbf{w}^*\mathbf{x}_s). \tag{7.20}$$

7.2 Soft margin SVM

In Figure 7.1, examples in the two classes are linearly separable. However, in more general cases in practice, the two classes of examples are linearly inseparable. That is, they cannot be separated using a straight line, and thus the hard margin SVM cannot be applied to the classification task as the constraints cannot be fully satisfied. An example is shown in Figure 7.3, where the two classes cannot be separated using a line as they are somewhat "mixed." One viable approach is to allow an SVM to make some mistakes in classification while keeping the margin as wide as possible such that the other points can still be classified correctly. This is the basic idea of a

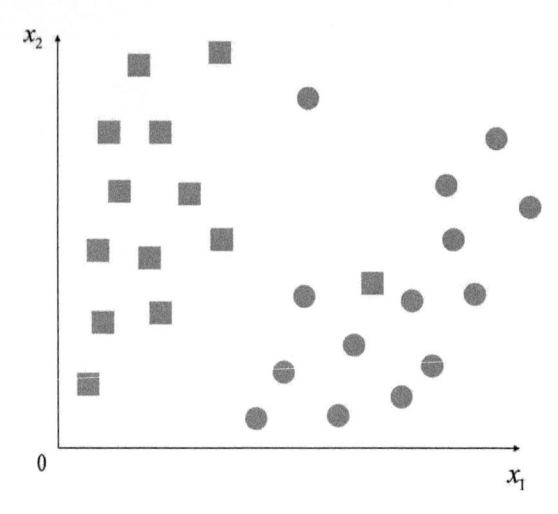

84 *Machine learning and data analytics for maritime studies*

Figure 7.3   Examples that are linearly inseparable

soft margin SVM. Then, hyperplane h_0 can still be used to separate the two classes, as presented in Figure 7.4.

Mathematically, the above approach can be realized by allowing some points (where the set of these points is denoted by D') to violate the constraint,
i.e.,

$$y'_i(\mathbf{w}^T\mathbf{x}_{i'} + b) < 1,\ i' \in D'. \tag{7.21}$$

We further denote the violation degree of example i' of constraint (7.7) by a slack variable ξ, $\xi > 0$, and thus Equation (7.21) can be written as

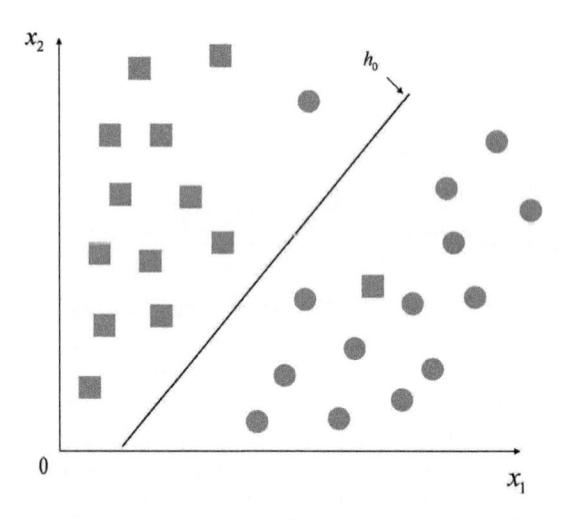

Figure 7.4   An example of a soft margin SVM

$$y_{i'}(\mathbf{w}^T\mathbf{x}_{i'} + b) \geq 1 - \xi_{i'}, \ i' \in D'. \tag{7.22}$$

To unify the format, we also attach the slack variable ξ to examples that do not violate constraint (7.7) while setting $\xi = 0$. Meanwhile, optimization model $[M_0]$ is changed to

$$[M_0'] \quad \min \frac{1}{2}\|\mathbf{w}\|^2 + C\sum_{i=1}^{n} \xi_i \tag{7.23}$$

subject to

$$\begin{aligned} y_i(\mathbf{w}^T\mathbf{x}_i + b) &\geq 1 - \xi_i, \ i = 1, \ldots, n \\ \xi_i &\geq 0, \ i = 1, \ldots, n. \end{aligned} \tag{7.24}$$

In objective function (7.23), C is a hyperparameter taking positive value and should be understood as the penalty on the examples that are wrongly classified. It decides a trade-off between maximizing the margin and minimizing the mistakes made. If $C \to \infty$, then $\xi_i \to 0$, which means that all the examples are expected to be classified as correctly as possible. If C takes a finite value, certain errors can be made. If C is small, less importance is given to classification mistake and more importance is given to maximizing the margin; otherwise, more importance is given to avoid misclassification at the expense of keeping the margin small. Similar to hard margin SVM, Lagrange multipliers can be introduced to optimization model $[M_0']$, and it is turned to optimization model $[M_1']$

$$[M_1'] \quad L(\mathbf{w}, b, \alpha, \boldsymbol{\xi}, \mu) = \frac{1}{2}\|\mathbf{w}\|^2 + C\sum_{i=1}^{n}\xi_i \\ + \sum_{i=1}^{n}\alpha_i(1 - \xi_i - y_i(\mathbf{w}^T\mathbf{x}_i + b)) - \sum_{i=1}^{n}\mu_i\xi_i \tag{7.25}$$

subject to

$$\begin{aligned} \alpha_i &\geq 0, i = 1, \ldots, n, \\ \mu_i &\geq 0, i = 1, \ldots, n, \end{aligned} \tag{7.26}$$

where α and μ are Lagrange multipliers and $\mathbf{w}$, b, and $\boldsymbol{\xi}$ are model parameters. Optimization model $[M_1']$ can be solved similarly to $[M_1]$.

Soft margin SVM can easily be implemented by the *scikit-learn* API [1] in Python. Here is an example of ship deficiency condition prediction (denoted by deficiency_state, which is a binary variable with 1 indicating a ship has six or more deficiencies and 0, otherwise) similar to Example 6.1 in Chapter 6.1, where a training set with 200 examples and five features, namely, ship age (age), type, last inspection time (l_ins_time), last deficiency number (l_def_no), and last inspection state (whether a ship is detained in the last inspection; l_ins_state) are used to construct a soft margin SVM model.

Example 7.1: As distance needs to be calculated in SVM, all the numerical features, i.e., age, l_ins_time, l_def_no, and l_ins_state, are processed by min–max scaler. For the categorical feature type, it is discretized to six features using one-hot encoding: type_bulk_carrier, type_container_ship, type_general_cargo/ multipurpose, type_passenger_ship, type_tanker, and type_other. The target variable deficiency_state is encoded to a binary variable, where "1" indicates a ship has six or more deficiencies and "0" otherwise.

We use fivefold cross validation to select the optimal value of hyperparameter C. The search space is set to [0.001, 0.01, 0.1, 1, 10, 100, 1000]. Note that the metric used for parameter selection is $F1score$, as the data set is slightly imbalanced: 46 out of the 200 ships are with six or more deficiencies. The core code is as follows:

```
from sklearn.model_selection import GridSearchCV
model = SVC(kernel='linear')
param_grid = {'C': [1e-3, 1e-2, 1e-1, 1, 10, 100, 1000]}
grid_search = GridSearchCV(model, param_grid, n_jobs = -1,
    cv=5, scoring='f1')
grid_search.fit(X_train_scaled, y_train)
best_parameters = grid_search.best_estimator_.get_params()
# output: the optimal C is 10
```

Then, the optimal value of C is used to train and test the final soft margin SVM model using the following code:

```
final_model = SVC(kernel='linear',C=best_parameters['C'])
final_model.fit(X_train_scaled, y_train)
y_pred = final_model.predict(X_test_scaled)
```

The *Precision*, *Secall*, and F_1 score are 0.57, 0.53, and 0.51, respectively.

7.3 Kernel trick

Soft margin SVM is more suitable when the linearly inseparable issue is caused by only a few examples in classification problems. When the examples in different classes are interwoven with each other, soft margin SVM might be ineffective. Another approach to address the problem of linearly inseparable classification problem in SVM is using kernel trick. The basic idea of kernel trick is to use a transform function to map the features from the original space to a higher space where they can be linearly separated. The dot product used in SVM models (i.e., $\mathbf{w}^T\mathbf{x}$) constructed so far on the original features of each two examples can be viewed as a linear kernel. Moreover, it has been proven that if the original feature space is of limited dimension, there must be a feature space of a higher dimension where the examples are linearly separable.

Mathematically, a kernel function can be written as

$$K(\mathbf{x}_i, \mathbf{x}_j) = \langle \phi(\mathbf{x}_i), \phi(\mathbf{x}_j) \rangle \\ = \phi(\mathbf{x}_i)^T \phi(\mathbf{x}_j). \quad (7.27)$$

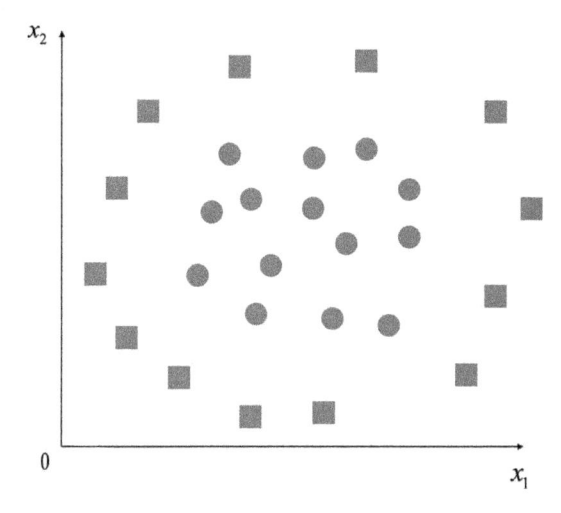

Figure 7.5   A linearly inseparable case

The kernel function in Equation (7.27) can be understood as taking a dot product (which measures the similarity of the two terms) of the transformed input vectors by function $\phi(\cdot)$. The kernel function can easily be incorporated into the original hard margin SVM model by replacing $\mathbf{x}_i$ by $\phi(\mathbf{x}_i)$ in optimization models [M_0], [M_1], and [M_2].

An intuitive example to use the kernel trick in SVM to tackle the linearly inseparable problem is as follows. Suppose we need to separate examples of two classes presented in Figure 7.5. It is obvious that a straight line cannot separate the two classes of examples perfectly or nearly perfectly. Instead, a circle can do the job as shown in Figure 7.6.

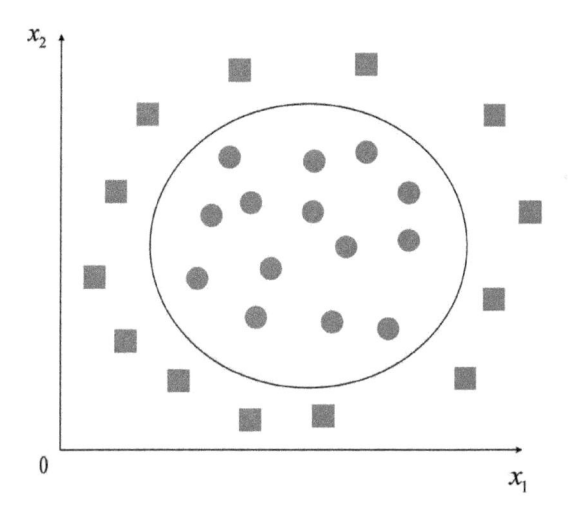

Figure 7.6   Using a circle to separate examples in the linearly inseparable case

Then, a kernel function measuring the similarity of two terms based on whether the related points are within a circle can be considered. For a point $(x_{1'}, x_{2'})$ in Figure 7.6, we define a transform function:

$$\phi(x_{1'}, x_{2'}) = (x_{1'}^2, x_{2'}^2, \sqrt{2}x_{1'}x_{2'}). \tag{7.28}$$

Then, the kernel function takes the following form:

$$\begin{aligned} K((x_{1'}, x_{2'}), (x_{1''}, x_{2''})) &= \langle \phi(x_{1'}, x_{2'}), \phi((x_{1''}, x_{2''})) \rangle \\ &= x_{1'}^2 x_{1''}^2 + x_{2'}^2 x_{2''}^2 + 2x_{1'}x_{2'}x_{1''}x_{2''}, \end{aligned} \tag{7.29}$$

which is actually the function form of a circle that can transform the examples into a 3D space. By doing so, the way to measure the similarity of two terms is changed from using a dot product (i.e., a linear kernel) to judging whether points are within a circle by using the new decision boundary generated by a kernel function as is shown in Figure 7.6. In this way, a linear hyperplane in the 3D space (which is equivalent to a circle in the 2D space) can perfectly separate the two classes, while the linear form of the classifier can also be reserved.

Having a closer examination of Equation (7.29), we can find that it is actually the square of the dot product in the 2D space, i.e.,

$$\begin{aligned} K((x_{1'}, x_{2'}), (x_{1''}, x_{2''})) &= x_{1'}^2 x_{1''}^2 + x_{2'}^2 x_{2''}^2 + 2x_{1'}x_{2'}x_{1''}x_{2''} \\ &= (x_{1'}x_{1''} + x_{2'}x_{2''})^2 \\ &= \langle (x_{1'}, x_{2'}), (x_{1''}, x_{2''}) \rangle^2. \end{aligned} \tag{7.30}$$

Therefore, the value of the selected kernel function can be calculated directly from the dot product of two points in the 2D space, and thus the computation burden can be reduced. The kernel function presented in Equation (7.30) is called a polynomial kernel, whose degree is 2. The above SVM with kernel trick can be used to address the problem in Example 7.1 by the *scikit-learn* API, which is presented in Example 7.2.

Example 7.2: The features and their processing methods are the same as those in Example 7.1.1. We also use fivefold cross validation to select the optimal value of parameter C with $F1score$ as the metric. The search space is set to $[0.001, 0.01, 0.1, 1, 10, 100, 1000]$. The core code is as follows:

```
from sklearn.model_selection import GridSearchCV
model = SVC(kernel='poly',degree=2) # This means that the
    'poly' kernel with degree 2 is used
param_grid = {'C': [1e-3, 1e-2, 1e-1, 1, 10, 100, 1000]}
grid_search = GridSearchCV(model, param_grid, n_jobs = -1,
    cv=5,scoring='f1')
grid_search.fit(X_train_scaled, y_train)
best_parameters = grid_search.best_estimator_.get_params()
# output: the optimal C is 1000
```

It can be found that the optimal value of the penalty term C in this example is much larger than that in Example 7.1 when linear SVM is used (C = 10 in this case). This is generally because using polynomial kernel maps the data to a higher dimension where they are more linearly separable. Then, the tolerance of an example that is wrongly classified gets smaller, and thus the penalty gets larger. The optimal value of C is then used to train and test the final soft margin SVM model using the following code:

```
final_model =
    SVC(kernel='poly',degree=2,C=best_parameters['C'])
final_model.fit(X_train_scaled, y_train)
y_pred = final_model.predict(X_test_scaled)
```

The *Precision*, *Recall*, and F_1 score are 0.57, 0.59, and 0.58, respectively. Recall that these three metrics for Example 7.1 are 0.57, 0.53, and 0.51, respectively. Therefore, introducing a polynomial kernel into an SVM increases the prediction accuracy.

The degree of a polynomial kernel can be any positive integer. The larger value the degree is, the more complex model will be generated. Therefore, it can be seen that selecting a proper kernel function is of vital importance to guarantee the accuracy of an SVM model. This can be a tricky task, as how to accurately map the original data to a higher dimensional space where they are linearly separable is unknown. In addition to the linear kernel and the polynomial kernel, some other common kernel functions are given in Table 7.1.

Table 7.1   Common kernel functions

Kernel function name	Format	parameters
Linear kernel function	$K(\mathbf{x}_i, \mathbf{x}_j) = \mathbf{x}_i^T \mathbf{x}_j$	None
Polynomial kernel function	$K(\mathbf{x}_i, \mathbf{x}_j) = (\mathbf{x}_i^T \mathbf{x}_j)^d$	$d \geq 1$, which is the degree of the polynomial. A larger d leads to a more complex SVM model.
Gaussian kernel	$K(\mathbf{x}_i, \mathbf{x}_j) = e^{-\frac{\|\mathbf{x}_i - \mathbf{x}_j\|^2}{2\sigma^2}}$	$\sigma > 0$, which is the width of the Gaussian kernel. A smaller σ leads to an SVM model that is more sensitive to the distance between each pair of points, and thus is more likely to overfit the data.
Laplacian Kernel	$K(\mathbf{x}_i, \mathbf{x}_j) = e^{-\frac{\|\mathbf{x}_i - \mathbf{x}_j\|}{\sigma}}$	The same as Gaussian kernel.
Sigmoid kernel	$K(\mathbf{x}_i, \mathbf{x}_j) = \tanh(\beta \mathbf{x}_i^T \mathbf{x}_j + \theta)$	tanh is the hyperbolic tangent function, $\beta > 0, \theta < 0$.

7.4 Support vector regression

SVM is designed for classification problems. It can be extended to address regression problems by conducting proper revisions, and the model is called support vector regression or SVR for short. In standard ML models for regression tasks, if the predicted output and the real output are not equal, their difference will be counted in the loss function such as MSE and MAE. Nevertheless, SVR aims to find a line with a certain width to fit the data, such that the error of the examples within the line with certain width is not considered when calculating the value of the loss function, while only the examples outside the line with certain width will be considered. An illustration of SVR where only one feature is considered is shown in Figure 7.7. The line of the SVR model developed to predict the output of the examples is $f(x) = wx + b$, and the width is 2ϵ. Thus, the error of the examples falling within the area bounded by lines $f(x) + \epsilon$ and $f(x) - \epsilon$ will not be accounted into the calculation of the loss function, while the others will be accounted for.

Mathematically, an SVR model can be formulated as follows:

$$[M_3] \quad \min_{w,b} \frac{1}{2}\|w\|^2 + C \sum_{i=1}^{n} l_\epsilon(f(x_i) - y_i). \tag{7.31}$$

where C is the penalty term, and $l_\epsilon(f(x_i) - y_i)$ is the loss function given the value of ϵ which takes the following form:

$$l_\epsilon(f(x_i) - y_i) = \begin{cases} 0, & \text{if } |f(x_i) - y_i| \leq \epsilon \\ |f(x_i) - y_i| - \epsilon, & \text{otherwise.} \end{cases}$$

Then, two slack variables, ξ_{i1} and ξ_{i2}, are introduced to linearize $l_\epsilon(f(x_i) - y_i)$, and model $[M_3]$ is turned to

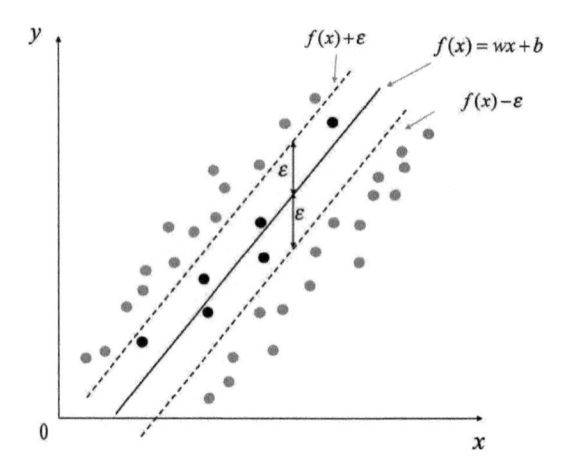

Figure 7.7   An illustration of an SVR model

$$[M_3'] \quad \min_{w,b,\xi_{i1},\xi_{i2},i=1,...,n} \frac{1}{2}\|w\|^2 + C\sum_{i=1}^{n}(\xi_{i1}+\xi_{i2}) \quad (7.32)$$

subject to

$$\begin{aligned} f(x_i) - y_i &\leq \epsilon + \xi_{i1}, i=1,...,n, \\ y_i - f(x_i) &\leq \epsilon + \xi_{i2}, i=1,...,n, \\ \xi_{i1} &\geq 0, i=1,...,n, \\ \xi_{i2} &\geq 0, i=1,...,n. \end{aligned} \quad (7.33)$$

Similar to optimization model $[M_1']$, Lagrange multipliers can be introduced to model $[M_3']$ so as to obtain its dual problem. Then, the format of an SVR model can be written as

$$f(x) = \sum_{i=1}^{n}(\alpha_{i1} - \alpha_{i2})x_i x + b. \quad (7.34)$$

where α_{i1} and α_{i2} are the Lagrange multipliers for example i, $i = 1, ..., n$, for the first two constraints in Equation (7.33). Examples with $\alpha_{i1} - \alpha_{i2} \neq 0$ are the support vectors of the SVR, which are the examples lying outside of the line with width 2ϵ. Obviously, they are a subset of the whole training set. Similar to SVM, kernel trick can also be applied in SVR to improve model performance.

SVR is also implemented in the *scikit-learn* API. The following example is to predict the number of deficiencies of a ship (i.e., a continuous output) using the same features as well as their decoding methods in Example 7.2.

Example 7.3: We first use fivefold cross validation to select the most suitable kernel function from "linear," "poly" (i.e., polynomial kernel), "rbf" (i.e., Gaussian kernel), "sigmoid," the degree value if "poly" kernel is used, and the penalty term C. The core code is as follows:

```
param_grid = {'C': [1e-3, 1e-2, 1e-1, 1, 10, 100, 1000],
  'kernel':['linear', 'poly', 'rbf', 'sigmoid'],
  'degree':[2,3,4,5],}
model = SVR()
grid_search = GridSearchCV(model, param_grid, n_jobs=-1,
    cv=5,scoring='neg_mean_squared_error')
grid_search.fit(X_train_scaled, y_train)
best_parameters = grid_search.best_estimator_.get_params()
print(best_parameters)
# output: the optimal kernel is poly, degree is 4, and C is
    1
```

Then, the optimal kernel and its degree and the value of the penalty term C are used to train and test the final SVR model using the following code:

```
final_model = SVR(kernel='poly',degree=4,C=1)
final_model.fit(X_train_scaled, y_train)
y_pred = final_model.predict(X_test_scaled)
```

The *MSE*, *MAE*, and *MAPE* are 11.1886, 2.3153, and 0.7085, respectively.

Reference

[1] Pedregosa F., Varoquaux G., Gramfort A., *et al*. 'Scikit-learn: machine learning in python'. *Journal of Machine Learning Research*. 2011;**12**:2825–30.

Chapter 8

Artificial neural network

This chapter aims to introduce a widely-used machine learning model for both regression and classification tasks called artificial neural network (ANN). Basic concepts of ANNs are first covered, and then typical algorithms and tricks for ANN training are introduced. Finally, deep learning models, as an important type of neural network, are briefly discussed.

8.1 The structure and basic concepts of an ANN

Suppose we have a data set $D = \{(\mathbf{x}_i, y_i), i = 1, ..., n\}$, where $\mathbf{x}_i \in R^m$ is the vector of features and $y_i \in R$ is the output. The structure of a typical ANN model to predict y_i (denoted by $\hat{y}_i$, $i = 1, ..., n$) is presented in Figure 8.1. It contains three layers: an input layer, a hidden layer, and an output layer. Especially, the input layer contains m nodes to receive the m feature values of the examples, which can either be categorical or numerical, in addition to a bias node. The output layer gives the final predicted target $\hat{y}_i$. For a specific problem, the structure of the input and output layers is fixed. In contrast, the hidden layer can be much more flexible: an ANN can have one or more hidden layers, and the number of nodes in one hidden layer is also flexible. In particular, the number of hidden layers and the number of nodes contained by each of them depend on the specific problem and the data structure. In Figure 8.1 one hidden layer with $K + 1$ nodes is contained. Weight is attached to each arrow connecting two nodes in consecutive layers in Figure 8.1. The direction of the arrows shows the data flow direction in an ANN model, which is opposite to the direction of the error flow which will be covered later.

One node in the neural network is also called a neuron. The details of a neuron in Figure 8.1 are shown in Figure 8.2. A neuron is an elementary unit of an ANN model, which can be viewed as a mathematical function receiving one or more inputs from an example or from the neurons in the preceding layer. For the neuron shown in Figure 8.2, it receives inputs from a_1, a_2, and a_3 with weights w_1, w_2, and w_3 attached. The weighted sum of the inputs, which is denoted by t, is first calculated by $t = w_1 a_1 + w_2 a_2 + w_3 a_3 + b$, where b is the bias. Then, t is passed to an activation function f to generate the output of this neuron, which is denoted by z, i.e., $z = f(t)$. In particular, activation functions deciding whether the neurons should be activated

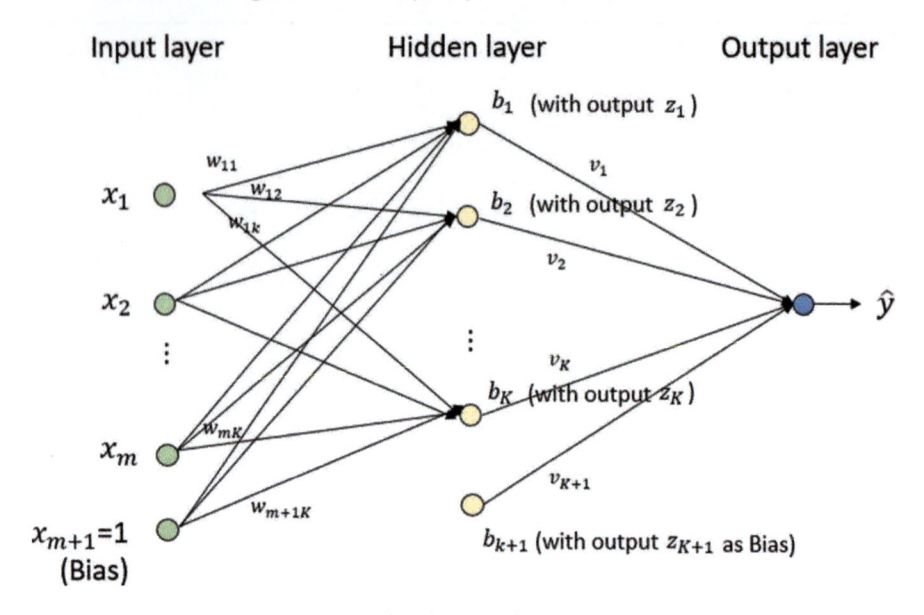

94 *Machine learning and data analytics for maritime studies*

Figure 8.1   An illustration of a typical ANN model

play a fundamental role in ANNs as they introduce nonlinearity (i.e., nonlinear transformation) into the network. Nonlinearity guarantees the universal approximation property of ANNs: "Standard multilayer feedforward networks with as few as one hidden layer using arbitrary squashing functions are capable of approximating any Borel measurable function from one finite dimensional space to another to any desired degree of accuracy, provided sufficiently many hidden units are available" [1]. Popular activation functions are presented in Figures 8.3–8.6.

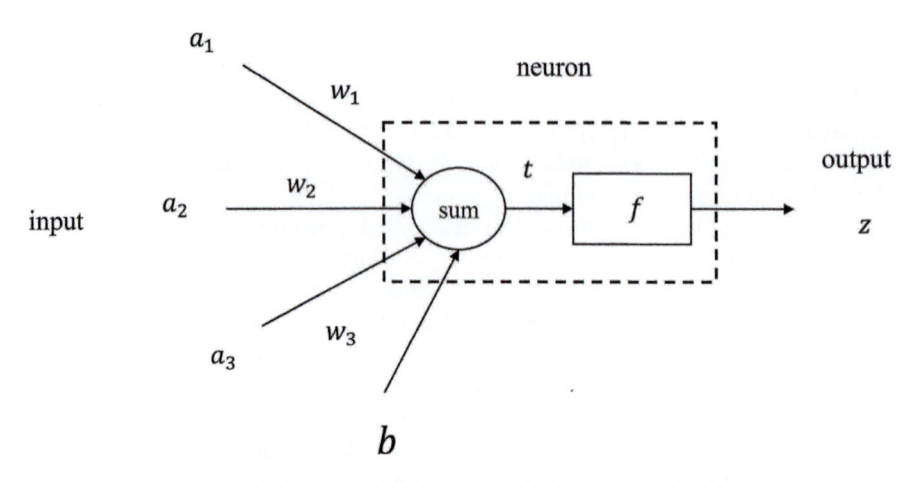

Figure 8.2   An illustration of a neuron in ANNs

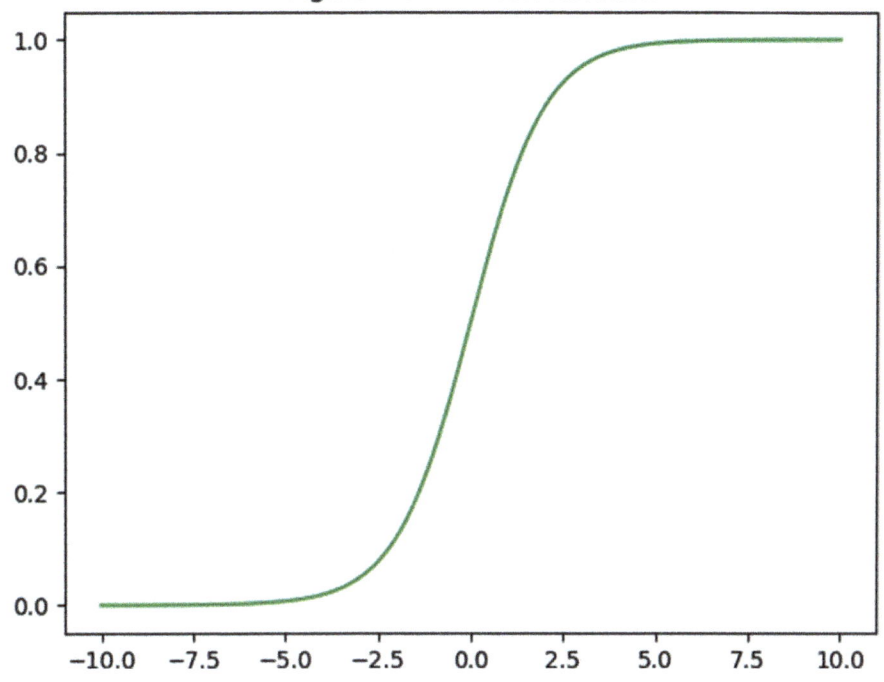

Figure 8.3   An illustration of Sigmoid activation function: $f(x) = \frac{1}{1+e^{-x}}$

Based on the structure and mathematical representation of a neuron, the layers in an ANN shown in Figure 8.1 can be mathematically presented as follows. As stated above, the input layer of an ANN model receives the input from the feature values. Then, they serve as the input to the hidden layers, and the weighted sum of the inputs to neuron $b_k, k = 1, ..., K$ denoted by t_k, is calculated by

$$t_k = \sum_{m'=1}^{m} w_{m'k} x_{m'}, k = 1, \ldots, K. \tag{8.1}$$

Then, the output of node $k, k = 1, ..., K + 1$ is

$$\begin{aligned} z_k &= f(t_k), \quad k = 1, \ldots, K, \\ z_{K+1} &= 1. \end{aligned} \tag{8.2}$$

Note that, for simplicity, we set the bias terms $x_{m+1} = 1$ in the input layer and $z_{K+1} = 1$ in the hidden layer. As the bias term is connected by its weights, which can be adjusted, to the neuron(s) in the consecutive layer, the value of the bias term itself can be set to an arbitrary nonzero constant and is usually set to be 1. The outputs of nodes in the hidden layer also serve as the input to the output layer, which is

$$t_y = \sum_{k=1}^{k+1} v_k z_K. \tag{8.3}$$

Tanh activation function

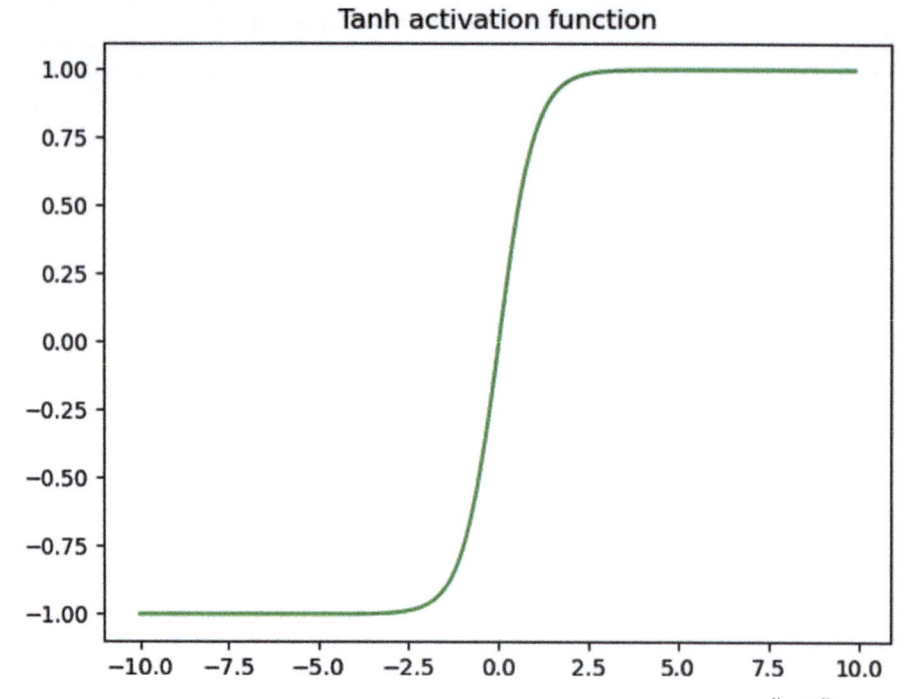

Figure 8.4 An illustration of Tanh activation function: $f(x) = \frac{e^x - e^{-x}}{e^x + e^{-x}}$

In regression tasks where the output is numerical, the weighted sum of the outputs given by the hidden layer is used directly as the final prediction in the output layer, that is

$$\hat{y} = t_y. \tag{8.4}$$

In classification tasks where the output is categorical, the continuous weighted sum input to the output layer is further discretized. In binary classification problems, sigmoid function is often used, where the predicted value given by the output layer is the probability of being a positive class:

$$\hat{y} = \frac{1}{1 + e^{-t_y}}. \tag{8.5}$$

In multiclass classification problems, the output is multi-dimensional, and a Softmax function is often used to process the outputs, which turns absolute values into probabilities. We do not discuss the details of this case here. Then, the predicted output of the output layer can be uniformly expressed as

$$\hat{y} = g(t_y). \tag{8.6}$$

Therefore, the overall relationship between the input features $x_{m'}$, $m' = 1, ..., m$, and the output $\hat{y}$ can be presented by

ReLU activation function

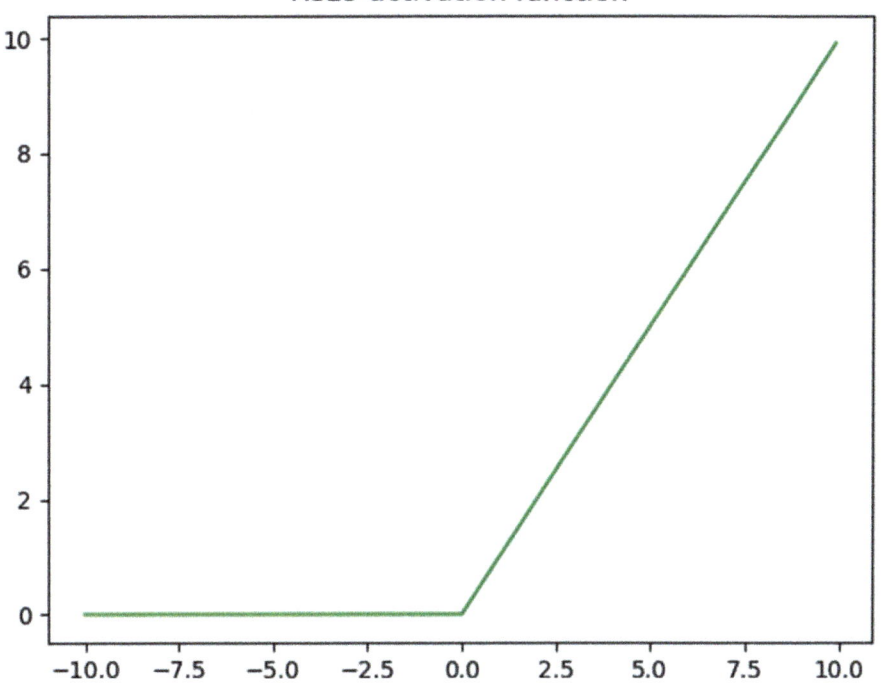

Figure 8.5 An illustration of ReLU (short for rectified linear units) activation function: $f(x) = 0$ if $x < 0$ and $f(x) = x$, otherwise

$$\begin{aligned}
\hat{y} &= g(t_y) \\
&= g(\sum_{k=1}^{k+1} v_k z_k) \\
&= g(\sum_{k=1}^{k} v_k f(t_k) + v_{K+1}) \\
&= g(\sum_{k=1}^{k} v_k f(\sum_{m'=1}^{m+1} w_{m'k} x_{m'}) + v_{K+1}).
\end{aligned} \quad (8.7)$$

8.1.1 Training of an ANN model

In the above analysis, the output of one layer serves as the input to its directly consecutive layer. Then, the output of the last layer, i.e., the output layer, gives the final prediction $\hat{y}$. The cost function is an evaluation of the loss brought about by the predicted output $\hat{y}$ with regard to the real output y, which is used to evaluate the prediction error. In regression problems, the cost function is usually half of the MSE between the real and the predicted output values. In binary classification problems, cross entropy is a popular cost function. That is, if y_i is numerical, the cost function of this example can be calculated by

Leaky ReLU activation function

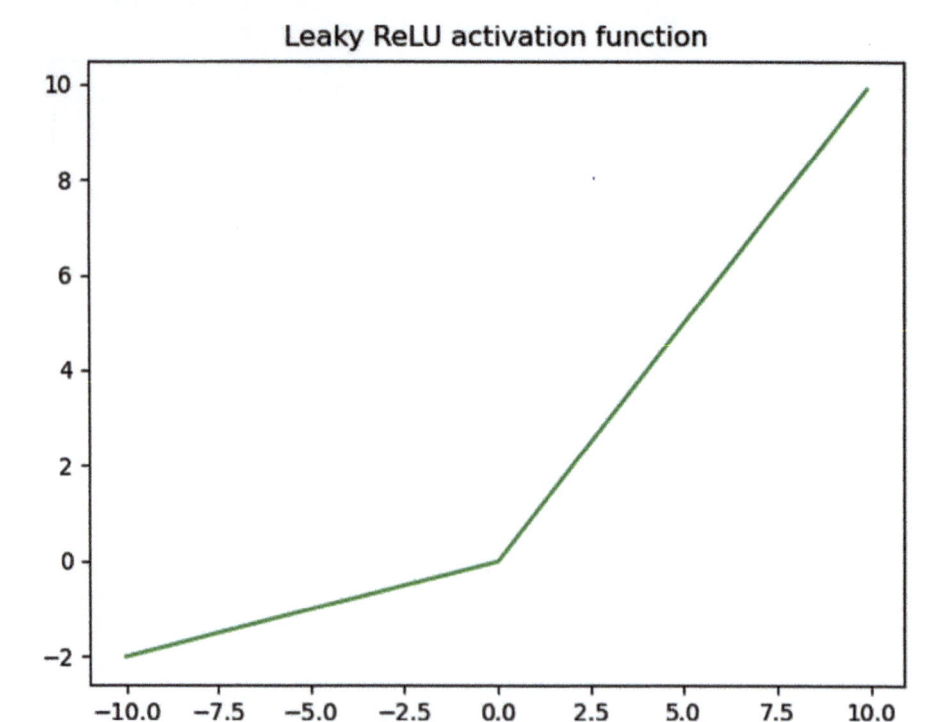

Figure 8.6 An illustration of leaky ReLU activation function: $f(x) = 0.2x$ if $x < 0$ and $f(x) = x$, otherwise

$$L(y_i, \hat{y}_i) = \frac{1}{2}(y_i - \hat{y}_i)^2. \tag{8.8}$$

If y_i is binary, the cost function of this example can be calculated by

$$L(y_i, \hat{y}_i) = -[y_i(\log(\hat{y}_i)) + (1 - y_i)(\log(1 - \hat{y}_i))]. \tag{8.9}$$

The training process of an ANN model in one round is to adjust the weights connecting the neurons, with the goal to minimize the cost function: the smaller the value of the cost function, the closer the predicted and actual outputs. The most popular ANN learning algorithm is called backpropagation, which is a way of computing gradients of expressions through recursive application of chain rule, such that the current error in the output layer can be reversely passed to each layer, and the weights can be adjusted accordingly to minimize the error. We use the case of regression problem as an example to show the working mechanism of the backpropagation algorithm.

We first calculate the gradients of the weights connecting the hidden layer and the output layer. Assume sigmoid function is used as the activation function, i.e., $f(x) = \frac{1}{1+e^{-x}}$. Then, its derivative can be calculated by $f'(x) = f(x)(1 - f(x))$. Denote the current training example by (x_i, y_i). For neuron k, $k = 1, ..., K$ in the hidden layer whose weight is denoted by v_k, its connection with the output layer can be represented by

$$\begin{cases} L = \dfrac{1}{2}(y_i - \hat{y}_i)^2 \\ \hat{y}_i = t_y \\ t_y = \displaystyle\sum_{k=1}^{K+1} v_k z_k. \end{cases} \quad (8.10)$$

Then, the gradient of L with respect to v_k, denoted by δv_k for short, can be calculated by the following chain rule:

$$\begin{aligned} \delta v_k &= \dfrac{\partial L}{\partial v_k} \\ &= \dfrac{\partial L}{\partial \hat{y}_i}\dfrac{\partial \hat{y}_i}{\partial t_y}\dfrac{\partial t_y}{\partial v_k} \\ &= (\hat{y}_i - y_i)z_k. \end{aligned} \quad (8.11)$$

where all the terms involved are known. The gradient of weight $w_{m'k}$ connecting neuron m' in the input layer and neuron k in the hidden layer can be calculated in a similar way. The connection of neuron m' and the prediction error can be represented by

$$\begin{cases} L = \dfrac{1}{2}(y_i - \hat{y}_i)^2 \\ \hat{y}_i = t_y \\ t_y = \displaystyle\sum_{k=1}^{K+1} v_k z_k \\ z_k = f(t_k) \\ t_k = \displaystyle\sum_{m'=1}^{m+1} w_{m'k} x_{m'}. \end{cases} \quad (8.12)$$

Then, the gradient of L with respect to $w_{m'k}$, denoted by $\delta w_{m'k}$ for short, can be calculated by the following chain rule

$$\begin{aligned} \delta w_{m'k} &= \dfrac{\partial L}{\partial w_{m'k}} \\ &= \dfrac{\partial L}{\partial \hat{y}_i}\dfrac{\partial \hat{y}_i}{\partial t_y}\dfrac{\partial t_y}{\partial z_k}\dfrac{\partial z_k}{\partial t_k}\dfrac{\partial t_k}{\partial w_{m'k}} \\ &= (\hat{y}_i - y_i)v_k f(t_k)(1 - f(t_k))x_{m'}. \end{aligned} \quad (8.13)$$

where all the terms involved are known. δv_k and $\delta w_{m'k}$ are then used to update the weights connecting the hidden layer and the output layer and those connecting the input layer and the hidden layer, respectively. To reduce the problem of over-fitting, learning rate, which is denoted by η, is proposed to control the speed of weight updating. Then, after this round of learning, the weight connecting neuron k, $k = 1, ..., K$ and the output layer is updated to

$$\tilde{v}_k = v_k - \eta \delta v_k, \tag{8.14}$$

and the weight connecting neuron m', $m' = 1, ..., m$, and neuron k, $k = 1, ..., K$, is updated to

$$\tilde{w}_{m'k} = w_{m'k} - \eta \delta w_{m'k}. \tag{8.15}$$

By doing so, the current prediction error is fed back throughout the entire network, and the weights can be adjusted with the aim to minimize the prediction error. Note that in one round of the above updating using standard backpropagation, only one example, i.e., $(\mathbf{x}_i, y_i)$, is used to calculate the gradient of the weights. As only a single random example is used, standard backpropagation might lead to the problem of local minima, and the updating results might be unstable. An alternative is to use the whole training set in one round to calculate the gradients, which is also called accumulated error backpropagation. However, in many tasks, when the accumulated error is small, it can be very slow to further reduce it. Under this condition, standard backpropagation is more suitable. To reach a trade-off between standard backpropagation and accumulated error backpropagation, a few examples are selected randomly, instead of a single example or the entire training set, for each iteration for weight updating. The set of examples is called a mini batch, and the number of examples in one mini batch is called batch size. The process of weight updating is iterated several rounds, where the number of rounds is denoted by "*epoch.*"

8.1.2 Hyperparameters in an ANN model

Hyperparameters in an ANN model can be divided into two categories: one is related to network structure, and the other is related to learning algorithm. The number of hidden layers and the number of neurons contained in each of the hidden layers are two main types of hyperparameters to control the complexity of the network structure in ANNs. Generally speaking, more hidden layers or more neurons contained in one hidden layer increase model complexity, and thus might lead to the problem of over-fitting. One trick to reduce the problem of over-fitting is called dropout. It is a regularization approach that can be applied to input and hidden layers. Dropout means for an input or a hidden layer, temporarily removing a certain proportion (denoted by p) of neurons from the layer, including the neuron itself as well as the connections with its preceding (if any) and following layers. Then, the outputs of this hidden layer are scaled by multiplying each of the outputs by $(1-p)$. This approach is widely used in deep neural networks (DNNs). Besides, the types of activation functions of the neurons are also hyperparameters that need to be decided before ANN training, and they can have a large impact on model performance.

Hyperparameters regarding the learning algorithm mainly include the learning rate (i.e., η), the number of rounds (i.e., *epoch*), and batch size. In particular, η controls the speed of weight updating. If η is too small, the speed of learning would be slow and a larger value of *epoch* might be needed. If η is too large, the optimal values of the weights might be surpassed in the updating process. If the value of *epoch* is too large, the ANN model developed might learn the data too well, leading to the problem of over-fitting. In contrast, if *epoch* is too small, the problem of under-fitting might occur. To find a proper value for *epoch*, a validation set that is independent of the training set should be used to test the performance of the temporary ANN model constructed: if the cost function on the validation set decreases moderately or even increases, the training should be stopped as the problem of over-fitting is highly likely to occur. This trick is also called "early stopping." Finally, batch size is highly dependent on the size of the whole data set and the network structure. Common batch size is 1, 2, 4, 16, 32, 64, 128, and 256. In addition, to overcome the problem of over-fitting, regularization on the weights in the network is introduced to the cost function shown in Equation (8.8):

$$\tilde{L} = \lambda \frac{1}{n} \sum_{i=1}^{n} L(\hat{y}_i, y_i) + (1 - \lambda)(\sum_{m'=1}^{m} \sum_{k=1}^{k+1} w_m^2/k + \sum_{k=1}^{k+1} v_k^2), \qquad (8.16)$$

where λ is a hyperparameter that needs to be tuned, which balances the trade-off between model accuracy and network complexity. Note that regularization is not imposed on the bias term, as it is usually a parameter with a fixed and small value (e.g., 1). All the hyperparameters mentioned above should be tuned by a hold-out validation set or based on cross validation and grid search.

ANN models for regression tasks are implemented in *scikit-learn* API [2], which is named "MLPRegressor" under the class of "neural_network." We use the following example to show the procedure of developing an ANN model to predict ship deficiency number.

Example 8.1. We first randomly divide the whole data set containing 1 991 examples to a training set with 1 500 examples and a test set with 491 examples. The numerical features considered are GT, age, last_inspection_time, and last_deficiency_no. The categorical features considered are flag_performance, RO_performance, company_performance, last_inspection_state, and ship type. Especially, for flag_performance, white is encoded to 0, gray is encoded to 1, black is encoded to 2, and unknown is encoded to 3. For RO_performance, high is encoded to 0, medium is encoded to 1, and unknown is encoded to 2. For company_performance, high is encoded to 0, medium is encoded to 1, low is encoded to 2, very low is encoded to 3, and unknown is encoded to 4. For last_inspection_state, if a ship has no last inspection record in the database of Tokyo MoU, it is decoded to −1 (this also applies to numerical features last_inspection_time and last_deficiency_no). If a ship is detained in the last inspection, it is encoded to 1. If a ship is not detained in the last inspection, it is encoded to 0. Ship type is encoded using one-hot encoding the same as the examples in Chapter 7.

We fix the number of hidden layers as one and the number of neurons contained as 10. All the nodes in the hidden layer use ReLU function as the activation function. The following hyperparameters are tuned by fivefold cross validation on the training set: learning rate (learning_rate_init in *scikit-learn* API), batch size (batch_size

Table 8.1 Popular deep learning models

Model name	Main structure	Applications
Convolutional neural network (CNN)	Three main types of layers are contained: convolution, pooling, and fully-connected layers.	Supervised and unsupervised: image processing, object detection.
Recurrent neural network (RNN)	RNNs have cyclic connections in the structure that employ recurrent computations to sequentially process the input data. The edges in an RNN are fed into the next time step instead of into the next layer in the same time step.	Supervised: speech recognition, handwriting recognition, and text recognition where sequences and patterns are involved.
Long short term memory network (LSTM)	LSTM is a variant of RNN that can handle long-term dependencies. Therefore, an LSTM network has the same structure as an RNN, except that the neurons in the hidden layer of the RNN are replaced by memory blocks. The basic component of an LSTM network is called a memory block, which contains a cell and three gates.	Supervised: time-series prediction, speech or video recognition, music composition, and pharmaceutical development.
Deep belief network (DBN)	DBNs integrate restricted Boltzmann machines (RBMs) with deep feedforward neural networks (D-FFNNs), where RBMs form the input unit and D-FFNNs form the output unit.	Supervised and unsupervised: image recognition, video recognition, motion-capture data processing.
Generative adversarial network (GAN)	GANs have two components, namely, a generator and a discriminator, and they can create new data that resemble the training data by learning from them.	Supervised and unsupervised: Image generation, human face generation, cartoon character generation, image transformation, and astronomical image improvement.

in *scikit-learn* API), and epoch (max_iter in *scikit-learn* API). The core code is as follows:

```
param_grid = {'learning_rate_init': [0.001, 0.05, 0.01,
    0.05, 0.1],
'batch_size':[1,4,16,64,128,256],
'max_iter':[100,200,300,500,800,1000]}
model = MLPRegressor(hidden_layer_sizes=(10,),
activation='relu',solver='sgd',random_state=0)
grid_search = GridSearchCV(model, param_grid, n_jobs=-1,
    cv=5,scoring='neg_mean_squared_error')
grid_search.fit(X_train_scaled, y_train)
best_parameters = grid_search.best_estimator_.get_params()
print(best_parameters)
# output: the optimal learning_rate_init is 0.01,
    batch_size is 256, and max_iter is 100
```

Then, the optimal learning rate init, batch size, and max iter are used to train andtest the final ANN model using the following code:

```
final_model =
    MLPRegressor(hidden_layer_sizes=(10,),activation='relu',
    solver='sgd',random_state=0,
learning_rate_init=0.01, batch_size=256,max_iter=100)
final_model.fit(X_train_scaled, y_train)
y_pred = final_model.predict(X_test_scaled)
```

The *MSE*, *MAE*, and *MAPE* are 14.3317, 2.5591, and 0.6437, respectively.

8.2 Brief introduction of deep learning models

With the improvement of computation resources, training more complex ANN models where adequate storage and computational power are needed becomes possible. With the accessibility of big data, the problem of over-fitting in these complex models can be largely reduced. The most common way to improve the complexity of an ANN model is to increase the number of its hidden layers: the deeper a network, the more complex a model. The reasons are intuitive: as the number of hidden layers becomes larger, more weights between consecutive layers are added, and a larger number of activation functions associated with the neurons can be involved. As a result, their joint efforts can significantly increase the complexity and capacity of an ANN model. An ANN model with a large number of hidden layers is called a DNN, or a deep learning model. The term "deep learning" describes a family of learning algorithms rather than a single algorithm to develop DNN models. Popular deep learning models are summarized in Table 8.1.

References

[1] Hornik K., Stinchcombe M., White H. 'Multilayer feedforward networks are universal approximators'. *Neural Networks*. 1989;2(5):359–66.
[2] Pedregosa F., Varoquaux G., Gramfort A., *et al*. 'Scikit-learn: machine learning in python'. *Journal of Machine Learning Research*. 2011;12:2825–30.

Chapter 9
Tree-based models

This chapter aims to introduce several machine learning models based on a tree structure called the decision tree, which are widely believed to be among the most popular methods for both classification and regression tasks. The basic structure and concepts as well as the tree-growing algorithms of a single decision tree are first introduced. As a single decision tree is prone to over-fitting, ensemble models consisting of a certain number of decision trees are developed. Random forest-based on bagging and gradient boosting decision trees based on boosting will be introduced as the representatives of ensemble models on decision trees.

9.1 Basic concepts of a decision tree

As the name suggests, a decision tree, which we call DT for short, uses a tree-like model for decision making. A DT consists of nodes and directed edges, where a set of examples is contained in a node, and a directed edge means splitting a node into consecutive nodes. An example of a DT is shown in Figure 9.1. The orange node at the top is a root node, which contains the whole data set used to construct the current DT. The green nodes are leaf nodes that are not split any more and give the final prediction. The blue node is the internal node, which is to be further split. The directed edges in the DT show the process of node splitting: a parent node at the tail of a directed edge is split into two child nodes at the head following some criteria, which is usually a selected feature or a (feature, value) pair. The term "Splitting" here means that the examples contained by the parent node are separated into two complementary and disjoint child nodes, and the splitting criterion is the feature or the (feature, value) pair. The examples with feature values less than or equal to the threshold value are split to the left child node, and others to the right child node. The splitting aim is to make the two child nodes become "purer." The term "purer" here means that the examples contained by one node are more similar to each other: in classification problems, the examples in one node are expected to be in the same class; in regression problems, the targets of examples in the same node should be as close as possible. Finally, the output of a leaf node is also determined by the problem. In general, in classification problems, the class of the majority of the examples contained within it is used as the output; in regression problems, the average target of all the examples contained is the output. The depth of a tree is one more than the number of splits from the root node to the deepest leaf node. The depth of the tree in Figure 9.1 is 3.

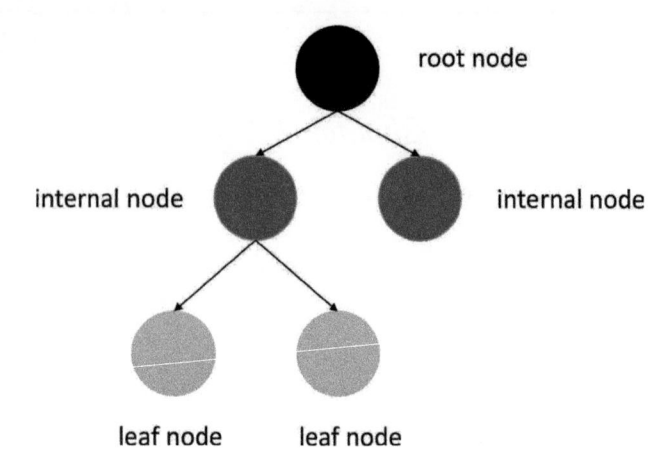

Figure 9.1 The structure of an example DT

Figure 9.1 shows that a DT is constructed in a recursive manner: from the root node, the nodes are split following the same rule until one node cannot be split any more (e.g., only one example is contained by it or the targets of all the examples contained by it are identical) or some preset criteria are reached (e.g., the tree has reached the maximum depth, or the number of examples contained by a leaf node of the current node, if it is to be split, will be less than the minimum number of examples required to be in a leaf node). Then, when using a constructed decision tree to predict the output of an unseen example, the prediction process follows a set of "if-then" rules from the root node to a certain leaf node. In the process of decision tree construction, the most important task is how to select the optimal splitting criteria to split a node. The strategies used to find the optimal splitting criteria are different in classification and regression tasks and in different tree construction algorithms, which will be introduced in the following sections in detail.

9.2 Node splitting in classification trees

Three algorithms are popular to construct classification DTs: iterative dichotomizer 3 (ID3), C4.5, and classification and regression tree (CART).

9.2.1 Iterative dichotomizer 3 (ID3)

In ID3, a node in a decision tree is repeatedly (iteratively) divided (dichotomize) into two or more child nodes from top to bottom in a greedy manner. The term "greedy" means that for each node, the best splitting criteria at the present moment for the current node is selected, while it may not be the optimal of the whole tree. ID3 is based on entropy and information gain in information theory. Especially, entropy is used to measure the degree of randomness in the information being processed. The higher the entropy, the harder to draw any conclusion from the information. Denote by X a discrete random variable with I values which has the following probability distribution

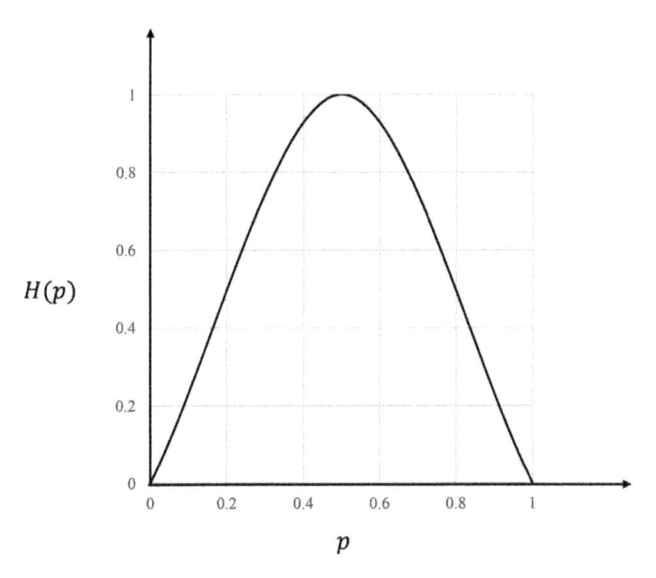

Figure 9.2 The relationship between entropy and probability in binary classification problems

$$P(X = x_i) = p_i, i = 1, ..., I. \tag{9.1}$$

Then, the entropy of random variable X is

$$H(X) = -\sum_{i=1}^{I} p_i \log_2 p_i. \tag{9.2}$$

If $p_i = 0$, $p_i \log p_i$ is set to 0. The entropy of X only depends on its distribution, and thus Equation (9.2) can be written as Equation (9.3):

$$H(p) = -\sum_{i=1}^{I} p_i \log_2 p_i. \tag{9.3}$$

In particular, in binary classification problems where $I = 2$, the probability of one class is p and that of the other class is $1 - p$. The relationship between entropy $H(p)$ and probability p can be shown as follows:

Figure 9.2 shows that when $p = 0$ or $p = 1$ we have $H(p) = 0$, which means that there is no uncertainty in the random variable. When $p = 0.5$ we have $H(p) = 1$, which means that the uncertainty in the random variable is at the maximum. For two random variables X and Y, conditional entropy $H(X, Y)$ is the uncertainty of Y when X is known, and can be presented as follows

$$H(Y|X) = \sum_{i=1}^{I} p_i H(Y|X = x_i), \tag{9.4}$$

where $p_i = P(X = x_i), i = 1, ..., I$. Based on the concepts of entropy and conditional entropy, information gain (IG) is proposed as a statistical property that measures how much uncertainty can be reduced by introducing another random variable as the condition, that is,

$$G(Y,X) = H(Y) - H(Y|X). \tag{9.5}$$

The concept of IG can be used as the criterion to find the optimal split of a node in a DT for classification, where Equation (9.5) can be understood as how much uncertainty regarding the categorical target in the training set can be reduced if a feature is introduced as the condition. The larger the IG value of a feature, the more suitable the feature is for splitting the node, as the subsequent nodes are purer. This is the basic idea of the ID3 algorithm for DT construction.

ID3 is applicable to classification problems where all the features are categorical. In the process of DT construction using ID3, from splitting the root node, the IG of each candidate feature for splitting the current node (where each feature value results in a child node) is calculated, and the feature leading to the largest IG is used as the feature to split the node. Especially, the child nodes of the current node are obtained by using all the values of the selected feature to split the current node, and the number of resulting child nodes is equal to the number of values the selected feature has. Note that for a numerical feature, the values of the feature may be infinite, while there might be only one example with a specific feature value. Consequently, it is believed that ID3 is not able to deal with numerical features.

The above process can be mathematically presented as follows. Given training set $D = \{(x_i, y_i), i = 1, ..., n\}$ for a classification problem, where the target has a total of K possible values and one value is denoted by c_k, $k = 1, ..., K$. For example, in the ship detention prediction problem, $K = 2$ and $c_1 = 0$ and $c_2 = 1$. 1 indicates that a ship is detained and 0, otherwise. C_k is the number of examples in class c_k. For a feature x (e.g., ship type) with J candidate values, i.e., the set of values of x is $\{x_1, ..., x_j, ..., x_J\}$, at the root node, D can be divided into J subsets with each value of x as a child node (i.e., examples with the same value of x are divided into one child node), which are denoted by $D_1, ..., D_j, ..., D_J$. Denote the set of examples belonging to class c_k in D_j by D_{jk}. For a node other than the root node, denote the set of examples contained in it by D', $D' \in D$, the set of examples in subset j by D'_j, the number of examples in D' belonging to class c_k by C'_k, and the set of examples belonging to class c_k in D'_j by D'_{jk}. The entropy of D' is

$$H(D') = -\sum_{k=1}^{K} \frac{C'_k}{|D'|} \log_2 \frac{C'_k}{|D'|}. \tag{9.6}$$

Then, the conditional entropy of feature x over data set D' is

$$H(D'|x) = \sum_{j=1}^{J} \frac{|D'_j|}{|D'|} H(D'_j) = -\sum_{j=1}^{J} \frac{|D'_j|}{|D'|} \sum_{k=1}^{K} \frac{|D'_{jk}|}{|D'_j|} \log_2 \frac{|D'_{jk}|}{|D'_j|}. \tag{9.7}$$

Then, the IG can be calculated by

$$G(D',x) = H(D') - H(D'|x). \tag{9.8}$$

Feature x^* leading to the maximum value of among all candidate features is used as the optimal feature to split the current node, where the examples of the same value for x^* are split to one child node. This means that a total of J child nodes are generated. Then, the child nodes generated are split following the above steps. A node will not be split any more if its entropy is below a preset threshold.

ID3 is an intuitive algorithm based on information theory, which is easy to understand. However, it has some obvious drawbacks, which are listed as follows:

- ID3 cannot properly process continuous feature, as a continuous feature usually has too many possible values and each of the subsequent child nodes may only contain one example. Therefore, continuous features cannot be directly used by the ID3 algorithm.
- IG gives an absolute value which is highly related to the number of feature values: IG gets large accordingly if the number of feature values increases. For example, ship type has many possible values such as container ship, bulk carrier, heavy lift cargo ship, tanker ship, chemical or LNG ship, multipurpose ship, passenger ship, fishing vessel, tug, liquefied natural gas tanker, and nuclear fuel vessel, among others. In contrast, ship RO performance only has three possible values, namely high, medium, not listed in our data set. If IG is used as the criteria to select the best feature for splitting, it would be biased toward feature ship type as this feature has more values, and thus it is more likely to be selected.
- Missing values of the features cannot be processed by ID3.
- There is no pruning process in a DT constructed by ID3 when it was first proposed, and thus might lead to the problem of over-fitting.

9.2.2 C4.5

The above drawbacks of ID3 can be addressed by C4.5 algorithm in the following ways, which is developed by Quinlan [1]:

- For a continuous feature x with J values (where J can be close to n especially for non-integer values), dichotomy is used to discretize the features. First, rank the values of feature x in ascending order denoted by $\{x^1, ..., x^j, ..., x^J\}$. Then, there will be a total of $J-1$ split points that can divide the values of feature x into two categories. One candidate split point is denoted by x^t, and $x^t = \frac{x^j + x^{j+1}}{2}, j = 1, ..., J-1$. Based on x^t, the current data set D' can be divided into two categories: D'_{x^t-} with examples with feature x less than or equal to x^t, and D'_{x^t+} with feature x larger than x^t. Then, the continuous feature x can be treated as a binary feature.
- Mainly aimed for a categorical feature x with J values, information gain ratio, or IGR, is used as the criteria to find the best split feature in C4.5 algorithm. The IGR of feature x to data set D' can be calculated by

$$G_R(D', x) = \frac{G(D', x)}{H_x(D')}, \qquad (9.9)$$

where $H_x(D') = -\sum_{j=1}^{J} \frac{|D'_j|}{|D'|} \log_2 \frac{|D'_j|}{|D'|}$ is the intrinsic value of feature x. The larger number of values of J, the larger value of $H_x(D')$. Hence, it can be used to reduce the influence of the number of feature values on the IG value calculated.
- Two issues regarding missing values need to be addressed: one is how to find the best split feature when some features have missing values, and the other is after selecting the best split feature, how to deal with the examples with missing feature values. The basic idea to address the above two issues is to attach different weights to the examples. Readers are referred to Reference 1 for more detailed information.
- Two tree pruning strategies can be adopted: prepruning and postpruning. In prepruning, some limitations called tree growing termination conditions are set to restrict tree growing, which aim to prevent the tree from learning the training data too well by limiting its size. Common tree growing termination conditions include:

1. max_tree_depth, which is the maximum depth of the tree.
2. min_samples_leaf, which is the minimum number of examples that should be contained by a leaf node of the tree.
3. min_gain, which is the minimum IG or IGR required for splitting a node.

In postpruning, a complete tree is first constructed, and then two leaf nodes from the same parent node are "drawn back" to their parent node, i.e., assume that the parent node is not split. Then, the performance of the DT with leaf nodes drawn back is tested on a hold-out validation set. If the prediction performance can be improved, the two leaf nodes involved are to be pruned, and they are reduced to their parent node.

9.2.3 Classification and regression tree (CART)

CART is a binary tree where each node is split into two child nodes following a given splitting criterion: examples satisfying the criterion are split to the left child node, and the other examples are split to the right child node. As the name suggests, CART can be used for both classification and regression tasks. In this section, we introduce how to apply CART algorithm to construct DT for classification tasks. In a classification tree constructed by CART, Gini index is used as the criterion to find the best split. For data set $D = \{(x_i, y_i), i = 1, ..., n\}$ where $y_i \in \{c_1, ..., c_k, ..., c_K\}$ and c_k is a possible target value. The number of examples in class c_k is denoted by C_k. The probability of an example to be of class c_k is p_k. Then, the Gini index of the probability distribution is

$$Gini(p) = \sum_{k=1}^{K} p_k(1-p_k) = 1 - \sum_{k=1}^{K} p_k^2. \tag{9.10}$$

For binary classification problem, denote the probability of an example to be of one class is p, then the Gini index of the probability distribution can be calculated by

$$Gini(p) = 2p(1-p). \tag{9.11}$$

The Gini index of data set D can be calculated by

$$Gini(D) = 1 - \sum_{k=1}^{K}\left(\frac{C_k}{|D|}\right)^2. \tag{9.12}$$

In CART classification tree where the nodes are divided in a binary manner, feature x as well as its value x^j is used to divide the current node containing data set D' into two mutually exclusive nodes containing data set D'_1 and D'_2, respectively. If feature x is continuous, D'_1 contains examples with feature value less than or equal to x^j and D'_2 contains examples with feature value larger than x^j. If feature x is categorical, D'_1 contains examples with feature value $x = x^j$ and D'_2 contains examples with feature value $x \neq x^j$. Then, the Gini index of data set D' if feature value pair (x, x^j) is used for node splitting can be calculated by

$$Gini(D', x, x^j) = \frac{|D'_1|}{|D'|}Gini(D'_1) + \frac{|D'_2|}{|D'|}Gini(D'_2). \tag{9.13}$$

Similar to the entropy of data set D', the Gini index of D' shows its degree of uncertainty: the larger value of $Gini(D')$, the higher degree of uncertainty in D'. Therefore, to split a node containing data set D' in a classification tree based on CART, the feature value pair leading to the minimum $Gini(D', x, x^j)$ is selected for node splitting.

CART is able to process missing values in features, and it is based on finding surrogate splits. We do not cover the details here. Readers are referred to [2] for more illustration. A DT constructed using CART can be pruned to reduce over-fitting. The prune process starts from pruning the leaf nodes to their parent node until coming to the root node. Then, a set of sub-trees can be formulated in the pruning process, which is denoted by $\{T_0, T_1, ..., T_N\}$. Then, a hold-out validation set or cross validation should be used to find the best sub-tree leading to the minimum loss function:

$$C_\alpha(T_n) = C(T_n) + \alpha|T_n|, \tag{9.14}$$

where $C(T_n)$ is the prediction error of sub-tree T_n on the validation set or in cross validation, $|T_n|$ is the number of leaf nodes in the sub-tree used to represent tree complexity, and α is a hyperparameter taking non-negative value to achieve a trade-off between tree prediction error and tree complexity.

The overall procedure of constructing a classification tree is shown in Algorithm 1.

Algorithm 1 Construction of a DT for classification

Input: Training set D, the algorithm (e.g., ID3, C4.5, or CART) to find the best split criterion for each node, the set of tree growing termination conditions.
Output: A classification DT $f_{DT}^c(x)$.
1: Find the best split criterion of the current node according to the given tree training algorithm.
2: Split the current node using the best split criterion.
3: Repeat the above two steps in a depth-first manner until any of the tree growing termination conditions is reached. Then, this node becomes a leaf node and a new node for splitting is found by backtracking.
4: Repeat Step 3 until there is no more nodes that can be split. Then, the growing process of the classification tree terminates.

5: Conducting post-pruning of the tree if it is needed. Then, the structure of the classification DT $f_{DT}^c(x)$ is determined.
6: Finally, the total training set is separated into M mutually exclusive sub-sets (i.e., leaf nodes in the DT constructed) D_1 to D_M, with one sub-set denoted by D_m, $m = 1, ..., M$. The output of examples in leaf node D_m is the majority class of the examples contained. Then, the outputs of classification DT $f_{DT}^c(x)$ are determined.
7: **return** $f_{DT}^c(x)$.

An improved version of CART is implemented by *scikit-learn* API [3] to deal with classification problems. We use the same features as those in Example 9.1. Chapter 9 to predict whether a ship has more than 5 deficiencies in the following Example 9.1.

Example 9.1. *Recall that the whole data set containing 1991 examples is randomly divided into a training set with 1500 examples and a test set with 491 examples. The numerical features are GT, age, last_inspection_time, and last_deficiency_no. The categorical features are flag_performance, RO_performance, company_performance, last_inspection_state, and ship type. The feature encoding method is the same as that in Example LABEL:eg:ANN_reg. Note that as the features as well as their values are used to find the best split directly without calculating distance or similarity, there is no need for feature scaling.*

We use Gini index as the node splitting criteria, and we tune the values of two hyperparameters to control tree complexity: max_depth and min_samples_leaf. As the training set is unbalanced, where only 381 ships have more than 5 deficiencies among all the 1500 ships contained, $F1$ *score* is used as the metric for hyperparameter tuning. The core code is as follows:

```
from sklearn.model_selection import GridSearchCV
param_grid = {'max_depth': range(3,16,3),
    # This means that the candidate values of 'max_depth'
        is 3,6,9,12,15
    'min_samples_leaf':range(1,10,2)}
model =
    DecisionTreeClassifier(criterion='gini',random_state=0)
grid_search = GridSearchCV(model, param_grid, n_jobs=-1,
    cv=5,scoring='f1')
grid_search.fit(X_train, y_train)
best_parameters = grid_search.best_estimator_.get_params()
print(best_parameters)
# output: the optimal max_depth is 9 and min_samples_leaf
    is 7
```

Then, the optimal max_depth and min_samples_leaf are used to train and test the final classification DT model using the following code:

```
final_model =
    DecisionTreeClassifier(criterion='gini',random_state=0,
max_depth=9,min_samples_leaf=7)
final_model.fit(X_train, y_train)
y_pred = final_model.predict(X_test)
```

The Precision, Recall, and F1 score are 0.66, 0.64, and 0.65, respectively.

9.2.4 Node splitting in regression trees

When applying DTs to address regression tasks, CART is the most popular algorithm for tree construction, and we only cover CART in this section. Given a training data set $D = \{(\mathbf{x}_i, y_i), i = 1, ..., n\}$ where y_i is continuous, MSE is used as the criterion to split one node in a regression tree in CART. Starting from the root node, the tree is constructed in a greedy manner: to split one node containing data set D', all features as well as their values are enumerated to form the feature value pair denoted by (feature, value) or (x, x^j). Then, examples with the feature less than or equal to the threshold value are split to the left child node, and the other examples are split to the right child node. That is, the example set of the left child node is $D'_1(x, x^j) = \{i = 1, ..., n | x_{ij} \leq x^j\}$, and the example set of the right child node is $D'_2(x, x^j) = \{i = 1, ..., n | x_{ij}^j\}$. The output of one child node is the average targets of the examples contained in that node. That is, the output of $D'_1(x, x^j)$ is $c_1 = \frac{1}{|D'_1(x, x^j)|} \sum_{i \in D'_1(x, x^j)} y_i$, and the output of $D'_2(x, x^j)$ is $c_2 = \frac{1}{|D'_2(x, x^j)|} \sum_{i \in D'_2(x, x^j)} y_i$. The best split pair is the one that leads to the minimum sum of MSE of the two child nodes, that is

$$(x^*, x^{j*}) = \arg\min_{x \in \{x_1, ..., x_J\}, x^j \in \{x^1, ..., x^J\}} \left[\sum_{i \in D'_1(x, x^j)} (y_i - c_1)^2 + \sum_{i \in D'_2(x, x^j)} (y_i - c_2)^2 \right]. \quad (9.15)$$

After a regression tree based on CART is constructed, tree pruning can be conducted similar to the classification tree based on CART to reduce over-fitting. The overall procedure of constructing a regression tree using CART is shown in Algorithm 2.

Algorithm 2 Construction of a DT for regression based on CART

Input: Training set D, the set of tree growing termination conditions.
Output: A regression DT $f_{DT}^r(x)$.
1: Find the best split feature value pair (x^*, x^{j*}) of the current node based on Eq. (9.15).
2: Split the current node using (x^*, x^{j*}).
3: Repeat the above two steps in a depth-first manner until any of the tree growing termination conditions is reached. Then, this node becomes a leaf node and a new node for splitting is found by backtracking.

4: Repeat Step 3 until there is no more nodes that can be split. Then, the growing process of the regression tree terminates.
5: Conduct post-pruning of the tree if it is needed. Then, the structure of the regression DT $f_{DT}^r(x)$ is determined.
6: Finally, the total training set is separated into M mutually exclusive sub-sets (i.e., leaf nodes in the DT constructed) D_1 to D_M, with one sub-set denoted by D_m. The output of examples in leaf node D_m is the average target values of the examples contained in D_m, $m = 1, ..., M$. Then, the outputs of the regression DT $f_{DT}^r(x)$ are determined.
7: **return** $f_{DT}^r(x)$.

An improved version of CART is implemented by *scikit-learn* API [3] to deal with regression problems. We use the same features and prediction target as those in Example 9.1 of this chapter to predict the number of deficiencies of a ship in the following Example 9.2.

Example 9.2. We use MSE as the node splitting criteria, and we tune the values of two hyperparameters to control tree complexity: max_depth and min_samples_leaf, similar to that in classification DT. The core code is as follows:

```
from sklearn.model_selection import GridSearchCV
param_grid = {'max_depth': range(3,16,3),
    'min_samples_leaf':range(1,10,2)}
model =
    DecisionTreeRegressor(criterion='mse',random_state=0)
grid_search = GridSearchCV(model, param_grid, n_jobs=-1,
    cv=5,scoring='neg_mean_squared_error')
grid_search.fit(X_train, y_train)
best_parameters = grid_search.best_estimator_.get_params()
print(best_parameters)
# output: the optimal max_depth is 6 and min_samples_leaf
    is 7
```

Then, the optimal max_depth and min_samples_leaf are used to train and test the final regression DT model using the following code:

```
final_model =
    DecisionTreeRegressor(criterion='mse',random_state=0,
max_depth=6,min_samples_leaf=7)
final_model.fit(X_train, y_train)
y_pred = final_model.predict(X_test)
```

The *MSE*, *MAE*, and *MAPE* are 18.4986, 2.8218, and 0.6946, respectively.

DT is a popular machine learning method to address both classification and regression problems. The advantages and disadvantages of DTs are summarized as follows.

Advantages:

- DTs are easy to understand and intuitive. They can also be visualized. Decision rules can be summarized from a constructed DT.
- Not many data as well as feature preprocessing procedures are needed. For example, feature normalization and scaling are not needed, and some DT construction algorithms can deal with missing values.
- DTs can easily be extended to deal with multi-output problems.

Disadvantages:

- DTs can easily be over-fitting, especially when the tree constructed is very complex.
- DTs can be unstable, which means that they can easily be influenced by small variations in the data, especially some extreme feature values.
- The prediction given by DTs for regression problems is not smooth or continuous.
- Learning an optimal DT is known to be NP-hard. Therefore, all the above popular DT construction methods are in a heuristic manner, where optimal decisions in a local manner (i.e., for each node in the tree) are made[*].

We summarize the features and properties of the common algorithms for DT construction as follows in Table 9.1.

9.3 Ensemble learning on tree-based models

The above disadvantages of DTs can be mitigated by ensemble learning, a technique in machine learning model to combine a certain number of individual learners to reach a final decision, where the basic structure is shown in Figure 9.3.

The individual learners contained by an ensemble learning model can be identical in terms of model type (i.e., homogeneous) or distinct (i.e., heterogeneous). In the former case, the individual learners are called base learners, and

---

[*]Constructing global optimal DTs is also studied by researchers, where readers are refered to [4–6]

Table 9.1 Common DT construction algorithms

Name of the algorithm	Output data	Input data	Splitting criterion	Tree structure	Processing of missing values	Pruning
ID3	Categorical data	Categorical data	Information gain	Polytree (i.e., a tree node can have more than two child nodes)	No	None
C4.5	Categorical data	Categorical and numerical data	Information gain ratio	Polytree	Yes, based on assigning weights to examples	Prepruning and postpruning based on nodes drawn back (i.e., pessimistic error pruning)
CART	Categorical and numerical data	Categorical and numerical data	Classification: Gini index; Regression: MSE (mean squared error) or MAE (mean absolute error)	Binary tree (i.e., a tree node only has two child nodes)	Yes, based on surrogate splits	Prepruning and postpruning which is first generating all possible sub-trees and then finding the best one (i.e., cost-complexity pruning)

in the latter case, they are called component learners. In this section, we mainly focus on how to develop an ensemble model consisting of a certain number of DTs constructed by CART algorithm as base learners to address classification or regression tasks. Actually, based on Hoeffding inequality, as the number of base learners contained by the ensemble model, i.e., T, increases, the error rate decreases dramatically and will be close to zero eventually [7]. Meanwhile, it should also be noted that the above analysis is based on an important assumption: the error of the base learners should be independent with each other. However, "good" while "independent" base learners are hard to find, because all the base learners are developed to address the same problem, and thus they cannot be independent, especially under the condition that the data set size is limited while certain accuracy should be achieved. Therefore, the core steps of developing a strong ensemble model include how to develop "divergent and good" base learners that can capture different aspects of the data accurately, and how to produce a strong ensemble model based on these base learners. To achieve this, two strategies are widely used, namely bagging and boosting.

9.3.1 Bagging

As mentioned in section 9.2.4, one disadvantage of DTs is that they are very sensitive to the training set: a slightly different training set can result in a totally different DT, especially for training sets with extreme points. Therefore, to produce divergent base learners in an ensemble model, different training sets generated from the original training set can be used. At the same time, as the dimension of the original training set is fixed and limited, to ensure that the performance of a base learner is not too bad, the number of examples contained by a training set for a base learner should be the same as the original training set. One viable approach is to draw samples of the same size as the original training set from the whole training set and construct machine learning models on these samples. Bootstrapping aggregating (bagging) is such a widely used approach to develop ensemble learning method in a parallel manner. Bagging is based on bootstrap sampling: for a training set D containing n examples, to formulate a sample training set $\hat{D}_t$, a random example is first selected and place to $\hat{D}_t$, and then it is placed back to D. The above process is repeated n times, and then the sample training set $\hat{D}_t$ with n examples can be formulated. As the examples are selected with replacement, it is highly likely that there are duplicate examples in $\hat{D}_t$ while some examples in D are not contained by $\hat{D}_t$. Actually, only about 63.2% of the examples in D are chosen in $\hat{D}_t$ when the number of examples contained in training set D is sufficiently large [8]. After repeating the above process T times, T sample training sets, $\hat{D}_1$ to $\hat{D}_T$, can be formulated, and can be used to train T DTs. To ensemble the outputs of the T DTs in the final ensemble model, voting strategy is adopted in classification problems, i.e., the target of the majority DTs is used as the final output. In case of tie, a random output in tie is chosen. In regression problems, the average of the targets of all the DTs is used as the final output.

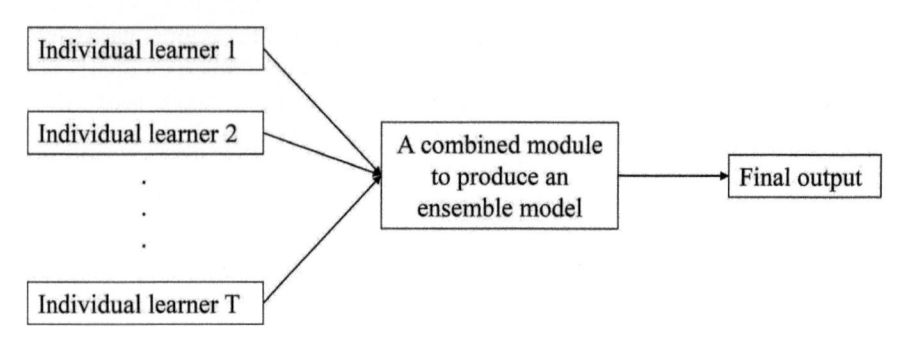

118 *Machine learning and data analytics for maritime studies*

Figure 9.3 An illustration of ensemble learning model

A famous tree ensemble method, called random forest or RF for short, is an extension of bagging method based on DTs where one more layer of randomness is introduced to the training process of a DT. The only difference between bagging based on DTs and RF is the candidate features used to split a node in a DT: all the features are considered to find the optimal feature value pair in bagging, while only a subset of all the features are considered to find the optimal feature value pair in RF. Denote the number of features used to split a node of a DT in the RF by d and recall that data set D has m features, the recommended value for d is $\log_2 m$ [9]. Therefore, the randomness in RF comes from two aspects: samples from bootstrap sampling used to construct each DT and a subset of features used to split each node. The characteristics and working mechanism of an RF model is shown in Figure 9.4.

RF is implemented by *scikit-learn* API [3] to deal with both regression and classification tasks. We use the following Example 9.3 and Example 9.4 to show how to call *scikit-learn* API to predict ship detention and ship deficiency

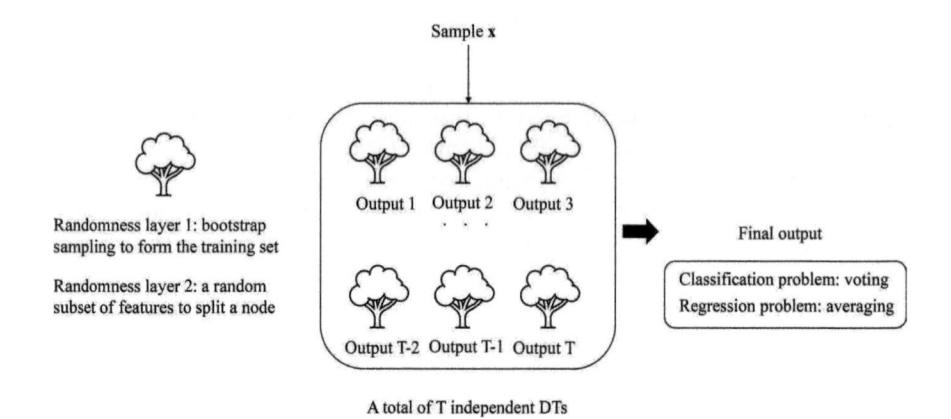

Figure 9.4 An illustration of an RF model

number in PSC. The features and target of Example 9.3 for classification are the same as Example 9.1, and those of Example 9.4 for regression are the same as Example 9.2.

Example 9.3. *In addition to the hyperparameters in DT, RF has more hyperparameters to control the whole structure of the model. The number of DTs contained, i.e., T, is called n_estimators in scikit-learn, the number of features to consider when looking for the best split for a node is called max_features. Similar to the DT constructed to predict ship detention, we also use Gini index as the node splitting criteria of each DT. We tune a total of four hyperparameters: two from the DT to control tree complexity: max_depth and min_samples_leaf, and the other two from the ensemble model n_estimators and max_features. As the training set is unbalanced, where only 381 ships have more than 5 deficiencies among all the 1500 ships contained, F1 score is used as the metric for hyperparameter tuning. The core code is as follows*:

```
from sklearn.model_selection import GridSearchCV
param_grid = {'n_estimators':range(100,501,100),
 'max_features':[0.2,0.4,0.6,0.8],
 'max_depth': range(3,16,3),
 'min_samples_leaf':range(1,10,2)}

model =
    RandomForestClassifier(criterion='gini',random_state=0)
grid_search = GridSearchCV(model, param_grid, n_jobs=-1,
    cv=5,scoring='f1')
grid_search.fit(X_train, y_train)
best_parameters = grid_search.best_estimator_.get_params()
print(best_parameters)
# output: the optimal n_estimators is 300, max_features is
    0.6, max_depth is 15 and min_samples_leaf is 9
```

Then, the optimal n_estimators, max_features, max_depth, and min_samples_leaf are used to train and test the final classification DT model using the following code:

```
final_model =
    RandomForestClassifier(criterion='gini',random_state=0,
n_estimators=300, max_features=0.6, max_depth=15,
    min_samples_leaf=9)
final_model.fit(X_train, y_train)
y_pred = final_model.predict(X_test)
```

The *Precision, Recall*, and F_1 score are 0.68, 0.63, and 0.65, respectively.

Example 9.4. *We use MSE as the node splitting criteria, and we tune the values of four hyperparameters to control the structure of RF and tree complexity: n_estimators, max_features, max_depth, and min_samples_leaf, similar to that in classification RF. The core code is as follows:*

```
from sklearn.model_selection import GridSearchCV
param_grid = {'n_estimators':range(100,501,100),
 'max_features':[0.2,0.4,0.6,0.8],
 'max_depth': range(3,16,3),
 'min_samples_leaf':range(1,10,2)}

model =
    RandomForestRegressor(criterion='mse',random_state=0)
grid_search = GridSearchCV(model, param_grid, n_jobs=-1,
    cv=5,scoring='neg_mean_squared_error')
grid_search.fit(X_train, y_train)
best_parameters = grid_search.best_estimator_.get_params()
# output: the optimal n_estimators is 200, max_features is
    0.6, max_depth is 9 and min_samples_leaf is 3
```

Then, the optimal n_estimators, max_features, max_depth, and min_samples_leaf are used to train and test the final regression RF model using the following code:

```
final_model =
    RandomForestRegressor(criterion='mse',random_state=0,n_estimators=20(
final_model.fit(X_train, y_train)
y_pred = final_model.predict(X_test)
```

The *MSE*, *MAE*, and *MAPE* are 14.2134, 2.5567, and 0.6212, respectively.

9.3.2 Boosting

The base learners are independently constructed and are ideally independent of each other in bagging; therefore, they can be constructed in parallel. In contrast, the base learners, which are also called weak learners, are strongly dependent on each other in boosting. The weak learners are constructed one by one, which aims to constantly improve (boost) the performance of the current model developed by paying more attention to the examples with larger prediction errors by attaching higher weights to them or changing the prediction target. Then, after developing T weak learners for the boosting model, they are combined linearly to produce the final output. In boosting, the number of weak learners T is also called the maximum number of iterations. In this section, we introduce two boosting methods: adaptive boosting or Adaboost and boosting tree, where the former is based on changing the examples' weights and the latter is to learn the error of the base models constructed so far.

9.3.2.1 Adaboost

The weak learner of Adaboost is the decision stump, which is a DT with only one split on the root node. In each round of weak learner construction, the weights of the examples are changed according to their prediction error in Adaboost, and the examples with weights attached are used to train the weak learner in the next round. To be more specific, consider a binary classification problem with a training set denoted by $D = \{(\mathbf{x}_i, y_i), i = 1, ..., n\}$, where $y_i \in \{-1, 1\}$. The procedure of constructing an Adaboost model is shown in Algorithm 3.

Algorithm 3 Construction of an Adaboost model

Input: Training set D, the maximum number of iterations T.
Output: An Adaboost model for classification $f^c_{Adaboost}(x)$.
1: Initialize the weight of each example in D for the first DT in Adaboost: $D_1 = (w_{t1}, w_{t2}, ..., w_{tn}), w_{ti} = \frac{1}{n}, i = 1, ..., n$.
2: Construct an initial Adaboost model $f^0_{Adaboost}(x) = 0$.
3: **for** $t = 1, ..., T$ **do**
4:    Use the training set to develop a weak learner and denote it by $G_t(x)$.
5:    Evaluate the performance of $G_t(x)$ considering the weights of the examples:

$$e_t = \sum_{i=1}^{n} w_{ti} I(G_t(x_i) \neq y_i), \tag{1.16}$$

and we have $e_t \in [0, 1]$.
6:    Calculate the weight of weak learner t:

$$\alpha_t = \frac{1}{2} \ln \frac{1 - e_t}{e_t}. \tag{1.17}$$

7:    The current model in round t can be presented by

$$f^t_{Adaboost}(x) = f^{t-1}_{Adaboost}(x) + \alpha_t G_t(x). \tag{1.18}$$

8:    Update the weight of the examples in D: $D_{t+1} = (w_{t+1,1}, w_{t+1,2}, ..., w_{t+1,n})$, where $w_{t+1,i} = \frac{w_{ti}}{Z_t} \exp(-\alpha_t y_i G_t(x_i)), i = 1, ..., n,$ and $Z_t = \sum_{i=1}^{n} w_{ti} \exp(-\alpha_t y_i G_t(x_i))$ is the normalizer used to turn D_{t+1} to a probability distribution.
9: **end for**
10: After developing T weak learners, they are combined in a linear way by

$$f^{combined}_{Adaboost}(x) = f^{T-1}_{Adaboost}(x) + \alpha_T G_T(x). \tag{1.19}$$

Then, the final Adaboost model for regression is

$$f^c_{Adaboost}(x) = sign(f^{combined}_{Adaboost}(x)), \tag{1.20}$$

where

$$sign(x) = \begin{cases} +1 & \text{if } x \geq 0, \\ -1 & \text{if } x < 0. \end{cases}$$

11: **return** $f^c_{Adaboost}(x)$.

There are some further explanations of Algorithm 3.

- In step 1, all the examples are assigned identical weights to train the first (an initial) weak learner. Then, the weights of the examples will be updated according to the performance of the current weak learner.
- Examples that are wrongly predicted will be assigned higher weights as shown by how $w_{t+1,i}$ is calculated. Therefore, these examples need more attention in the next round of learning. Therefore, it can be seen that in the learning process of Adaboost, the examples are not changed while their weights, i.e., the distribution of the examples, are changed.
- From Equation (1.16), it can be seen that the error of round t is the sum of weights of the examples that are wrongly classified, which indicates the relationship between distribution D_t of the data set and the weak learner $G_t(x)$.
- The weight of weak learner $G_t(x)$, i.e., α_t is important of this weak learner in the final classification model. It is shown in Equation (1.17) that when $e_t \leq \frac{1}{2}$, $\alpha_t \geq 0$, and as e_t decreases α_t increases. This shows that the more accurate a weak learner is, the more important role it plays in the final prediction model. In contrast, when $e_t > \frac{1}{2}$, we have $\alpha_t < 0$. It is not reasonable to have a weak learner whose weight is less than zero. To address this issue, one approach is to change the predicted outputs to their inverse numbers, and the error rate should become smaller than $\frac{1}{2}$. Another approach is to discard the current weak learner as its performance is too bad.
- The sign of $f_{Adaboost}^{combined}(x)$ determines the predicted class of example x, and the absolute value of $f_{Adaboost}^{combined}(x)$ shows the reliability of the prediction.
- Regularization can be applied in Adaboost in Equation (1.18) of Step 7 by introducing learning rate denoted by v, $v \in (0, 1)$, which is used to control the model updating speed. Then, Equation (1.19) is changed to $f_{Adaboost}^{t}(x) = f_{Adaboost}^{t-1}(x) + v\alpha_t G_t(x)$. The larger value v is, the faster the model learns.

The following Example 9.5 is an example to use Adaboost for ship risk prediction using the same features and target as those in Example 9.3.

Example 9.5. *As decision stump is used as the weak learner, there is no hyperparameter used to control the complexity of a DT contained by an Adaboost. Only two hyperparameters are tuned to control the learning performance of the whole Adaboost model: n_estimators, i.e., T, which is the number of weak learners contained, and learning_rate, i.e., v to control model learning speed. As the training set is unbalanced, where only 381 ships have more than five deficiencies among all the 1500 ships contained, F1 score is used as the metric for hyperparameter tuning. The core code is as follows:*

```
from sklearn.model_selection import GridSearchCV
param_grid = {'n_estimators':range(50,301,50),

 'learning_rate':[0.1,0.2,0.5,1,0.5,2]}

model = AdaBoostClassifier(random_state=0)
grid_search = GridSearchCV(model, param_grid, n_jobs=-1,
    cv=5,scoring='f1')
grid_search.fit(X_train, y_train)
best_parameters = grid_search.best_estimator_.get_params()
print(best_parameters)
# output: the optimal n_estimators is 100, learning_rate is
    0.5
```

Then, the optimal n_estimators and learning_rate are used to train and test the final classification DT model using the following code:

```
final_model =
    AdaBoostClassifier(random_state=0,n_estimators=100,
    learning_rate=0.5)
final_model.fit(X_train, y_train)
y_pred = final_model.predict(X_test)
```

The *Precision*, *Recall*, and F_1 score are 0.69, 0.66, and 0.67, respectively.

Adaboost is widely used for classification tasks, and it can also be used for regression tasks. When used for regression, the main idea and procedure are identical to the situation when it is used for classification, except for the equations used to calculate the performance of a weak learner in Equation (9.16) of Algorithm 3 (which is determined by the regression algorithm adopted), the weight of the examples in *D*, and the weight of each weak learner in Equation (9.17). In addition, theoretically, any ML model can be used as weak learner under the framework of Adaboost, and the most popular ones are DTs and ANNs. It should also be noted that Adaboost can be very sensitive to outliers or extreme examples, as these examples will be attached with large weights in model constructing as they are more likely to be badly predicted. Consequently, the performance of Adaboost is badly influenced.

9.3.2.2 Boosting tree

Boosting tree is also called gradient boosting decision tree, or GBDT for short, which can be used for classification and regression tasks. It is based on CARTs and uses forward-stagewise additive modeling where the model is fit by minimizing a loss function averaged over the training data, i.e., to use the prediction error in the previous round of learning as the new target in the next round of weak learner learning. The above idea can be summarized as "learn from the prediction error so far," and it is achieved by changing the prediction target in the current round to the negative gradient value of the loss function in the last round. The procedure of using GBDT for a regression task is shown in Algorithm 4.

Algorithm 4 Construction of a GBDT model for regression

Input: Training set D, the maximum number of iterations T.
Output: A GBDT model for regression $f^r_{GBDT}(x)$.
1: Construct an initial weak learner $G_0(x) = 0$.
2: **for** $t = 1, ..., T$ **do**
3:      Calculate the negative gradient of the current prediction error of each example by

$$r_{ti} = -\frac{\partial L(y_i, G_{t-1}(x))}{\partial G_{t-1}(x)}, i = 1, ..., n. \quad (1.21)$$

4:      Use a new training set $D_m = (x_i, r_{ti})$ to fit a new regression tree G_t with J_t leaf nodes denoted by $R_{tj}, j = 1, ..., J_t$. For leaf node j, its best output is

$$c_{tj} = \arg\min_c \sum_{x_i \in R_{tj}} L(y_i, G_{t-1}(x) + c). \quad (1.22)$$

5:      The current GBDT model is updated to

$$G_t(x) \leftarrow G_{t-1}(x) + G_t(x)$$
$$= G_{t-1}(x) + \sum_{j=1}^{J_t} c_{tj} I(x \in R_{tj}), \quad (1.23)$$

     where J_t is the number of leave nodes in tree t.
6: **end for**
7: The final GBDT model for regression can be presented by

$$f^r_{GBDT}(x) = G_T(x) = \sum_{t=1}^{T} \sum_{j=1}^{J_t} c_{tj} I(x \in R_{tj}). \quad (1.24)$$

8: **return** $f^r_{GBDT}(x)$.

When using MSE as the loss function in this above task, the negative gradient calculated in Equation (1.21) is $y_i - G_{t-1}(x_i), i = 1, ..., n$, which is the difference between the real and current predictions for example i in the training set. This difference is also called residual and is denoted by r_i. Therefore, it can be seen that the learning objective of each round in GBDT for regression is to minimize the residual of the current prediction, and the final output is calculated by summing the outputs of all rounds of learning. The following Example 9.6 is used to show how to use the *scikit-learn* API [3] to develop a GBDT model for regression to predict the number of deficiencies a ship has.

Example 9.6. *In addition to the hyperparameters in DT (i.e., max_depth and min_samples_leaf), GBDT for regression has more hyperparameters to control the* construction of the whole ensemble model. Similar to Adaboost, it has n_estimators and learning_rate to control the learning process of the whole model. We use MSE as the node splitting criteria, and we tune the values of four hyperparameters to control the structure and complexity of GBDT: n_estimators, learning_rate, max_depth and min_samples_leaf. The core code is as follows:

```
from sklearn.model_selection import GridSearchCV
param_grid = {'n_estimators':range(100,501,100),
 'learning_rate':[0.2,0.4,0.6,0.8],
 'max_depth': range(3,16,3),
 'min_samples_leaf':range(1,10,2)}

model =
    GradientBoostingRegressor(criterion='mse',random_state=0)
grid_search = GridSearchCV(model, param_grid, n_jobs=-1,
    cv=5,scoring='neg_mean_squared_error')
grid_search.fit(X_train, y_train)
best_parameters = grid_search.best_estimator_.get_params()
print(best_parameters)
# output: the optimal n_estimators is 100, learning_rate is
    0.2, max_depth is 3 and min_samples_leaf is 9
```

Then, the optimal n_estimators, learning_rate, max_depth and min_samples_leaf are used to train and test the final regression RF model using the following code:

```
final_model =
    GradientBoostingRegressor(criterion='mse',random_state=0,
    n_estimators=100,learning_rate=0.2,max_depth=3,min_samples_leaf=9)
final_model.fit(X_train, y_train)
y_pred = final_model.predict(X_test)
```

The *MSE*, *MAE*, and *MAPE* are 16.4603, 2.6489, and 0.6671, respectively.

The basic idea of using GBDT for classification is similar to applying GBDT for regression. Consider a GBDT model for binary classification. The main difference is that the output of GBDT for classification is discrete, and thus the prediction error cannot be directly used as the target in the next round of learning like GBDT used for regression. This problem can be solved by using exponential loss, which is similar to the case of Adaboost, or using log loss, which is similar to the case of logistic regression. In this section, we discuss the case of using log loss for classification tasks by GBDT. Log loss aims to minimize the difference between the predicted probability and the real probability of the examples. The log loss function for classification can be presented by

$$L(y,\hat{y}) = \ln(1 + \exp(-y\hat{y})), \tag{9.25}$$

Table 9.2 Comparison of boosting and bagging based on DTs

Name of algorithm	Representative models	Type of base learners	Relationship of base learners	Learning objective of each round of learning	Combining weak learners	Learning rate
Bagging	RF	CART	Independent, and can be paralleled	Real target	Averaging in regression and voting in classification	No
Boosting	Adaboost and GBDT	Adaboost: decision stump; GBDT: CART	Dependent, and cannot be paralleled	Adaboost: real target; GBDT: negative gradient	Adaboost: weights; GBDT: additive manner	Yes

where $y \in \{-1, 1\}$. The procedure of using GBDT for classification task is shown in Algorithm 5.

Algorithm 5 Construction of a GBDT model for classification

Input: Training set D, the maximum number of iterations T.
Output: A GBDT model for classification $f_{GBDT}^c(x)$.
1: Construct an initial weak learner $G_0(x) = 0$.
2: **for** $t = 1, ..., T$ **do**
3:   Calculate the negative gradient of the current error of each example by

$$r_{ti} = -\frac{\partial L(y_i, G_{t-1}(x))}{\partial G_{t-1}(x)} = \frac{y_i}{1 + \exp(y_i G_{t-1}(x_i))}, i = 1, ..., n. \tag{9.26}$$

4:   Use a new training set $D_m = (x_i, r_{ti})$ to fit a new classification tree G_t with J_t leaf nodes denoted by $R_{tj}, j = 1, ..., J_t$. For leaf node j, its best output is

$$c_{tj} = \arg\min_c \sum_{x_i \in R_{tj}} \ln(1 + \exp(-y_i(G_{t-1}(x_i) + c))). \tag{9.27}$$

As the optimal c in Eq. (9.27) is hard to find, an approximate value shown below is used for c:

$$c_{tj} = \frac{\sum\limits_{x_i \in R_{tj}} r_{ti}}{\sum\limits_{x_i \in R_{tj}} |r_{ti}|(1 - |r_{ti}|)}. \tag{9.28}$$

5:   The current GBDT model is updated to

$$G_t(x) \leftarrow G_{t-1}(x) + G_t(x)$$
$$= G_{t-1}(x) + \sum_{j=1}^{J_t} c_{tj} I(x \in R_{tj}), \tag{9.29}$$

where J_t is the number of leave nodes in tree t.
6: **end for**
7: The final GBDT model for classification can be presented by

$$G_T(x) = \sum_{t=1}^{T} \sum_{j=1}^{J_t} c_{tj} I(x \in R_{tj}). \tag{9.30}$$

The probability of example x to be in class 1 is calculated by

$$f_{GBDT}^c(x) = \frac{\exp(2G_T(x))}{1 + \exp(2G_T(x))} = \frac{1}{1 + \exp(-2G_T(x))}. \tag{9.31}$$

8: **return** $f_{GBDT}^c(x)$.

Regularization can be used in GBDT to reduce the problem of over-fitting. For each base learner based on CART, pruning of each tree is a viable approach. Another way is to use sub-sampling without replacement to construct one weak learner, with the aim to reduce the over-fitting of each weak learner. For the whole GBDT model, one possible way is to introduce learning rate v, $v \in (0, 1)$ like that in Adaboost to control the model learning speed, and then Equation (9.23) is changed to $G_t(x) = G_{t-1}(x) + v \sum_{j=1}^{J_t} c_{tj} I(x \in R_{tj})$ in Algorithm 4 and (9.29) is changed to $G_t(x) = G_{t-1}(x) + v \sum_{j=1}^{J_t} c_{tj} I(x \in R_{tj})$ in Algorithm 5.

We compare boosting and bagging models based on DTs in Table 9.2.

References

[1] Quinlan J.R. *C4.5: programs for machine learning*. San Mateo, CA: Morgen Kaufmann Publishers; 2014.
[2] Hastie T., Tibshirani R., Friedman J.H. 'The elements of statistical learning' in *The Elements of Statistical Learning: Data Mining, Inference, and Prediction*. Vol. 2. New York, NY: Springer; 2009.
[3] Pedregosa F., Varoquaux G., Gramfort A, *et al.* 'Scikit-learn: machine learning in python'. *Journal of Machine Learning Research*. 2011;12:28–2825.
[4] Bertsimas D., Dunn J. 'Optimal classification trees'. *Machine Learning*. 2017;106(7):82–1039.
[5] Aghaei S., Gómez A., Vayanos P. 'Strong optimal classification trees'. [arXiv preprint arXiv:210315965] 2021.
[6] Verwer S., Zhang Y. 'Learning optimal classification trees using a binary linear program formulation'. *Proceedings of the AAAI Conference on Artificial Intelligence*. 2019;33:32–1625.
[7] Hoeffding W. 'Springer'. [In: The Collected Works of Wassily Hoeffding] *Probability Inequalities for Sums of Bounded Random Variables*. 1994::26–409.
[8] Breiman L. 'Bagging predictors'. *Machine Learning*. 1996;24(2):40–123.
[9] Breiman L. 'Random forests'. *Machine Learning*. 2001;45(1):5–32.

Chapter 10
Association rule learning

Association rule learning is a rule-based and unsupervised machine learning method, which is widely used to discover interesting relations between items (i.e., records) in a database and also to explore how and why these items are connected. A widely known application of association rule learning is to analyze the purchased items in market basket transactions, which aims to identify what goods are often bought together in one transaction so as to adjust sales strategies to increase sales. For example, a famous story is that on Friday nights, the sales of diapers and beer were correlated in Walmart: a bottle of beer was often bought when diapers were bought. The explanation was that working men were asked to pick up diapers on their way home from work, and they would also buy a bottle of beer for themselves at the same time. Based on this finding, Walmart put the shelves of these two goods close to each other on Friday nights, and the sales volume of both increased greatly. In the above example, diapers and beer are correlated in the following way: diapers → beer, meaning that if diapers are bought, beer is highly like to be bought in the same transaction. This is a basic form of association rule we are going to cover in this section. Actually, association rule learning is widely used to mine the relations of items from transaction databases, and the rules generated are used to guide the activities of personalized product recommendation, shelf placement, combined coupon dispatching, and bundle sales in the retailing industry.

10.1 Large item sets

To introduce how association rules are generated from a given database, we first introduce some basic concepts. Denote a set of N items by $I = \{i_1, ..., i_N\}$. An item set is a subset of I, and an item set containing n' items (where $n' \in [1, ..., N]$) is also called an n' item set and is denoted by $I_{n'}$. A database consisting of M records, where each record is an item set, is denoted by $D = \{t_1, ..., t_M\}$. Define the event of observing the occurrence of a particular item set $I_{n'}$ by $E(I_{n'})$, which means that all the items in $I_{n'}$ are observed in one record. We further define $P(E(I_{n'}))$ as the proportion of the M records that have all the items in $I_{n'}$, which can also be interpreted as the probability of the occurrence of event $E(I_{n'})$. It should also be noted that a record that includes all the items in item set $I_{n'}$ can also include items not in $I_{n'}$ but in I. The probability of $E(I_{n'})$ is also called the *Support* of item set $I_{n'}$, that is, $P(E(I_{n'})) = Sup(I_{n'})$, and $P(E(I_{n'})) \in [0, 1]$. A larger value of $P(E(I_{n'}))$ indicates that the more frequently item set $I_{n'}$ occurs in D. In order to define a large item set denoted by $I_{n'}^*$, which frequently occurs in the database, we define the minimum threshold of *Support* for an item set to be a large item set by *min_Sup*. That is, if and only if $I_{n'}^* \subseteq I$ and $I_{n'}^* \neq \emptyset$ is a large item set, we have $Sup(I_{n'}^*) \geq min\_Sup$.

A rule is generated by dividing a large n' item set, i.e., $I_{n'}^*$ and $n' \geq 2$, into two mutually exclusive and non-empty item sets I_j and I_k, with $I_j \cup I_k = I_{n'}^*$. A rule can be generated from I_j to I_k in form $I_j \to I_k$. To determine whether rule $I_j \to I_k$ is an association rule denoted by $I_j \Rightarrow I_k$, two indicators are further introduced: *Confidence* and *Lift* of $I_j \to I_k$ is calculated by

$$Conf(I_j \to I_k) = \frac{P(E(I_j) \cap E(I_k))}{P(E(I_j))} = P(E(I_k)|E(I_j)), \qquad (10.1)$$

and $Conf(I_j \to I_k) \in [0, 1]$. The larger value *Confidence* is, the more likely that the items in I_k appear given that the items in item set I_j appear. *Lift* of $I_j \to I_k$ is calculated by

$$Lift(I_j \to I_k) = \frac{P(E(I_j) \cap E(I_k))}{P(E(I_j)) \times P(E(I_k))} = \frac{P(E(I_k)|E(I_j))}{P(E(I_k))}, \qquad (10.2)$$

and $Lift(I_j \to I_k) \in [0, +\infty)$ presents the influence of the occurrence of event $E(I_j)$ on event $E(I_k)$, which is the ratio of the probability of the occurrence of event $E(I_k)$ under the condition that event $E(I_j)$ occurs and the probability that event $E(I_k)$ occurs unconditionally in the database. This can be interpreted as how the occurrence of event $E(I_j)$ can increase/decrease (i.e., lift) the occurrence of $E(I_k)$. To be more specific, if $lift(I_j \to I_k) \in [0, 1)$, the occurrence of $E(I_j)$ decreases the probability of the occurrence of $E(I_k)$. If $Lift(I_j \to I_k) \in (1, +\infty)$, the occurrence of $E(I_j)$ increases the probability of the occurrence of $E(I_k)$. If $Lift(I_j \to I_k) = 1$, the occurrence of $E(I_j)$ has no influence on the occurrence of $E(I_k)$, that is, $E(I_j)$ and $E(I_k)$ are independent. It is also interesting to find that as event $E(I_k)$ acts as the denominator to calculate *Lift* of rule $I_j \to I_k$, if $P(E(I_k))$ is large, meaning that the occurrence probability of event $E(I_k)$ is high, the value of $Lift(I_j \to I_k)$ would be reduced. This shows that a frequently occurring event would have less contribution to generating association rules compared to rare events.

The concept of association rule is defined based on *Confidence* and *Lift* of a rule as follows:

Definition 10.1. *Given a large item set $I_{n'}^*$ with $n' \geq 2$ and two non-empty, mutually exclusive, and complementary item sets I_j and I_k, i.e., $I_j \neq \emptyset$, $I_k \neq \emptyset$, $I_j \cap I_k = \emptyset$, and $I_j \cup I_k = I_{n'}^*$, and given the minimum thresholds of Confidence and Lift denoted by min_Conf and min_Lift, respectively, the rule $I_j \to I_k$ is an association rule if and only if $Conf(I_j \to I_k) \geq min\_Conf$ and $Lift(I_j \to I_k) \geq min\_Lift$.*

In association rule $I_j \Rightarrow I_k$, I_j is called antecedent or left-hand-side and I_k consequent or right-hand-side, and it can be interpreted by "if I_j then I_k." The implication of this association rule is that if items contained by I_j occur, there is a high (which is guaranteed by *min_Conf*) and also higher (which is guaranteed by *min_Lift*) probability that the items in I_k can be detected. The thresholds of *Confidence* and *Lift* show the basic idea of association rule generation: the threshold of *Confidence* guarantees that the ratio of $Sup(I_j \cap I_k)$ and $Sup(I_j)$ is above a certain level, i.e., the rule is meaningful, and the threshold of *Lift* guarantees that the association rule is strong enough to be regarded as an effective "rule." In the following sections, the most popular algorithm to generate association rules, namely, Apriori, will be introduced in detail first. Then, as an improvement of Apriori, FP-growth, which uses a more efficient data structure to store the data and realize more efficient searching, is then briefly covered.

10.2 Apriori algorithm

Before presenting the details of the Apriori algorithm, we first present the following two properties of large item sets:

Property 1: Any non-empty and strict subset of a large item set is large.

Property 2: Any super-set of a non-large item set cannot be large.

The above properties are intuitive. For Property 1, suppose we have a large item set I^*, and it has two mutually exclusive and non-empty item sets I_j and I_k where $I_j \cup I_k = I^*$. The number of times where I^* occurs in the database should be no less than *min_Sup* because it is a large item set. As the individual occurrence time of I_j in the database should be no less than the occurrence time of I^* in the database, I_j is also a large item set. This also applies to item set I_k. For Property 2, for a non-large item set I', its occurrence time in the database is less than *min_Sup*, and thus the occurrence time of any super-set of I' is also less than *min_Sup*. Therefore, it cannot be a large item set.

The Apriori algorithm is proposed by Agrawal et al. in 1993 [1]. It is developed based on the above properties to improve the computational efficiency of large item set generation. Denote a large item set containing n' items as a large n'-item set. Denote $L_{n'}$ as the set of all large n'-item sets. Denote $N(I_{n'})$ as the occurrence times of item set $I_{n'}$ ($I_{n'} \in L_{n'}$) in database D. The algorithm to generate the set of all large n'-item sets, i.e., $L_{n'}$, is shown in Algorithm 1.

Algorithm 1 Generation of $L_{n'}, n' = 1, ..., N$

Input: The whole database D, *min_Sup*, which is the threshold of *Support*.
Output: $L_{n'}$, which is the set of all large n'- item sets.
1: **Step 1:**
2: $n' = 1$;
3: $L_{n'} = \emptyset$;
4: **for** all $i_n \in I$ **do**
5: $\quad Sup(i_n) = 0$;
6: $\quad N(i_n) = 0$;
7: $\quad$ **for** all $t_m \in D$ **do**
8: $\quad\quad$ **if** i_n is contained in t_m **then**

9:         $N(i_n) = N(i_n) + 1$;
10:       **end if**
11:    **end for**
12:    $Sup(i_n) = \frac{N(i_n)}{M}$;
13:    **if** $Sup(i_n) \geq min\_Sup$ **then**
14:       $L_1 = L_1 \cup \{\{i_n\}\}$;   ▷ where $\{i_n\}$ is a large 1-item set, and L_1 is the set of all large 1-item sets.
15:    **end if**
16: **end for**
17: **Step 2**:
18: **for** $n' = 2; L_{n'-1} \neq \emptyset$ and $n' \leq N; n'++$ **do**
19:    $C_{n'} = $ generate_candidate$(L_{n'-1})$;   ▷ see Algorithm 2
20:    $L_{n'} = \emptyset$;
21:    **for each** $c \in C_{n'}$ **do**
22:       $N(c) = 0$;
23:       $Sup(c) = 0$;
24:       **for all** $t_m \in D$ **do**
25:          **if** c is contained in t_m **then**
26:             $N(c) = N(c) + 1$;
27:          **end if**
28:       **end for**
29:       $Sup(c) = \frac{N(c)}{M}$;
30:       **if** $Sup(c) \geq min\_Sup$ **then**
31:          $L_{n'} = L_{n'} \cup \{c\}$
32:       **end if**
33:    **end for**
34: **end for**

In Algorithm 1, Step 1 aims to find all large 1-item sets by scanning the whole database D and examining each item it contains. When $n' \geq 2$, by iteration, candidate large n'-item sets are first found by combining the large item sets in $L_{n'-1}$ where the first $(n'-2)$ items are the same while the last item is different, assuming that the items are ordered in alphabet. Then, subsets containing $(n'-1)$ items of each combination are checked to ensure that all of these subsets are large item sets. Otherwise, the item set containing n' items should be removed from the set of candidate large item sets. The detailed steps are shown in Algorithm 2. Then, *Support* of each of the candidate large n'-item sets generated by Algorithm 2 is checked, and the sets with *Support* no less than *min_Sup* are finally retained as the large n'-item sets, i.e., $L_{n'}$.

Algorithm 2 generate_candidate$(L_{n'-1})$

1: Denote a pair of large item sets in $L_{n'-1}$ by $I^{*'}_{n'-1} = \{i'_1, i'_2, ..., i'_{n'-2}, i'_{n'-1}\}$ and $I^{*''}_{n'-1} = \{i''_1, i''_2, ..., i''_{n'-2}, i''_{n'-1}\}$. "<" is used to indicate that the left-hand side item precedes the right-hand side item in the alphabet.
2: **Step 1**: Joining step

3: $C_{n'} = \emptyset$;
4: **for** all pairs of item sets in $L_{n'-1}$ where $n' \geq 2$ **do**
5:     **if** $n' = 2$ or $i'_1 = i''_1, i'_2 = i''_2, \ldots, i'_{n'-2} = i'''_{n'-2}$ **then**
6:         $C_{n'} = C_{n'} \cup \{\{i'_1, i'_2, \ldots, i'_{n'-2}, i'_{n'-1}, i''_{n'-1}\}\}$;
7:     **end if**
8: **end for**
9: **Step 2**: Pruning step
10: **for** all item sets $c \in C_{n'}$ **do**
11:     **for** all subsets s containing $(n'-1)$ items of c **do**
12:         **if** $s \notin L_{n'-1}$ **then**
13:             delete c from $C_{n'}$;
14:             break;
15:         **end if**
16:     **end for**
17: **end for**
18: Return $C_{n'}$.

The following toy example is used to show the procedure of using Apriori algorithm to generate large item sets from a given database D. We require that $min\_Sup = 0.5$. The set of records is shown in Table 10.1.

The candidate large 1-item sets and final large 1-item sets can be generated by Step 1 of Algorithm 1. To be more specific, the candidate large 1-item sets are shown in Table 10.2, and the large 1-item sets are shown in Table 10.3.

Then, all the large n'-item sets, where $n' \geq 2$, are generated by Step 2 of Algorithm 1, which combines the identified large $(n'-1)$-item sets one by one if they have the same first $(n'-2)$ items. Then, subsets of each candidate large item set are evaluated. The candidate large 2-item sets and the final large 2-item sets are shown in Tables 10.4 and 10.5, respectively.

Based on the large 2-item sets generated, only one candidate large 3-item set $\{2, 3, 5\}$ can be formulated, which is to combine $\{2, 3\}$ and $\{2, 5\}$ as they have the same first item 2. The *Support* of $\{2, 3, 5\}$ is 2/4 and is equal to the threshold, and

Table 10.1   Records in the database

Records	Items
t_1	$\{1, 3, 4\}$
t_2	$\{2, 3, 5\}$
t_3	$\{1, 2, 3, 5\}$
t_4	$\{2, 5\}$

Table 10.2   The candidate large 1-item sets

Candidate large 1-item set	Support
{1}	2/4
{2}	3/4
{3}	3/4
{4}	1/4
{5}	3/4

thus it is a large 3-item set. The process of large item set generation terminates, as no more large item sets can be formulated.

After generating all large n'-item sets, association rules can be generated from them. Thresholds for *Confidence* and *Lift* denoted by *min_Conf* and *min_Lift*, respectively, should be specified before association rule generating. In Apriori algorithm, Property 3 is used to guide the generation of association rules:

Property 3: Partitioning a large item set I^* into two mutually exclusive and complementary subsets I_j and I_k. The rule from I_j to I_k is denoted by $I_j \rightarrow I_k$. Assume that $Conf(I_j \rightarrow I_k)\_Conf$. Then, for any non-empty and strict subset of I_j, denoted by I'_j, and the superset of I_k, denoted by $I'_k = I_k - I'_j$, the rule from I'_j to I'_k is called a sub-rule of $I_j \rightarrow I_k$, and $Conf(I'_j \rightarrow I'_k)\_Conf$. The proof of Property 3 is as follows.

Proof: Denote the events of observing I_j and I_k by $E(I_j)$ and $E(I_k)$, respectively, and the events of observing I'_j and I'_k by $E(I'_j)$ and $E(I'_k)$, respectively. The *Confidence* of rule $I_j \rightarrow I_k$ can be calculated by

$$Conf(I_j \rightarrow I_k) = \frac{P(E(I_j) \cap E(I_k))}{P(E(I_j))} = \frac{P(E(I^*))}{P(E(I_j))}, \quad (10.3)$$

Table 10.3   The large 1-item sets

Large 1-item set	Support
{1}	2/4
{2}	3/4
{3}	3/4
{5}	3/4

Table 10.4   The candidate large 2-item sets

Candidate large 2-item set	Support
{1, 2}	1/4
{1, 3}	2/4
{1, 5}	1/4
{2, 3}	2/4
{2, 5}	3/4
{3, 5}	2/4

and the *Confidence* of rule $I'_j \rightarrow I'_k$ can be calculated by

$$\mathit{Conf}(I'_j \rightarrow I'_k) = \frac{P(E(I'_j) \cap E(I'_k))}{P(E(I'_j))} = \frac{P(E(I^*))}{P(E(I'_j))}. \tag{10.4}$$

As $I'_j \subset I_j$, we have $P(E(I'_j))(E(I_j))$ and thus $\mathit{Conf}(I'_j \rightarrow I'_k)(I_j \rightarrow I_k) < \mathit{min\_Conf}$. Therefore, it can be concluded that $\mathit{Conf}(I'_j \rightarrow I'_k) < \mathit{min\_Conf}$. □

Property 3 shows that a sub-rule of a rule whose *Confidence* is less than *min_Conf* has a smaller *Confidence* value, and thus, it cannot be an association rule. This property is used in Apriori algorithm to simplify the process of generating all association rules from a large item set by not examining sub-rules of a rule with *Confidence* less than *min_Conf*. Given a large n'-item set $I^*_{n'}$, a consequent with m $(1 \leq m')$ items is denoted by h_m and the set of all h_m derived from $I^*_{n'}$ is denoted by H_m. In Apriori, all rules with *Confidence* no smaller than *min_Conf* are first generated, and then a rule is deleted if its *Lift* is less than *min_Lift*. The Algorithm of association rule generation is presented in Algorithm 3.

Especially, Line 10 of Algorithm 4 is based on Property 3, which deletes the rules with *Confidence* less than *min_Conf* and thus all the sub-rules of this rule will not be considered, so as to reduce the computational burden. Line 14 of Algorithm 4 calls Algorithm 2 to generate H_m from all pairs of item sets in H_{m-1}.

Table 10.5   The large 2-item sets

Large 2-item set	Support
{1, 3}	2/4
{2, 3}	2/4
{2, 5}	3/4
{3, 5}	2/4

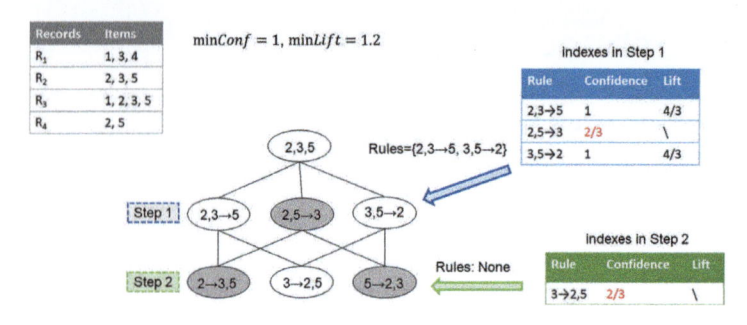

Figure 10.1 Generation of association rules from large 3-item set $\{2, 3, 5\}$

The procedure of generating association rules from the large 3-item set $\{2, 3, 5\}$ found in the toy example is shown in Figure 10.1. Suppose that we set $min\_Conf = 1$ and $min\_Lift = 1.2$.

As shown in Figure 10.1, in Step 1, three candidate rules can be generated: $2, 3 \rightarrow 5$, $2, 5 \rightarrow 3$, and $3, 5 \rightarrow 2$. Their *Confidence* values are first calculated, and it is found that *Confidence* of rule $2, 5 \rightarrow 3$ is less than $min\_Conf = 1$. Hence, its *Lift* value is not calculated, and its sub-rules are not considered, which are shown by the gray nodes. Then, the *Lift* values of the other two rules are checked, which are larger than the threshold of *Lift*, and thus these two rules are association rules. Then, in Step 2, only one candidate rule needs to be further investigated, namely, $3 \rightarrow 2, 5$. This is because the other two candidate association rules: $2 \rightarrow 3, 5$ and $5 \rightarrow 2, 3$ are sub-rules of rule $2, 5 \rightarrow 3$ whose *Confidence* is smaller than $min\_Conf = 1$. As $3 \rightarrow 2, 5$ has *Confidence* = 2/3, which is less than $min\_Conf = 1$, it cannot be an association rule. To conclude, two association rules can be generated from the large 3-item set $\{2, 3, 5\}$ in the toy example: $2, 3 \rightarrow 5$ and $3, 5 \rightarrow 2$.

Algorithm 3 Generation of association rules

1: **Step 1**: Generating step
2: $Rules = \emptyset$, which is the set of association rules to be generated;
3: **for** each large n'-item set $I_{n'}^*, n' \geq 2$ **do**
4: $\quad Rules = $ Ap$\_$AssRule$(I_{n'}^*)$;
5: **end for**
6: **Step 2**: Pruning step
7: **for** each *rule* in *Rules* **do**
8: $\quad$ Calculate $Lift(rule)$
9: $\quad$ **if** $Lift(rule) < min\_Lift$ **then**
10: $\quad\quad$ delete *rule* from *Rules*;
11: $\quad$ **end if**
12: **end for**
13: **return** *Rules*

Algorithm 4 Ap_AssRule($I_{n'}^*$)

1: $m = 1$;
2: $H_m = \{h_m | h_m \subset I_{n'}^*\}$;
3: **while** $m \leq n' - 1$ **do**
4:     **for** each $h_m \in H_m$ **do**
5:         $rule = I_{n'}^* - h_m \rightarrow h_m$;
6:         Calculate $Conf(rule)$;
7:         **if** $Conf(rule) \geq min\_Conf$ **then**
8:             $Rules = Rules \cup \{rule\}$;
9:         **else**
10:            Delete h_m from H_m;
11:         **end if**
12:     **end for**
13:     $m = m + 1$;
14:     $H_m = \text{generate\_candidate}(H_{m-1})$, where generate_candidate($\cdot$) is defined by Algorithm 2;
15: **end while**
16: **return** $Rules$

Association rule learning is a popular approach to analyzing the relations between deficiency codes as well as ship characteristics and the inspection results including deficiencies and detention in ship PSC inspection in the existing literature. Example 10.1 is to mine the relations between deficiency codes in ship inspection records using a sub-dataset of the whole inspection data set with 200 inspection records. Weka, which is an open source software providing various tools for data preprocessing, data mining and machine learning algorithms implementation, and data visualization developed by researchers from The University of Waikato [2], is used as the tool to mine the association rules regarding the deficiency codes from the 200 inspection records.

Example 10.1: According to the Tokyo MoU [3], there are a total of 17 deficiency codes, which are *the* inspection targets of a PSC inspection conducted by each member state of the Tokyo MoU as listed in Table 10.6. Our aim is to figure out the correlations between the deficiency codes except for D99-Other, as this code does not have a specific meaning. Especially, the correlation is presented by the large item sets found and the association rules mined from them.

The deficiency information in the initial data collected online is specific deficiency items, e.g., "01103" representing "Passenger Ship Safety (including exemption)," "07101" representing "Fire prevention structural integrity," and "12101" representing "Stowage/segregation/packaging dangerous goods." Therefore, the initial data are first pre-processed by generating 16 features with each feature representing a deficiency code, i.e., D01, D02, ..., D15, D18, and the value of each record for each feature is the number of deficiencies of the record under this deficiency code. To enable the Weka software to generate association rules from the data set

138 Machine learning and data analytics for maritime studies

Table 10.6   Deficiency codes in the Tokyo MoU [3]

Deficiency code	Content	Deficiency code	Content
D01	Certificate and documentation	D10	Safety of navigation
D02	Structural conditions	D11	Life-saving appliances
D03	Water/weathertight conditions	D12	Dangerous goods
D04	Emergency systems	D13	Propulsion and auxiliary machinery
D05	Radio communications	D14	Pollution prevention
D06	Cargo operations including equipment	D15	International safety management
D07	Fire safety	D18	Labor Conditions
D08	Alarms	D99	Other
D09	Working and living conditions		

containing deficiency codes of each record, all non-zero feature values in the data set are encoded to "T", and zero feature values are encoded to "?". Then, the initial .csv file is converted to .arff file. The thresholds of *Support*, *Confidence*, and *Lift* are set to $min\_Sup = 0.1$, $min\_Conf = 0.7$, and $min\_Lift = 1.5$. The large item sets are shown in Tables 10.7–10.9, and the association rules mined from the large item sets are shown in Table 10.10.

Table 10.7 shows that the most frequently detected deficiency items are D07: fire safety, which is identified in more than half of the 200 inspections of concern, followed by D10: safety of navigation, which is found in 43.5% of the inspections. From Table 10.8, it can be seen that deficiency codes D07 and D10 are often detected together, which can be found in nearly 25% of the inspection records. In addition, D04 and D07 as well as D07 and D11 are often identified in one inspection, which

Table 10.7   Large 1-item sets with $min\_Sup = 0.1$

Large item set	Frequency (*support*)
D07	102(0.510)
D10	87(0.435)
D04	50(0.250)
D18	48(0.240)
D14	46(0.230)
D11	43(0.215)
D03	41(0.205)
D13	29(0.145)
D09	28(0.140)

Table 10.8   Large 2-item sets with min_Sup = 0.1

Large item set	Frequency (*Support*)
D07, D10	48(0.240)
D04, D07	32(0.160)
D07, D11	30(0.150)
D07, D18	28(0.140)
D07, D14	26(0.130)
D10, D11	26(0.130)
D03, D07	25(0.125)
D03, D10	25(0.125)
D04, D10	25(0.125)
D10, D14	24(0.120)
D04, D11	21(0.105)
D10, D18	21(0.105)

can be seen in more than 15% of the inspection records. Finally, Table 10.9 shows that D07, D10, and D11 together with D04, D07, and D10 are detected together in more than 10% of the inspection records. The association rules shown in Table 10.10 are all generated from the large 3-item sets and can be interpreted as follows. For example, for the association rule D10, D11 ⇒ D07, if D10 and D11 are detected on one ship, the probability that the ship has deficiency code D07 is 81%. Compared to the fact that the probability that a ship has deficiency D07 is 51% if no prior information is known as shown in Table 10.7, the *Lift* value of this association rule is about 1.58, showing that identifying deficiency codes D10 and D11 can increase the probability of identifying deficiency code D07 by 1.58 times. Therefore, this association rule can be used to guide onboard ship inspection: if deficiency codes D10 and D11 are detected on a ship, it is suggested that the PSC inspector should pay more attention to deficiency items under deficiency code D07 as there is a high probability that the ship has a deficiency/deficiencies under this code.

10.3   FP-growth algorithm

The main difference between the FP-growth algorithm and the Apriori algorithm is in the context of generating all large item sets, that is, finding all item sets with

Table 10.9   Large 3-item sets with min_Sup = 0.1

Large item set	Frequency (*Support*)
D07, D10, D11	21(0.105)
D04, D07, D10	20(0.100)

Table 10.10 Association rules mined with min_Conf = 0.7 and min_Lift = 1.5

Association rule	Confidence	Lift
D10, D11 $\Rightarrow$ D07	0.81	1.58
D04, D10 $\Rightarrow$ D07	0.80	1.57
D07, D11 $\Rightarrow$ D10	0.70	1.61

$Sup(I_{n'}) \geq min\_Sup$, while the procedure of generating all association rules is the same in both algorithms. FP-growth algorithm is proposed by Han et al. in 2000 [4] and is believed to be an improvement of Apriori as it is much faster. In Apriori, data are stored in set structure, and candidate large item sets should be generated, the number of which can be large. Moreover, the whole database needs to be traversed multiple times to check the *Support* values of the candidate item sets in a brute-force manner, and traversing the whole data set heavily reduces the computational efficiency. In the FP-growth algorithm, data are stored in a tree structure called a frequent pattern tree or FP tree. FP tree allows for faster scanning, while no candidate large item sets are to be generated. In an FP tree, each node represents a single item in the original database, and the links between the nodes represent their co-occurrence in the database. We do not show the detailed steps to construct an FP tree and the large item sets in this section, and we only present the rough steps as follows. Readers are referred to Han et al. (2001) [4] and Borgelt [5] for more details.

1. Scan the database to calculate the frequency (i.e., *Support*) of individual items, and remove the items with *Support* values less than *min_Sup*.
2. Start to construct the FP tree by creating the root of the tree that represents null.
3. Scan the database again to exam each of the records and add them to the tree one by one. Especially, the branch of the tree is constructed with items in descending order of count.
4. After constructing the FP tree, the tree is mined from the lowest nodes as well as the links to the nodes. The lowest nodes represent the frequency pattern length 1. From each of the lowest nodes, traverse the path in the FP Tree, and the path or paths are called a conditional pattern base.
5. Construct a conditional FP tree based on the conditional pattern bases of the nodes.
6. Large item sets are generated from the conditional FP tree constructed.

References

[1] Agrawal R., Imieliński T., Swami A. 'Mining association rules between sets of items in large databases'. *The 1993 ACM SIGMOD International Conference*

[online]; Washington, DC, New York, NY, 1993. pp. 207–16. Available from http://portal.acm.org/citation.cfm?doid=170035

[2] Witten I.H., Frank E. 'Data mining'. *ACM SIGMOD Record*. 2002;31(1):76–77.

[3] The Tokyo MOU; 2017 *List of tokyo mou deficiency codes* [online]]]. Available from http://www.tokyo-mou.org/doc/Tokyo%20MOU%20deficiency%20codes%20(December%202017).pdf [Accessed 9 Aug 2021].

[4] Han J., Pei J., Yin Y. 'Mining frequent patterns without candidate generation'. *ACM SIGMOD Record*. 2000;29(2):1–12.

[5] Borgelt C. *An implementation of the FP-growth algorithm* [online]. 2005:1–5. Available from http://portal.acm.org/citation.cfm?doid=1133905

Chapter 11

Cluster analysis

Cluster analysis, or clustering, is a general task aiming to group a given set of examples into several groups (i.e., clusters) following given criteria, such that the examples in the same group are as close as to each other, and the examples in different groups are as different as from each other. Generally, clustering works in the context of unsupervised learning, where only the features of the examples are known while there is no target defined or targets are unknown. It aims to divide the whole data set into several mutually exclusive and complementary clusters, so as to mine the properties of the clusters formulated where such properties are represented by cluster labels. Mathematically, suppose that we have an unlabeled data set $D = \{\mathbf{x}_i, i = 1, ..., n\}$, where $\mathbf{x}_i \in R^m$ is the feature vector that can be represented by $\mathbf{x}_i = (x_{i1}, ..., x_{im})$. A clustering algorithm aims to divide D into K groups $\{C_k | k = 1, ..., K\}$ such that $C_{k'} \cap C_{k''} = \emptyset, k' \neq k''$, $k' \in \{1, ..., K\}$ and $k'' \in \{1, ..., K\}$, and $D = \cup_{k=1}^{K} C_k$. The cluster label of cluster C_k is denoted by λ_k, which needs to be decided and summarized by model user. Based on the cluster labels, the label of each example, which we call it by example cluster label, can be obtained and is denoted by λ'_i, e.g., $\mathbf{x}_i$. One needs to distinguish between clustering and classification, although both aim to separate examples in the data set. In classification, labels of the examples are known, and the model training process aims to separate the data according to the features while minimizing a given loss function measured by the examples' labels, so as to obtain a classifier to predict the target of the new examples. In contrast, labels are unavailable in clustering, and the model training process aims to put the examples similar to each other regarding their features in the same cluster, and then to summarize the cluster labels of the clusters generated.

In the following sections, distance measures of two examples in the data set and of two clusters are first introduced, as distance measure is the most important concept in cluster analysis. Then, metrics to evaluate the performance of cluster analysis are introduced. Finally, popular clustering algorithms are discussed with examples of ship inspection by PSC provided.

11.1 Distance measure in clustering

11.1.1 Distance measure of examples

The key point in cluster analysis is how to measure the "similarity" between two examples in the data set, and this is usually achieved by the calculation of "distance" of these two examples. Distance measure is an objective score used to measure the relative difference/dissimilarity between two examples in the problem of concern. The distance between two examples $\mathbf{x}_i$ and $\mathbf{x}_j$ is denoted by $dist(\mathbf{x}_i, \mathbf{x}_j)$, which satisfies the following properties:

1. non-negativity: $dist(\mathbf{x}_i, \mathbf{x}_j) \geq 0$;
2. identity: If and only if $\mathbf{x}_i = \mathbf{x}_j$, $dist(\mathbf{x}_i, \mathbf{x}_j) = 0$;
3. symmetry: $dist(\mathbf{x}_i, \mathbf{x}_j) = dist(\mathbf{x}_j, \mathbf{x}_i)$; and
4. triangle inequality: $dist(\mathbf{x}_i, \mathbf{x}_j) \leq dist(\mathbf{x}_i, \mathbf{x}_k) + dist(\mathbf{x}_k, \mathbf{x}_j)$.

Features of an example can be numerical and categorical. Numerical features are ordinal, where the relative feature values are comparable. For example, ship age is a numerical feature, where a ship of age 5 is younger than a ship of age 10 by 5 years. Categorical features can be either ordinal, where the relative feature values are comparable like numerical features (e.g., low, medium, and high for ship company performance, where a ship company with high performance is better than a ship company with medium performance, and is much better than a ship company with low performance), or nominal, where the feature values only indicate the categories and cannot be compared (e.g., container ship, bulk carrier, and passenger ship belonging to the feature of ship type, and they cannot be compared directly with each other). As feature values are comparable in ordinal features and noncomparable in nominal features, different means of distance measure should be used in these two types of features. For data set D with m features, denote the number of its ordinal features by m_1 and the number of its nominal features by m_2, where $m = m_1 + m_2$. For ordinal features, Minkowski distance taking the following form is the most popular one:

$$dist_{mk}(\mathbf{x}_i, \mathbf{x}_j) = \left(\sum_{m'=1}^{m_1} \left| x_{im'} - x_{jm'} \right|^p \right)^{\frac{1}{p}}, \tag{11.1}$$

where the subscript is mk short for Minkowski, and p should be no less than 1, such that the properties of distance measure can be satisfied. Common values of p are 1 and 2. When $p = 1$, Equation (11.1) is also called Manhattan distance, and can be written as

$$dist_{man}(\mathbf{x}_i, \mathbf{x}_j) = \left\| \mathbf{x}_i - \mathbf{x}_j \right\|_1 = \sum_{m'=1}^{m_1} \left| x_{im'} - x_{jm'} \right|. \tag{11.2}$$

Manhattan distance is the sum of the absolute difference between each pair of components of the coordinates of two examples whose calculation is related to the L1 vector norm. It was first proposed to calculate the distance of driving from one crossroad to another crossroad in Manhattan. Therefore, Equation (11.2) is also referred to as city block distance.

When $p = 2$, Equation (11.1) is also called Euclidean distance, and can be written as

Cluster analysis 145

$$dist_{ed}(\mathbf{x}_i, \mathbf{x}_j) = \|\mathbf{x}_i - \mathbf{x}_j\|_2 = \sqrt{\sum_{m'=1}^{m_1} \left|x_{im'} - x_{jm'}\right|^2}. \tag{11.3}$$

Euclidean distance is the most widely used distance measure of two examples whose calculation is related to the L2 vector norm, and can be interpreted as the linear distance between two examples.

To measure the similarity of two examples regarding the angle between them, one can use cosine similarity, which is calculated as follows:

$$\cos(\mathbf{x}_i, \mathbf{x}_j) = \frac{\mathbf{x}_i \cdot \mathbf{x}_j}{|\mathbf{x}_i| \times |\mathbf{x}_j|} = \frac{\sum_{m'=1}^{m_1} x_{im'} x_{jm'}}{\sqrt{\sum_{m'=1}^{m_1} x_{im'}^2} \sqrt{\sum_{m'=1}^{m_1} x_{jm'}^2}}. \tag{11.4}$$

The range of the cosine similarity of two examples is $[-1, 1]$. Then, the cosine distance of two examples can be calculated from the cosine similarity of them by

$$dist_{cos}(\mathbf{x}_i, \mathbf{x}_j) = 1 - \cos(\mathbf{x}_i, \mathbf{x}_j). \tag{11.5}$$

One should also note that, for two vectors that are normalized by L2 normalization, the squared Euclidean norm of their difference is equivalent to twice of their cosine distance, whose proof is as follows:

Proof: For two vectors $A = (x_{1A}, x_{2A}, ..., x_{m_1 A})$ and $B = (x_{1B}, x_{2B}, ..., x_{m_1 B})$, they are first normalized by L2 normalization, respectively, and we have

$$A' = \left(\frac{x_{1A}}{\sqrt{x_{1A}^2 + x_{2A}^2 + ... + x_{m_1 A}^2}}, \frac{x_{2A}}{\sqrt{x_{1A}^2 + x_{2A}^2 + ... + x_{m_1 A}^2}}, ..., \frac{x_{m_1 A}}{\sqrt{x_{1A}^2 + x_{2A}^2 + ... + x_{m_1 A}^2}}\right) = (x_{1A'}, x_{2A'}, ..., x_{m_1 A'})$$

and

$$B' = \left(\frac{x_{1B}}{\sqrt{x_{1B}^2 + x_{2B}^2 + ... + x_{m_1 B}^2}}, \frac{x_{2B}}{\sqrt{x_{1B}^2 + x_{2B}^2 + ... + x_{m_1 B}^2}}, ..., \frac{x_{m_1 B}}{\sqrt{x_{1B}^2 + x_{2B}^2 + ... + x_{m_1 B}^2}}\right) = (x_{1B'}, x_{2B'}, ..., x_{m_1 B'})$$

Then, the cosine distance between A' and B' can be calculated by

$$\cos(A', B') = \frac{\sum_{m'=1}^{m_1} x_{m'A'} x_{m'B'}}{\sqrt{\sum_{m'=1}^{m_1} x_{m'A'}^2} \sqrt{\sum_{m'=1}^{m_1} x_{m'B'}^2}}$$
$$= \sum_{m'=1}^{m_1} x_{m'A'} x_{m'B'}. \tag{11.6}$$

The squared Euclidean norm of the difference of A' and B' has the following form

$$|A' - B'|^2 = (x_{1A'} - x_{1B'})^2 + (x_{2A'} - x_{2B'})^2 + ... + (x_{m_1 A'} - x_{m_1 B'})^2$$
$$= 2 - 2 \sum_{m'=1}^{m_1} x_{m'A'} x_{m'B'} \tag{11.7}$$
$$= 2 - 2 \cos(A', B')$$

Then, the equivalence of the squared Euclidean norm and twice of the cosine distance of two vectors is proven.□

Modified cosine similarity aims to reduce the influence of the absolute values of the components of a feature vector. It is calculated by subtracting the mean of components of the vector for an example from each component of the vector. To be more specific, the mean of $\mathbf{x}_i$ (which is of m_1 dimensions) is calculated by

$$\mu_{\mathbf{x}_i} = \frac{\sum_{m'=1}^{m_1} x_{im'}}{m_1}, \qquad (11.8)$$

and the mean of $\mathbf{x}_j$ is calculated by

$$\mu_{\mathbf{x}_j} = \frac{\sum_{m'=1}^{m_1} x_{jm'}}{m_1}. \qquad (11.9)$$

Modified cosine similarity can be calculated by

$$\cos'(\mathbf{x}_i, \mathbf{x}_j) = \frac{(\mathbf{x}_i - \mu_{\mathbf{x}_i}) \cdot (\mathbf{x}_j - \mu_{\mathbf{x}_j})}{|\mathbf{x}_i - \mu_{\mathbf{x}_i}| \times |\mathbf{x}_j - \mu_{\mathbf{x}_j}|}$$

$$= \frac{\sum_{m'=1}^{m_1} (x_{im'} - \mu_{\mathbf{x}_i})(x_{jm'} - \mu_{\mathbf{x}_j})}{\sqrt{\sum_{m'=1}^{m_1} (x_{im'} - \mu_{\mathbf{x}_i})^2} \sqrt{\sum_{m'=1}^{m_1} (x_{jm'} - \mu_{\mathbf{x}_j})^2}}. \qquad (11.10)$$

Similarly, the revised cosine distance can be calculated by

$$dist'_{cos}(\mathbf{x}_i, \mathbf{x}_j) = 1 - \cos'(\mathbf{x}_i, \mathbf{x}_j). \qquad (11.11)$$

Measuring distance of examples containing nominal features can be much more difficult, as their distance can hardly be measured because feature values are noncomparable. To measure the distance of two examples containing nominal features, cosine distance shown in Equation (11.5) and modified cosine distance shown in Equation (11.11) can be used, as they only consider the angle of the feature vector instead of the absolute values. If the nominal features are encoded into binary features (e.g., the three values for feature ship type, namely, container ship, bulk carrier, and passenger ship are encoded to three new features: is_container_ship, is_bulk_carrier, and is_passenger_ship, and 100, 010, and 001 are used as the feature values of the three new features to indicate the three types of ships), Hamming distance can be used. Hamming distance is a measurement of the number of different values in two strings of equal length, and it can be calculated by

$$dist'_{ham}(\mathbf{x}_i, \mathbf{x}_j) = \sum_{m'=1}^{m_2} \mathbf{1}(x_{im'} \neq x_{jm'}), \qquad (11.12)$$

where $\mathbf{1}(x)$ is the indicator function, which takes value 1 if x is true and 0, otherwise.

11.1.2 Distance measure of clusters

In some cluster analysis methods (e.g., hierarchical clustering), distance between clusters needs to be calculated. The distance between two clusters is usually calculated based on the function of the pairwise distance between one example in one cluster and one example in the other cluster. Given two clusters $C_{k'}$ and $C_{k''}$ and a distance measure of two examples $\mathbf{x}_i$ and $\mathbf{x}_j$ denoted by $dist(\mathbf{x}_i, \mathbf{x}_j)$, their distance, denoted by $Dist(C_{k'}, C_{k''})$, can be measured by

1. Single-linkage clustering:
   $Dist_{min}(C_{k'}, C_{k''}) = \min_{x_i \in C_{k'}, x_j \in C_{k''}} dist(x_i, x_j)$, which is the minimum distance of two examples belonging to the two clusters, respectively.
2. Complete-linkage clustering:
   $Dist_{max}(C_{k'}, C_{k''}) = \max_{x_i \in C_{k'}, x_j \in C_{k''}} dist(x_i, x_j)$, which is the maximum distance of two examples belonging to the two clusters, respectively.
3. Unweighted average-linkage clustering:
   $Dist_{uavg}(C_{k'}, C_{k''}) = \frac{1}{|C_{k'}| \times |C_{k''}|} \sum_{x_i \in C_{k'}, x_j \in C_{k''}} dist(x_i, x_j)$, which is the average distance of all pairs of two examples belonging to the two clusters, respectively.
4. Centroid-linkage clustering:
   $Dist_{cen}(C_{k'}, C_{k''}) = dist(c_{k'}, c_{k''})$ where $c_{k'}$ and $c_{k''}$ are the centroids of clusters $C_{k'}$ and $C_{k''}$, respectively, and can be calculated by $c_{k'} = \frac{1}{|C_{k'}|} \sum_{x_i \in C_{k'}} x_i$ and $c_{k''} = \frac{1}{|C_{k''}|} \sum_{x_j \in C_{k''}} x_j$. Therefore, centroid-linkage clustering measures the distance between centroids of the two clusters.

11.2 Metrics for clustering algorithm performance evaluation

Metrics to evaluate the performance of supervised ML models (mainly referring to regression and classification models) mainly aim to calculate the difference between the predicted and actual target values, and smaller difference usually indicates a more accurate prediction model, which is quite intuitive. Nevertheless, data sets used in cluster analysis are not labeled, i.e., unsupervised. Therefore, metrics for supervised learning are not applicable. Considering the fact that the core idea of clustering is that examples in the same cluster should be close to each other, while examples in different clusters should be far from each other, it can be seen that clusters of examples divided by a good clustering model should have high intra-cluster similarity while low inter-cluster similarity. Especially, two indicators are used to measure the distance between examples in one cluster C_k, given a distance measure of two examples is denoted by

$$avg(C_k) = \frac{2}{|C_k|(|C_k|-1)} \sum_{1 \leq i \leq |C_k|} dist(x_i, x_j), \tag{11.13}$$

which is the average distance of each pair of the examples belonging to cluster C_k, and

$$max(C_k) = \max_{1 \leq i \leq |C_k|} dist(x_i, x_j), \tag{11.14}$$

which is the largest distance of each pair of examples belonging to cluster C_k.

One popular metric to evaluate the performance of a clustering algorithm is called Davies–Bouldin index, or DBI for short, which considers average intra-cluster and inter-cluster distances, is shown in Equation (11.15):

$$DBI = \frac{1}{K} \sum_{k'=1}^{K} \max_{k''=1,\ldots,k'-1,k'+1,\ldots,K} \left(\frac{avg(C_{k'}) + avg(C_{k''})}{Dist_{cen}(C_{k'}, C_{k''})} \right), \tag{11.15}$$

where a smaller value of the whole part in parentheses indicates better clustering performance (as a smaller numerator shows smaller intra-cluster distance and a larger denominator shows larger inter-cluster distance). Then, the worst case of all the clusters is identified by the max operator, and the DBI calculates the sum of ratio between intra-cluster distance and inter-cluster distance in the worst case of all the clusters generated. Therefore, the smaller the value of DBI, the better the performance of a clustering model. Another popular metric is called Dunn index (DI), which considers extreme intra-cluster and inter-cluster distances shown as follows:

$$DI = \frac{\min_{1 \leq k' < k'' \leq K} Dist_{min}(C_{k'}, C_{k''})}{\max_{1 \leq k \leq K} max(C_k)}, \tag{11.16}$$

where $max(C_k)$ is calculated in Equation (11.14). A larger value of DI indicates that a clustering algorithm performs better, as a larger minimum inter-cluster distance (i.e., $Dist_{min}(C_{k'}, C_{k''})$) or a smaller maximum intra-cluster distance (i.e., $\max_{1 \leq k \leq K} max(C_k)$) increases the value of DI.

The above two metrics are called internal index for clustering performance evaluating as they only consider the clusters generated by the clustering algorithm itself. If there is a reference model that the clustering algorithm can be compared with, external indexes can be used. For data set D containing n examples, denote the clusters given by the reference model by $\{C_l^* | l = 1, ..., S\}$ with example cluster label $\{\lambda_i'^* | i = 1, ..., n\}$. Recall that the clusters given by the current clustering model is $\{C_k | k = 1, ..., K\}$ with cluster label $\{\lambda_i' | i = 1, ..., n\}$. Note that S is not necessarily equal to K. Examples in D are divided into clusters using the reference model and the current clustering model to be evaluated. For each two examples $x_i \in D$ and $x_j \in D$ with i, cases of how they belong to the clusters of the two clustering models are as follows:

1. $SS = \{(x_i, x_j) | \lambda_i' = \lambda_j', \lambda_i'^* = \lambda_j'^*, i\}$, which means that x_i and x_j are divided into the same cluster in both the clustering model to be evaluated and the reference model. Denote the number of such pairs of examples by $a = |SS|$;
2. $SD = \{(x_i, x_j) | \lambda_i' = \lambda_j', \lambda_i'^* \neq \lambda_j'^*, i\}$, which means that x_i and x_j are divided into the same cluster in the clustering model to be evaluated but different clusters in the reference model. Denote the number of such pairs of examples by $b = |SD|$;
3. $DS = \{(x_i, x_j) | \lambda_i' \neq \lambda_j', \lambda_i'^* = \lambda_j'^*, i\}$, which means that x_i and x_j are divided into the same cluster in the reference model but different clusters in the clustering model to be evaluated. Denote the number of such pairs of examples by $c = |DS|$;
4. $DD = \{(x_i, x_j) | \lambda_i' \neq \lambda_j', \lambda_i'^* \neq \lambda_j'^*, i\}$, which means that x_i and x_j are divided into different clusters in both the clustering model to be evaluated and the reference model. Denote the number of such pairs of examples by $d = |DD|$.

As we require that $i < j$, each pair of examples only appears once in SS, SD, DS, and DD, and we have $a + b + c + d = n(n-1)/2$. Based on the above cases, three external indexes, namely, Jaccard coefficient (JC), Fowlkes and Mallows index (FMI), and rand index (RI), are defined as follows:

$$JC = \frac{a}{a+b+c}, \tag{11.17}$$

$$FMI = \sqrt{\frac{a}{a+b} \cdot \frac{a}{a+c}}, \tag{11.18}$$

and

$$RI = \frac{2(a+d)}{n(n-1)}. \tag{11.19}$$

The range of the above three indicators is [0, 1], and larger values indicate better clustering algorithm performance.

11.2.1 Clustering algorithms

Classic clustering algorithms can roughly be divided into three types: partition-based methods, density-based methods, and hierarchy-based methods. A summary of these algorithms is presented in Table 11.1.

In the following subsections, a typical algorithm under each type of clustering methods is introduced. In particular, we introduce K-means as the representative of partition-based methods, density-based spatial clustering of applications with noise (DBSCAN) as the representative of density-based methods, and agglomerative algorithm as the representative of hierarchy-based methods.

11.2.2 K-means (partition-based method)

K-means is one of the simplest and the most widely used clustering algorithms. Especially, K is the number of clusters to be generated and its value needs to be pre-set, and "means" refers to averaging of the data by finding the centroids of each of the K clusters. The basic step of K-means algorithm is to find a total of K centroids as the centers of the clusters and allocate the examples in the data set to them. The positions of the K centroids are constantly updated to improve the clustering performance. The procedure of the K-means algorithm is shown in Algorithm 1.

Algorithm 1 K-means algorithm

Input: A data set $D = \{\mathbf{x}_1, ..., \mathbf{x}_n\}$, the number of clusters to be generated K.
Output: K clusters of examples $\{C_k | k = 1, ..., K\}$.
1: Randomly select K examples from D as the initial centroids, denoted by $\mathbf{c}_1$ to $\mathbf{c}_K$, with one centroid denoted by $\mathbf{c}_k$;
2: $\lambda' = \{\lambda'_1, ..., \lambda'_n\}$ as the example cluster label
3: **repeat**
4:     $C_k = \emptyset, k = 1, ..., K$, as the set of examples in cluster k;
5:     **for** $i = 1, ..., n$ **do**
6:         Calculate the distance between example $\mathbf{x}_i$ to each centroid $\mathbf{c}_k, k = 1, ..., K$, which is denoted by $dist(\mathbf{x}_i, \mathbf{c}_k)$;
7:         Choose the cluster whose centroid has the smallest distance from $\mathbf{x}_i$ as the cluster where $\mathbf{x}_i$ belongs to. Set $\lambda'_i = \arg\min_{k=1,...,K} dist(\mathbf{x}_i, \mathbf{c}_k)$ and add $\mathbf{x}_i$ to the corresponding cluster C_k: $C_k = C_k \cup \{\mathbf{x}_i\}$;
8:     **end for**
9:     **for** $k = 1, ..., K$ **do**

Table 11.1 A summary of classic clustering algorithms

Type of clustering methods	Basic idea	Typical algorithms	Advantages	Disadvantages
Partition-based methods	Regard the center of examples belonging to a cluster as the center of the corresponding cluster.	K-means and K-medoids	In general, relatively low time complexity and high computing efficiency.	The number of clusters (e.g., the value of K in K-means and K-medoids) should be preset, and the clustering results are highly sensitive to the preset cluster number; highly sensitive to outliers as they contribute much to the distance calculated; easily fall to local optimum; not suitable for non-convex data.
Density-based methods	Data in the region with high data space density are considered to belong to the same cluster.	DBSCAN, OPTICS, and mean-shift.	High computational efficiency; suitable to deal with data set with arbitrary shape.	When the density of data space is uneven, the resulting clusters might have low quality; when the data set is large, a big size memory is needed; highly sensitive to model parameters.
Hierarchy-based methods	Construct hierarchical structure among examples for clustering. Each example in the data set is first regarded as a cluster, and the two closest clusters are merged to form a new cluster until there is only one cluster left. Or, a reversed process.	Agglomerative algorithm and divisive algorithm.	Can deal with data set with arbitrary shape and feature types; the hierarchical structure in the data set is easily detected.	Relatively high time complexity; the number of clusters needs to be preset.

10:    Based on the examples belonging to cluster k, calculate its revised centroid by $\mathbf{c}'_k = \frac{1}{|C_k|} \sum_{\mathbf{x}_i \in C_k} \mathbf{x}_i$;
11:    $\mathbf{c}_k = \mathbf{c}'_k$;
12: **end for**
13: **until** $\mathbf{c}_k$ is not updated, i.e, the generated $\mathbf{c}'_k$ equals $\mathbf{c}_k$ for all $k = 1, ..., K$.
14: **return** $\{C_k | k = 1, ..., K\}$

It is noted that the stopping criterion of K-means is that $\mathbf{c}_k$ is not updated, which means that the centroids of the clusters are not changed any more. This can take quite a long time or lead to too many iterations. To reduce the computational burden, the longest algorithm running time or the maximum number of updates can be set, and the algorithm terminates when the threshold of the running time or the number of updates is reached. The advantages of K-means are that:

1. Relatively simple and easy to understand and implement;
2. Can guarantee convergence and is applicable to large-scale data sets;
3. Can be easily generalized to clusters of different shapes and sizes, as well as adapted to new examples.

Meanwhile, the disadvantages of K-means are that:

1. The value of K should be preset in a manual manner, and can be prone to biased subjective judgment.
2. As the initial centroids are generated randomly, the algorithm suffers from uncertainties.
3. As the centroids are determined by their distance to examples in the data set, clustering performance can be heavily influenced by outliers.

11.2.3 DBSCAN (density-based method)

DBSCAN is short for density-based spatial clustering of applications with noise, whose main idea is to find core examples exhibiting high density and expand clusters from them. In the meantime, outliers in low-density region that do not belong to any cluster can be marked. The basic idea of DBSCAN shows that there are two main parameters in the DBSCAN algorithm:

1. ϵ, which is a threshold of distance between two examples for one to be considered as in the neighborhood of the other. For example, $\mathbf{x}_i \in D$, its ϵ-neighborhood, denoted by $N_\epsilon(\mathbf{x}_i)$, is constituted by the examples whose distance from $\mathbf{x}_i$ is no more than ϵ, i.e., $N_\epsilon(\mathbf{x}_i) = \{\mathbf{x}_j \in D | dist(\mathbf{x}_i, \mathbf{x}_j) \leq \epsilon\}$. This guarantees that examples in the same cluster are close enough to each other.
2. *Min_Pts*, which is the minimum number of examples in the neighborhood of a certain example (including the example itself) for this examples to be considered as a core example. This guarantees that a cluster contains a certain number of examples.

Clusters in DBSCAN are formed based on the following concepts:

1. Core example: for a given example $x_i \in D$, if the number of examples in its ϵ-neighborhood is no less than Min_Pts, i.e., $|N_\epsilon(x_i)| \geq Min\_Pts$, x_i is a core example.
2. Directly density reachable: if example x_i is a core example and $x_j \in N_\epsilon(x_i)$, then example x_j is directly density reachable from example x_i. It should be mentioned that x_i might not be directly density reachable from example x_j, unless x_j is a core example.
3. Density reachable: for examples x_i and x_j, there is a series of examples $x_{k_1}, ..., x_{k_l}$ such that $x_{k_1} = x_i$ and x_j is directly density reachable from x_{k_l}. If $x_{k_{l'}}$ and $x_{k_{l'+1}}$, where $1 \leq l'$, are directly density reachable, meaning that $x_{k_{l'}}$ is within the ϵ-neighborhood of $x_{k_{l'+1}}$, and $x_{k_{l'+1}}$ is also within the ϵ-neighborhood of $x_{k_{l'}}$, then x_j is density reachable from x_i. As the series of examples x_{k_1} (i.e., x_i), ..., x_{k_l} are directly density reachable to each other, they must be core examples. Meanwhile, x_j is not necessary a core example.
4. Density connected: for examples x_i and x_j, if there is another example x_k such that x_i is density reachable from x_k, and x_j is density reachable from x_k, then x_i and x_j are density connected. For these three examples x_i, x_j, and x_k, x_k must be a core example, while x_i and x_j are not necessarily core examples.

Figure 11.1 is used to further illustrate the above concepts. Given the condition that $Min\_Pts = 3$, examples in red are core examples as no less than three examples are in their ϵ-neighborhood, while those not in red are noncore examples. Examples that are directly density reachable from a core example are in the blue circle with the core example in red at the center. The green links connecting the core examples show that these core examples are density reachable with each other. While

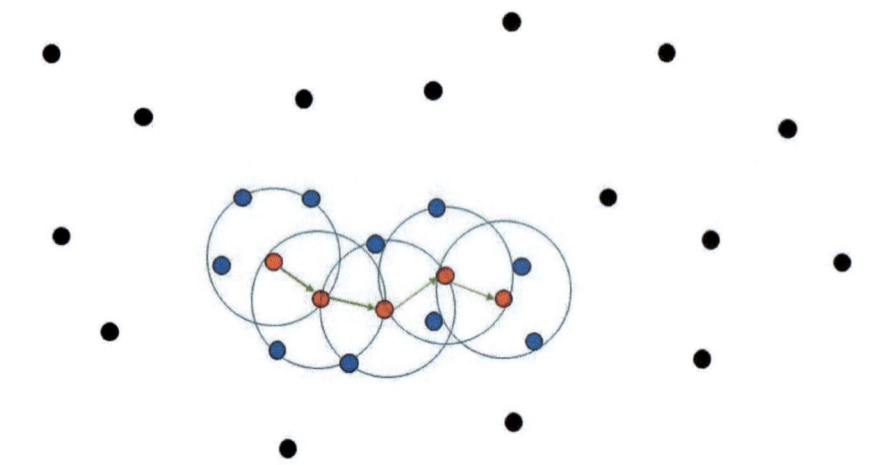

Figure 11.1 An example to illustrate the concepts in DBSCAN algorithm

Cluster analysis   153

if a core example is within the ϵ-neighborhood of another core example, they are directly reachable to each other. As examples in blue are within the ϵ-neighborhood of the core examples that are density reachable, all the examples in blue are density connected.

Based on the above concepts and explanation, a cluster in DBSCAN is defined as a maximum set of examples that are density connected from a core example. In other words, for a core example $\mathbf{x}_i^*$, a cluster expanded from it is denoted by $C_k = \{\mathbf{x} \in D | \mathbf{x} \text{ and } \mathbf{x}_i^* \text{ are density connected}\}$. The procedure of the DBSCAN algorithm is shown in Algorithm 2.

From lines 2 to 7 of Algorithm 2, the set of all core examples is first generated by traversing the whole data set. Then, a random core example is selected, and all the core examples that are density reachable from this core example are found. Especially, to ensure that the cluster is expanded from the initial random core example, queue data structure (i.e., Ω_{cur}) is used to store the core examples density reachable from this core example, so as to ensure that these core examples are visited one by one from the nearest to the farthest from the initial core example. Then, examples that are in the ϵ-neighborhood of all the core examples are added to the current cluster. This is shown from lines 14 to 18. Therefore, the initial (starting) core example is density connected with the other examples in this cluster. The above process is repeated until all the core examples are processed (i.e., added to clusters), which is shown by the while loop from lines 10 to 22 of Algorithm 2.

Algorithm 2 DBSCAN algorithm

Input: A data set $D = \{\mathbf{x}_1, ..., \mathbf{x}_n\}$, ε, Min_Pts.
Output: K clusters of examples $\{C_k | k = 1, ..., K\}$.
1: $\Omega = \emptyset$, which is the set of core examples;
2: **for** $i = 1, ..., n$ **do**
3:     Calculate the ε-neighborhood of example $\mathbf{x}_i$ by $N_\varepsilon(\mathbf{x}_i) = \{\mathbf{x}_j \in D | dist(\mathbf{x}_i, \mathbf{x}_j) \leq \varepsilon\}$.
4:     **if** $|N_\varepsilon(\mathbf{x}_i)| \geq Min\_Pts$ **then**
5:        Add $\mathbf{x}_i$ to the set of core examples: $\Omega = \Omega \cup \{\mathbf{x}_i\}$;
6:     **end if**
7: **end for**
8: Initialize $K = 1$, which is the index of the cluster to be generated;
9: Initialize $\Theta = D$, which is the set of un-visited examples;
10: **while** $\Omega \neq \emptyset$ **do**
11:     Randomly select a core example $\omega \in \Omega$;
12:     Initialize Ω_{cur}, $\Omega_{cur} = <\omega>$, which is the queue of core examples in the current cluster;
13:     Set $C_K = \{\omega\}$, and update the set of un-visited examples to $\Theta = \Theta - \{\omega\}$;
14:     **while** $\Omega_{cur} \neq \emptyset$ **do**
15:        Pick out the first core example ω' from Ω_{cur}, and find the examples in its ε-neighborhood and denote the set of such examples by $N_\varepsilon(\omega')$. Denote the un-visited examples in $N_\varepsilon(\omega')$ by $\delta_{\omega'}$, and $\delta_{\omega'} = N_\varepsilon(\omega') \cap \Theta$.

16:      Update the current cluster to $C_K = C_K \cup \delta_{\omega'}$, the set of un-visited examples to $\Theta = \Theta - \delta_{\omega'}$;
17:      The set of core examples added to the current cluster in this round can be calculated by $\delta_{\omega'} \cap \Omega$, and these core examples are added to queue Ω_{cur} considering their distance to the current core example ω', i.e., core examples with a closer distance to ω' are added earlier to Ω_{cur};
18:   **end while**
19:   Update the set of un-visited core examples to $\Omega = \Omega - C_K$;
20:   $K = K + 1$;
21: **end while**
22: **return** $\{C_k | k = 1, ..., K\}$

Unlike K-means where all the examples are divided into clusters, there might be some examples that do not belong to any cluster in DBSCAN, which are referred to as noises. The advantages of DBSCAN are that:

1. It is nonsensitive to noises and can also distinguish noise examples from normal examples in the data set.
2. The number of clusters does not need to be preset.
3. It can generate clusters of arbitrary shapes, which means that the contour of the examples contained by a cluster can be of any shape, as long as the examples in one cluster are density connected from the initial core example.

Meanwhile, the disadvantages of DBSCAN are that:

1. It is ineffective when dealing with data sets with uneven density, which means that the distribution of the examples in the data set is uneven in the feature space.
2. When the size of the data set is large, DBSCAN would take a long time to converge.
3. The values of parameters ϵ and *Min_Pts* can have a very large influence on the clusters generated.
4. As a random unvisited core example is selected as the seed from which a cluster is to be generated, the stability of the algorithm is adversely influenced.

11.2.4 Agglomerative algorithm (hierarchy-based methods)

Hierarchy-based methods can be either bottom-up (i.e., agglomerative clustering) or top-down (i.e., divisive clustering), where examples in the former are viewed as individual clusters at the beginning, and then merged with each other based on cluster distance until the preset number of clusters is reached, while examples in the latter are in one cluster at the beginning, and are gradually divided into subclusters based on cluster distance until the preset number of clusters is reached. The calculation of cluster distance is introduced in section 11.1.2, and the first three cluster distance calculation methods, namely, single-linkage clustering, complete-linkage clustering, and unweighted average-linkage clustering, can be used for cluster distance measuring in hierarchy-based methods. In this section, we introduce the agglomerative algorithm using unweighted average-linkage clustering to measure cluster distance as the representative of hierarchy-based methods. The basic idea of agglomerative algorithm is intuitive. Each example in the data set is viewed as an individual cluster at the beginning. Then, in each round of iteration, it constantly merges two clusters with the smallest cluster distance, until the number of clusters generated reaches the preset number of clusters. The procedure of agglomerative algorithm is presented by Algorithm 3.

From lines 1 to 5 of Algorithm 3, the clusters as well as the cluster distance matrix are initialized, where each example in the data set is regarded as an individual cluster. Then, from lines 6 to 15, a pair of clusters with the smallest cluster distance are merged into one cluster one by one in each iteration, and the distance matrix is also updated accordingly, until the number of preset clusters is reached, and then the algorithm terminates.

Algorithm 3 Agglomerative algorithm

Input: A data set $D = \{\mathbf{x}_1, ..., \mathbf{x}_n\}$, the number of clusters to be generated K, the distance measure method of two examples $\mathbf{x}_i$ and $\mathbf{x}_j$ denoted by $dist(\mathbf{x}_i, \mathbf{x}_j)$.
Output: K clusters of examples $\{C_k | k = 1, ..., K\}$.
1: **for** $k = 1, ..., n$ **do**
2:     $C'_k = \{\mathbf{x}_k\}$;
3: **end for**
4: Initialize the cluster distance matrix $M(k', k'') = Dist_{uavg}(C'_{k'}, C'_{k''}), k' = 1, ..., n, k'' = 1, ..., n$;
5: Initialize the current number of clusters $q = n$;
6: **while** $q > K$ **do**
7:     From M, pick out the pair of clusters $C'_{k'*}$ and $C'_{k''*}$, where $k'* < k''*$, that has the smallest cluster distance $Dist_{uavg}(C'_{k'*}, C'_{k''*})$;
8:     Combine $C'_{k'*}$ and $C'_{k''*}$ into one cluster denoted by $C'_{k'*}$, i.e., $C'_{k'*} = C'_{k'*} \cup C'_{k''*}$;
9:     **for** $k = k''* + 1, ..., K$ **do**
10:        Change the index of cluster C'_k to C'_{k-1}
11:     **end for**
12:     Delete the $k''*$th row and $k''*$th column from the cluster distance matrix M;
13:     Update the cluster distance matrix to $M(k'*, k) = Dist_{uavg}(C'_{k'*}, C'_k), k = 1, ..., q - 1$;
14:     $q = q - 1$;
15: **end while**
16: $C_k = C'_k, k = 1, ..., K$; **return** $\{C_k | k = 1, ..., K\}$

We use Example 1 to show how to use cluster analysis to cluster ships in PSC inspection while generating explanations and implications from the results using

156 Machine learning and data analytics for maritime studies

K-means as an example. To show the clustering process visually, we use a data set containing 50 ships that are randomly chosen from the whole data set with two features: age and last_deficiency_no as a toy example.

Example 11.1: The details of the 50 selected ships are shown in Table 11.2. The distribution of the features of the ships is shown in Figure 11.2.

In particular, we choose $K = 2$, $K = 3$, and $K = 4$ in the K-means algorithm. K-means is implemented by *scikit-learn* API [1]. It can easily be called by using the following lines of codes:

```
estimator = KMeans(n_clusters=2) # Construct a K-means
    clustering model with K=2
estimator.fit(X) # Apply the K-means model to the data set
label_pred = estimator.labels_ # obtain the exapmle cluster
    labels of the data set
```

The corresponding clusters are shown in Figures 11.3–11.5.

Table 11.2 Ship information in the toy example

Ship ID	age	last_deficiency_no	Ship ID	age	last_deficiency_no
1	8	1	2	5	0
3	22	5	4	8	0
5	13	2	6	14	0
7	20	13	8	11	5
9	10	0	10	14	1
11	9	5	12	16	6
13	19	4	14	19	9
15	6	0	16	8	4
17	6	9	18	11	17
19	14	1	20	16	4
21	19	5	22	38	8
23	5	1	24	9	4
25	12	0	26	13	1
27	18	8	28	10	1
29	11	0	30	9	11
31	8	0	32	23	17
33	9	0	34	4	1
35	7	1	36	20	3
37	2	4	38	9	3
39	15	2	40	14	8
41	4	1	42	19	3
43	13	2	44	4	0
45	7	1	46	16	0
47	6	0	48	17	0
49	8	4	50	18	4

Cluster analysis 157

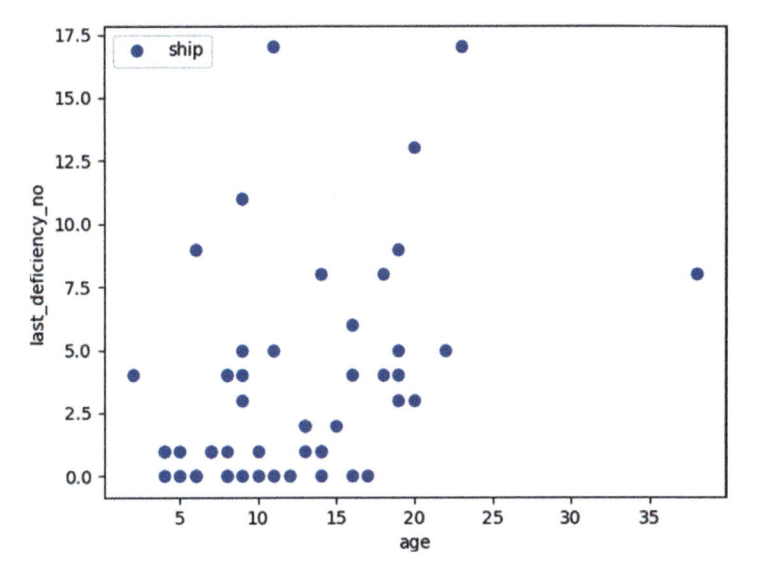

Figure 11.2   Distribution of the features of ships in the toy example

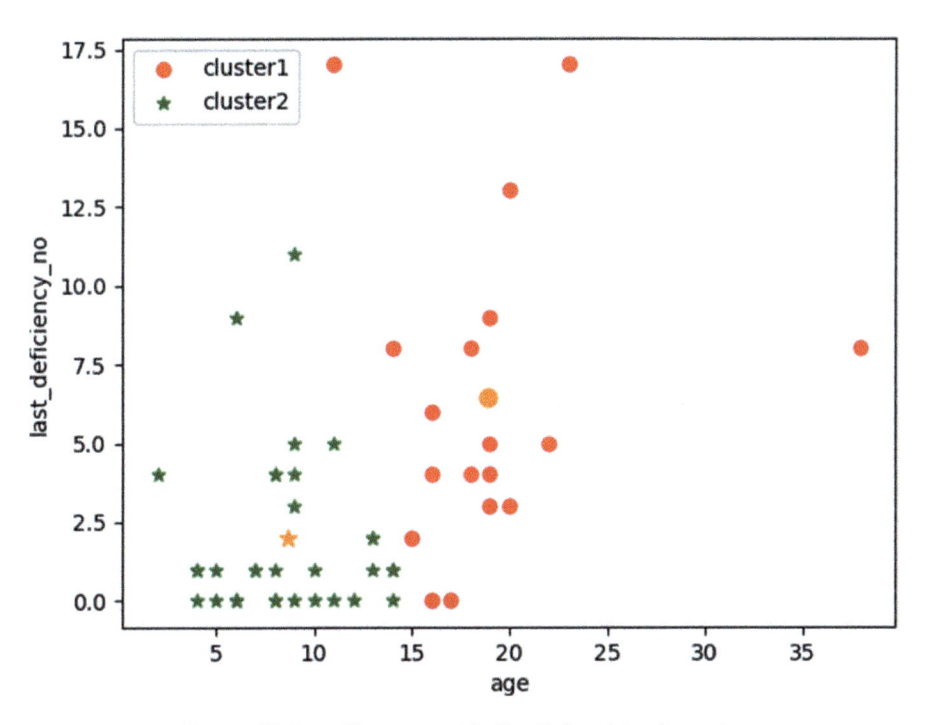

Figure 11.3   K-means with K = 2 for ship clustering

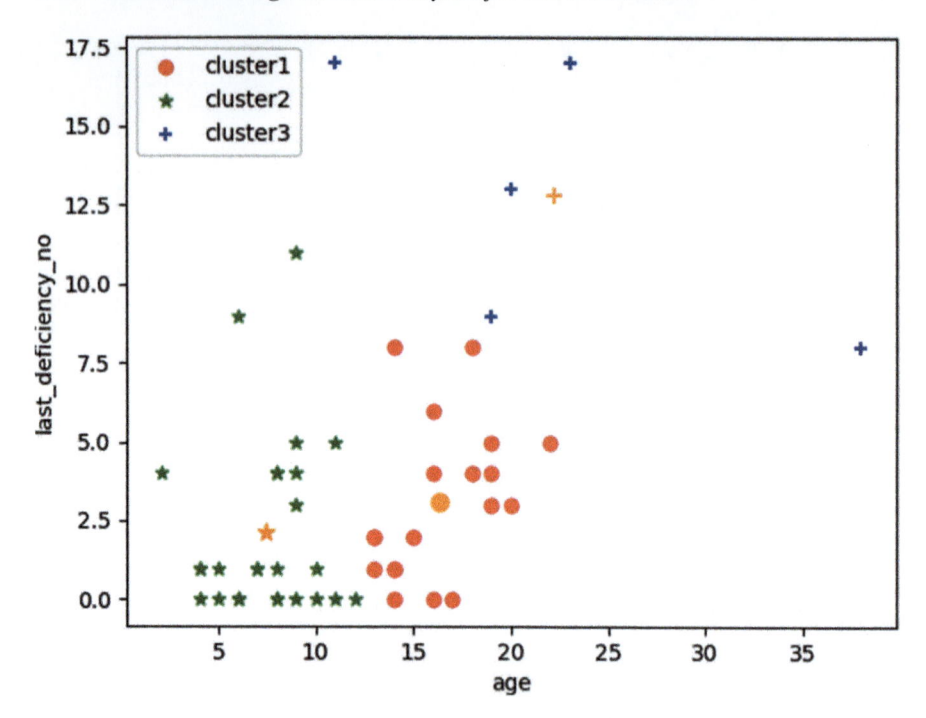

158 *Machine learning and data analytics for maritime studies*

Figure 11.4   K-means with K = 3 for ship clustering

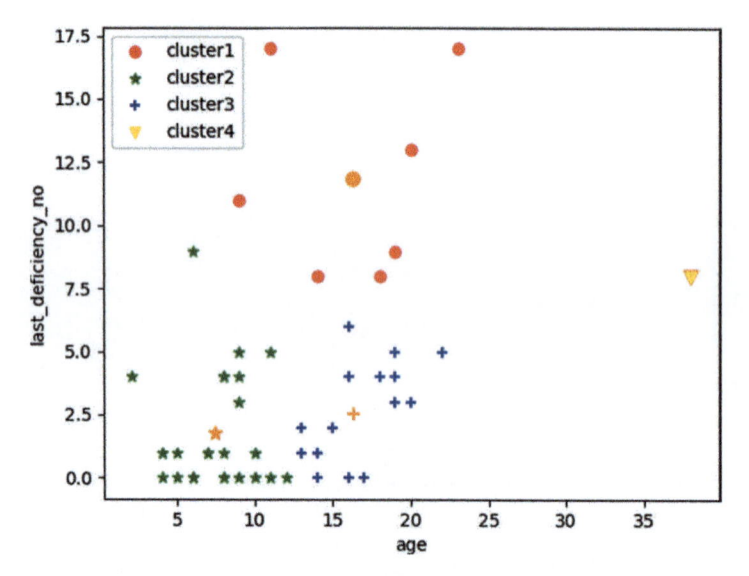

Figure 11.5   K-means with K = 4 for ship clustering

In Figures 11.3–11.5, ships in different clusters are represented by different symbols with different colors, and the centroids of the clusters are all in color orange and the same symbol as the examples in the corresponding cluster of a larger size. When $K = 2$, the centroid of cluster 1 is [18.9, 6.4], which means that the average age of the ships in this cluster is 18.9 and the average last deficiency number of these ships is 6.4. The centroid of cluster 2 is [8.6, 2.0]. It can be seen that the two clusters of ships are quite different in both ship age and the last inspection state, and it can also be anticipated that ships in cluster 1 have worse performance in the next PSC inspection, as both older ships and worse performance in the last PSC inspection are indicators of a higher ship risk.

When $K = 3$, the centroids of clusters 1–3 are [16.3, 3.1], [7.5, 2.2], and [22.2, 12.8], respectively, which also show a positively correlated relationship between ship age and the last deficiency number: as ship age increases from 7.5, to 16.3, and to 22.2, the last deficiency number increases from 2.2, to 3.1, and to 12.8. Moreover, as the last deficiency number of most ships is no more than 10 and ship age of most ships is less than 20, the centroids of clusters 1 and 2 are closer to each other compared to their distances from the centroid of cluster 3. Similar to the situation when $K = 2$, it can also be expected that ships in cluster 3 have the highest risk, followed by ships in cluster 1 and then by cluster 2.

When $K = 4$, the centroids of clusters 1–4 are [16.3, 11.9], [7.3, 1.9], [16.1, 2.4], and [38.0, 8.0], respectively. Cluster 4 only has one ship, which is very old and is far from the other ships. Then, it can be regarded as an outlier. For the other three clusters, the general relationship between ship age and the last deficiency number is similar to the situations when $K = 2$ and $K = 3$: these two features are positively correlated with each other. Furthermore, it can also be expected that ships in cluster 1 have the highest risk, while ships in cluster 2 have the lowest risk. In addition, as ship age is densely and nearly evenly distributed between 0 and 20, the difference of ship age value in the centroids of clusters 2 and 3 is much larger than that of clusters 3 and 1. Meanwhile, as there are several ships with last deficiency number more than 7, the difference of the last deficiency number of a ship in the centroids of clusters 2 and 3 is much smaller than that of clusters 3 and 1.

Based on the above three figures, it can also be seen that $K = 3$ is the most suitable value in K-means for ship clustering in this toy example, as the features of the ships are distinctive in different clusters, while the distribution of the clusters are relatively stable.

Reference

[1] Pedregosa F., Varoquaux G., Gramfort A., *et al.* 'Scikit-learn: machine learning in python'. *Journal of Machine Learning Research*. 2011;**12**:2825–30.

Chapter 12
Classic and emerging approaches to solving practical problems in maritime transport

This chapter aims to discuss classic and emerging approaches adopted in existing academic literature to address practical problems in maritime transport. First, widely studied practical problems in the maritime industry are summarized according to review papers. Then, classic methods that are widely adopted by the related studies are introduced. After that, data-driven methods, as a typical type of emerging approaches, are discussed. Especially, several examples of applying data-driven models to address prediction tasks in maritime studies are given. Finally, several issues regarding the application of data-driven models to address practical problems in maritime transport are presented with various examples in port state control (PSC) provided.

12.1 Topics in maritime transport research

Maritime transport is the transportation of cargo over the global water transportation network, where seaports serve as the nodes and waterways are the links of the network. Therefore, research topics in maritime transport academic research can be divided into "shipping part" and "port part" according to Talley [1]. There are several review papers on discussing the research themes, topics, and trends of the academic literature on maritime transportation, which we summarize as follows from the aspects of the number of papers involved, the research areas and topics covered, and main conclusions.

1. A series of review studies have been conducted by Woo *et al.* in 2011, 2012, and 2013 on how seaports have been studied from 1980s to 2000s from three perspectives: research themes and trends [2], research methodology [3], and research collaboration [4]. The authors found that in 1980s, about half of the papers in academic journals are published by *Maritime Policy & Management*. In 1980s, more than one-third of the related papers are still published by *Maritime Policy & Management*, while more journals, and some of them are newly established, started to publish papers on maritime transport, such as *Maritime Economics & Logistics*, *Journal of Transport Geography*, *Transportation Research Part E: Logistics and Transportation Review*, *Marine Policy*, and *Transport Policy*. The authors divide port research into two main

categories: government based and port authority/companies based, while the former can be further divided into port policy and governance and reform, and the latter can be further divided into management and strategy, competition and performance, and port in supply chains. Regarding the research strategies in port research, analytical and empirical strategies are the main research strategy, where analytical strategy includes conceptual, mathematical, and statistical methods, and empirical strategy includes experimental and case studies methods. In addition, the discipline bases in port research include economics, operations research, geography, strategic management, logistics, and marketing. Regarding the trend of research collaboration, the authors conclude that the research collaboration has substantially increased over the past three decades and in particular publication via collaboration has become dominant over the last decade.

2. Talley [1]: this study focuses on papers published in two leading maritime transport journals: *Maritime Policy & Management* and *Maritime Economics & Logistics*. Especially, papers published in *Maritime Policy & Management* from 2001 to 2012 are covered, and the most popular research topics of these papers are shipping performance/management, shipping finance, port governance/privatization, seafarers, and air emissions from shipping activities. Papers published in *Maritime Economics & Logistics* during 2001–2012 are covered, where the most popular topics are port performance, shipping performance/management, shipping finance, port choice, and supply chain.

3. Shi and Li [5]: this study reviews 1292 papers on maritime transport published in 19 transportation journals between 2000 and 2014. Among all the papers covered, more than one-third of them are published in *Maritime Policy & Management*, followed by *Maritime Economics & Logistics* that publishes about one-fifth of the papers, and then by *Transportation Research Part E*, *International Journal of Shipping and Transport Logistics*, and *Journal of Transport Geography*. Regarding the research areas covered, nearly 51% of the reviewed papers are in the area of shipping, about 43% are regarding port, and about 6% are related to maritime fleet. The authors then summarize the research topics under each of these three areas. Especially, the top five topics in shipping research include shipping market, industry, freight rate, and economic impact; operational management, mode choice, empty container; sailing speed, green shipping, and environment-related issues; shipping risk and maritime security; and corporate management, stock returns, and performance. The top four topics in port research include port management, service, performance, efficiency, and competitiveness; terminal studies and berth allocation; port planning, development, cluster, network, and economic impact, port governance, port policy, regulation, and legal issues; and ports in transport and supply chain. For maritime fleet, most of the related papers are regarding shipbuilding, demolition, new orders, and second-hand ships and fleet structure, deployment, ownership, and operation.

Classic and emerging approaches to solving practical problems 163

12.2 Research methods and their specific applications to maritime transport research

According to Shi and Li [5] and Yan *et al.* [6], seven methods are the most popular in academic research on maritime transportation, whose explanations are shown in Table 12.1.

From Table 12.1, it can be seen that the majority of the classic research methods are in a qualitative manner, including SIQO, case study, CCCQ, and literature review. Those in a quantitative manner mainly aim to construct models using mathematical modeling, economic and econometric theories, statistical modeling, and simulation methods to model or mimic the practical behavior of the system, with the aim to gain operational and managerial insights to reduce costs or increase benefits gained. The common point of the above models is that only when very specific instructions are given to them can they function well. After a long time of development and implementation, these classic methods as well as the results obtained from them are widely recognized and used by both academic researchers and industrial practitioners. However, it should also be mentioned that these methods are highly dependent on long-term practical experience and expert knowledge, while some of them might be biased, and they might not be adapted and updated to the ever-changing environment.

A typical example of using classic methods to address a practical problem in PSC is a ship selection scheme called ship risk profile (SRP) for high-risk ships selection introduced in Chapter 2, which is widely used by port states around the world for higher risk ship selection. It is heavily based on expert knowledge, which can be seen from the rough processing of the risk factors considered (e.g., only two states are considered for feature ship age: ships with age larger than 12, and ships with age no more than 12), the fixed and subjective weighting points assigned to the states of the features (e.g., ships with age larger than 12 are assigned with 2 weighting points, and 0, otherwise), and the simple weighted-sum manner for ship overall risk calculation. In addition, although ships can be quite different from each other, they are only divided into three risk profiles by the SRP, namely, HRS (high-risk ship), SRS (standard-risk ship), and LRS (low-risk ship) from the highest risk to the lowest risk, and a fixed (while different) time window is attached to each risk profile. The time window attached to HRS is longer than SRS, and is much longer than LRS. Then, a ship's inspection priority is highly dependent on the relationship between its last inspection time in the related MoU (Memorandum of Understanding) and the time window attached to the ship: if the period between the last inspection time and the current inspection is longer than the upper bound of the time window attached to the ship considering its SRP, the ship has high inspection priority; if the period is between the lower and upper bounds of the time window attached, the ship has medium inspection priority; if the period is shorter than the lower bound of the time window attached, the ship has low inspection priority. Besides, even if ship flag and RO (recognized organization) performances are updated annually and ship company performance is updated on a daily basis, feature processing methods, weighting points attached to the feature states, inspection time windows

Table 12.1 Summary of classic methods widely used in maritime transport research [6]

Research methods	Explanations
SIQO	"S" represents survey, which is to collect information from a selected sample of individuals by distributing and collecting their responses to certain questions. The whole procedure includes data collecting, aggregating, and analyzing. "I" represents interview, which is a conversation aiming to gather information and involves an interviewer, who is responsible to coordinate the whole process and ask questions, and an interviewee, who responds to the questions. "Q" represents questionnaire, which is a research instrument that consists of a set of questions with the aim to collect information from respondents. It should be noted that questionnaire and survey are related but different terms. Survey contains a relatively wider range of action from question design, data collection and analysis, and conclusion drawing. This shows that a survey usually involves questionnaires, while a single questionnaire is only one small part of survey. "O" represents observation, which is a study of nonexperimental situations where behavior is observed and recorded, so as to generate either qualitative or quantitative results.
Economic modeling	Using economic theories as well as quantitative and/or qualitative models and techniques to analytically evaluate the cause and effects of economic phenomenon.
MES	"M" represents mathematical analysis, which aims to model a nonmathematical situation, phenomenon, and the relationship between the situations mathematically, e.g., by developing mathematical models and proposing proper solution approaches. "E" represents econometric analysis, which aims to construct a set of equations to provide quantitative explanations and design tests to draw meaningful inferences from the samples to the whole population. "S" represents statistical analysis, which aims to collect and analyze data based on statistical assumptions and tests so as to draw meaningful inferences from the samples to the whole population.
Case study	An in-depth study of the research object from comprehensive perspectives to explore patterns and causes of its behaviors and states.
CCCQ	The first "C" represents conceptual analysis, which aims to break down and analyze concepts into their constituent parts to gain a better understanding of a particular issue. The second "C" represents content analysis, which aims to quantify and analyze the presence of certain words, themes, or concepts within some given quantitative data.

(Continues)

Table 12.1 Continued

Research methods	Explanations
	The third "C" represents comparative analysis: which aims to compare two or more processes, documents, data sets, and other objectives.
	"Q" represents qualitative analysis, which uses subjective judgments to analyze the research objective based on nonquantifiable information.
Literature review	An overview and evaluation of the available literature in a given subject of area.
Simulation	An imitation of the operation of a real-world process or system over time.

attached to the risk profiles, and ship inspection priority are not updated considering the changing conditions and factors, such as the evolution of ship conditions and the influences of the COVID-19 on ship behaviors and the PSC inspection mode.

To reduce the adverse influences of expert knowledge and subjective judgment on making assumptions and constructing models and to take the ever-changing environment into account, data-driven models as an emerging approach are rapidly developed and used to address practical problems in maritime transport research. As the name suggests, data-driven models are highly dependent on the data collected, so as to dynamically explore useful information from these data. ML (machine learning) models, which allow computer to learn from data facilitated by indirect programming, are a typical type of data-driven models. Especially, an overview of data-driven models is given in Chapter 3 and the key elements of data-driven models are summarized in Chapter 1 of this book. There are some initial trials of applying data-driven models to deal with practical issues in maritime transportation, which we briefly list as follows [6]:

1. Ship trajectory prediction (in the category of shipping research): it aims to predict the trajectory of a moving vessel in the near future by using historical time series trajectory points of the vessel, which is the foundation of vessel navigation risk analysis. Ship sailing trajectory is constructed from vessel automatic identification systems (AIS), where one sailing record comprising of vessel dynamic sailing information is proposed every tens of seconds to a few minutes. Popular data-driven models in existing literature for ship trajectory prediction include SVM (support vector machine) [7], extreme learning machine [8], k-nearest neighbors [9], autoencoder [10], and long short-term memory networks [11, 12].
2. Ship risk prediction and safety management (in the category of shipping research): two research topics are widely studied under this research stream: prediction of the occurrence probability of vessel accidents, and the analysis of regional and global ship accident statistics and accident reports. In the former research topic, AIS is the main data resource used to generate ship historical

trajectory and predict ship trajectory in the near future for risk assessment and analysis of ship collision [13–15], ship grounding accidents [16], fire accidents [17], and seafarers' nonfatal injuries [18]. In the latter research topic, maritime accident statistics and reports published by the IMO (Internation Maritime Organization) and local marine departments/institutions are the main data resource. BNs are the most popular model to analyze the relations among the risk factors and between the risk factors and the prediction target. The related studies can be found in References [19–22]. In addition, K-means is also applied to analyze maritime accidents [23, 24].

3. Ship inspection planning (in the category of shipping research): the related research topics are lying in four categories: ship risk prediction, onboard inspection sequence optimization, influencing factors on the inspection results of PSC, and PSC inspections' influence. Research data are mainly from the inspection data published by individual MoUs and ship specification data provided by online commercial databases. Widely used data-driven models in PSC-related research include BNs [25, 26], tree-based models [27–29], SVMs [30, 31], and association rule learning [32–34].

4. Ship energy efficiency prediction (in the category of shipping research): the related research aims to predict ship fuel consumption rate (i.e., hourly or daily fuel consumption amount) under various conditions using features of ship sailing information and the surrounding environments. Such ship- and environment-related features are collected from manually filled ship sailing records like noon reports or automatically collected from onboard sensors like fuel consumption sensors, global positioning system receivers, shaft power testers, wind speed sensors, and water depth sonars. ANNs are the most widely used fuel consumption rate prediction model, which can be found in References [35–38]. Additionally, tree-based models [39, 40], LASSO (least absolute shrinkage and selection operator) regression [41, 42], ridge regression [43, 44], and SVR (support vector regression) [44] are also popular for ship energy efficiency prediction. Then, the predicted ship energy efficiency indicators, with fuel consumption rate as the representative, are used to guide voyage management and optimization by means of speed optimization, trim optimization, weather routing, and their combinations.

5. Ocean freight market condition prediction (in the category of shipping research): two indicators of shipping market condition are widely used as the target of ocean freight market condition prediction, namely, baltic dry index (BDI), which is regarded as the leading indicator of economic activity and the barometer of the dry bulk shipping market, and sea freight rate, which is the requested price to transport cargo from the origin to the destination by sea, mainly determined by the weight and volume of the cargo and the distance between the origin and destination. For the prediction of BDI, ANNs (artifical neural networks) and SVRs are the most popular methods, and the adoption of ANNs can be seen in References [45–47], while the adoption of SVRs can be seen in References [48, 49]. For the prediction of sea freight rate, ANNs are also a popular method, which can be seen in References [50–52]. Other ML

models are also used and compared for sea freight rate prediction, as can be seen in [53, 54].
6. Ship destination port and arrival time at port prediction (in the category of port research): the related studies aim to predict the destination of a voyage and its arrival time. As such prediction is highly dependent on ship historical voyage information and the information of the current voyage, AIS is a widely used data source. Ship arrival time to the destination is also influenced by the sea and weather conditions along the way, as well as the port operating policies and conditions (e.g., the just-in-time arrival policy for ships to save fuel consumption and reduce greenhouse gas emissions). Therefore, marine weather forecast and port statistics are also important data resources. It should be mentioned that although the destination of the current voyage is required to be reported via the AIS by vessel captains, it is widely believed that such report is highly unreliable with complex causes as discussed by Yang *et al.* [55]. For ship destination prediction, the methods can be either turning point based or trajectory extraction based. As both tasks are nontrivial, especially when there are requirements on prediction accuracy, various ML-based models are developed, such as a framework based on anomaly detection and route prediction [56], DBSCAN (density-based spatial clustering of applications with noise) for turning region clustering and identification combined with ANN for next turning point prediction [57]. For literature in the second category, the prerequisite of arrival time prediction is to know the destination. Therefore, there are several studies that jointly predict vessel destination and the vessel arrival time, which can be found in References [58–60].
7. Port condition prediction (in the category of port research): this stream of research aims to predict port cargo volume and port traffic volume, where the related data are from port statistics and bill of lading. Port cargo volume is predicted by ANNs combined with time series methods [24] and by SVRs [61]. Port traffic volume at the Shanghai port is predicted by ANNs in Reference [62].

12.3 Issues of adopting data-driven models to address problems in maritime transportation

As introduced in Chapter 4, the procedure of adopting data-driven models to address a practical problem in maritime transportation starts from understanding the application scenarios and specification of the current problem as well as the feasibility of solving it. Then, related data are collected, which will be preprocessed and processed by various feature engineering techniques, including data cleaning, feature extraction, feature preprocessing, feature encoding and scaling, and feature selection, whose aim is to allow developing a proper and efficient data-driven model that can solve the problem more effectively. After that, a model/several models are constructed, evaluated, and refined while considering the properties and requirements of the practical problem to be solved, and then the best model on hand is finally selected and handed over to practical users to deal with the current problem. The users can be technicians who are familiar with the underlying of the model or practitioners who are unfamiliar with the theories of the model. During the practical usage of the model, ideally, its input, output, and settings should be constantly evaluated and updated, so as to better handle the ever-changing environment.

In the above process, four layers are involved and interwoven with each other, namely, data, model, user, and target. To develop proper and efficient data-driven models, there are issues within these four layers that should be paid close attention to which have been discussed by Yan *et al.* [6]. We discuss these issues here in detail with several new points added in the following subsections.

12.3.1 Data

Regarding the data used for model construction, one critical issue is that the data used, including the input features and output target(s), should be highly compatible with the focus of the current problem. To be specific, the features used should possess the properties of "relevance" and "comprehensiveness": "relevance" means that they are indeed related to/can contribute to the prediction of the target, and "comprehensiveness" means that the features used cover most of the indicators that affect the target.

Issues regarding data relevance can be assessed from qualitative perspective and quantitative perspective. Qualitatively, features selected should be of interest to model users, which is demonstrated in the following Example 12.1 of PSC.

Example 12.1: To predict the risk of a ship in a PSC inspection, it is common that ship characteristics and indicators of its historical inspection performance, such as ship dimensions (i.e., length, depth, beam, and draft), ship's flag, RO, and company performance, the date and performance of a ship's last PSC inspection, etc., are used as the input to develop prediction models, as can be seen in References [27, 28, 30, 31]. These features are relevant to ship's structural and operational conditions, navigation safety, and safe management, which are also key points in PSC. Therefore, it is justifiable to consider these factors when developing ship risk prediction models. In contrast, sailing information of the current voyage, such as the sailing route, average speed, freight rate, and charter mode, should deserve less attention since obtaining such information is costly or even impossible, while the port authorities cannot benefit much from the above information when evaluating ship risk levels.

Quantitatively, feature selection can be dealt with by relevance calculation and evaluation model development. The relevance calculation approach aims to evaluate the interrelationship between the features and the target. For example, in the step of feature selection in feature engineering, several approaches of feature selection can be used to identify the most relevant features from all candidate features by calculating the relation between the features and the target(s) as introduced in section 1.2.4 of Chapter 4. Such approaches include Pearson's correlation coefficient, mutual information, and recursive feature elimination. In addition, features can be selected based on regularization and feature importance, and both can be integrated in an ML algorithm. In particular, the former is by shrinking the coefficients, or weights, of the features that have very little impact on the prediction target to zero, so as to exclude this feature from the predicting process. The latter can be mainly achieved by tree-based models and permutation, as feature importance can be automatically generated by a tree-based model evaluated by the impurity reduction in node splitting by the features, while features that lead to higher prediction error in the permutation process should be more important in the permutation method.

In contrast, comprehensiveness of the features used can be much harder to evaluate, and it is highly dependent on and restricted by the available data sets that can be obtained. To consider more comprehensive features, data fusion is widely used to combine different data resources. Example 12.2 is provided as an example of data fusion in PSC inspection.

Example 12.2: In the Tokyo MoU where the port of Hong Kong belongs to, an online database called Asia Pacific Computerized Information System (APCIS) provides the detailed inspection records, where the record of an individual inspection includes ship basic information (i.e., ship name, ship type, IMO number, MMSI (maritime mobile service identity), call sign, classification society, flag, date keel laid, dead weight, and gross tonnage, ship's flag/RO/company performance) as well as the inspection details. The basic ship information can no doubt provide insights on the prediction of ship risk. Nevertheless, they cannot cover a full picture of ship conditions where the PSC inspectors might be interested in. For example, as shown by the econometric results of Cariou and Wolff [63], ships with worse historical PSC inspection performance are more likely to change their flag states. Therefore, it can be expected that taking ship flag change times between the period of keel laid date and the current inspection can increase prediction accuracy. Such information is provided by the World Register of Ship database, which can be combined with the APCIS database by ship IMO number. In addition, as ship inspection priority is highly related to its last inspection time as well as its performance in previous PSC inspections, taking factors regarding previous inspections, including but not limited to the last inspection time and the deficiency and detention conditions, is highly likely to improve the performance of the prediction model. Such information can be searched from APCIS by using a ship's IMO number, and can be combined with the original inspection records collected from APCIS by ships' IMO numbers.

12.3.2 Model

When developing data-driven models to solve a practical problem in maritime transportation, it is of utmost importance to understand and clarify the properties of the problem itself as well as the proper metrics to evaluate its performance. One basic step is to identify the nature of the current problem, i.e., to decide the current problem is a classification, regression, clustering, or association rule mining task. Then, models falling within certain category should be tried and evaluated. In this process, how to select an effective model as well as its evaluation metrics should be carefully planned. We use the following Example 12.3 to show the related issues.

Example 12.3: Suppose that we need to address the problem of ship detention prediction in PSC. The nature of this problem is classification, where a detained ship is predicted to be "1" and a non-detained ship is predicted to be "0." The main characteristic of this problem is that the data set is highly imbalanced regarding the prediction target: the number of non-detained ships is much larger than that of detained ships. For example, the annual detention rate at the Hong Kong port ranges from 2.96 to 3.67% between 2015 and 2019 as shown in the 2020 annual report of Tokyo MoU [64]. This means that among all 100 foreign visiting ships to the Hong Kong port that are inspected, no more than 3.67 of them are actually detained on average, while up to 97.04 of them are actually not detained on average. If standard classification models are directly applied, no matter how well they perform on balanced data sets, it can be anticipated that their performance for ship detention prediction would be relatively bad, or the prediction results cannot make sense (e.g., predicting all the inspected ships to be non-detained leading to a very high accuracy score, but the prediction is meaningless).

To take the imbalanced property of the data set into account, several strategies can be adopted. For example, oversampling (i.e., randomly selecting a certain number of detained ships from the original data set and then input the selected ships into the original data set to form a balanced one) or subsampling (i.e., randomly selecting a subset of non-detained ships from the original data set and then form a new data set containing all detained ships and the subset of the non-detained ships) can be applied before model construction. Moreover, specialized ML models suitable to address imbalanced classification tasks can also be used, such as balanced random forest and anomaly detection models.

Another major issue regarding data-driven models is that most of the ML models are of black-box nature, meaning that the model work mechanism and the prediction results cannot be understood by the developers and users. There are of course some exceptions, such as LR (linear regression), DT (decision tree), and BN (Bayesian network) models. Nevertheless, as these models are not sophisticated enough, they may have bad performance when addressing certain complex problems, and thus other sophisticated ML models of black-box nature have to be adopted. The black-box property of data-driven models would reduce their acceptance and credibility of the practitioners, and thus restrict their applicability. Furthermore, it is well known that when models are not conceived with self-explanatory characteristics, they may engender pitfalls. In recent years, a concept of "explainable artificial intelligence," or XAI for short, is proposed and rapidly developed, which aims to open up the black boxes of ML models and shed light on the model working mechanisms as well as the justification of the results generated. Details of the concepts, techniques, and applications of XAI will be covered in later sections of this part.

12.3.3 User

It is widely known that the maritime industry is relatively conservative and traditional, in which the decision support tools used by maritime practitioners are mainly based on expert knowledge or classic theories for a long time, and the stakeholders are reluctant to share the information and data with each other. Consequently, without the support of sufficient data, especially high-quality data, the digitization process of the maritime industry is relatively slow, where contemporary approaches, especially data-driven models, are not that popular compared to other industries such as financial technology and healthcare industry. Therefore, data-driven applications are still in their infancy, and it is also noted that applying data-driven models to maritime transport industry requires replacing the current (naive) decision support tools, which the decision makers are quite familiar with more sophisticated models that the decision makers may be unfamiliar with. Systems based on such models are more like black boxes to the users, with brand new graphical user interface (GUI) and obscure working mechanism. In this process, one guiding principle is to avoid imposing any unnecessary burden or pressure on the users during their usage.

However, this is not a trivial task as the developers and users of a data-driven-based model or system in the maritime industry usually have quite different backgrounds. On the one hand, model and system developers are usually engineering or researchers specializing in data analytics and with weak shipping background. The data-driven models developed usually have multifarious inputs and outputs as well as complex structures, which means that the data processing and model operation can be sophisticated. On the other hand, model and system users are usually practitioners of the maritime industry, like government officials, managers, and technicians in shipping companies and classification societies, and crew members, who may lack expertise in data analytics and system operation. It is therefore crucial to avoid requiring model users to input extra data or too much information to the models as well as conducting too many extra operations when using the systems. Ideally, a system with straightforward, concise, and friendly GUI as well as easy human–computer interactions should be developed. The following is an example of a high-risk ship selection decision support system in PSC developed for the Marine Department of Hong Kong by researchers from the Hong Kong Polytechnic University.

Example 12.4: This system is named by *AI for PSC at Hong Kong*[*], which automatically downloads the information of in port ships and due to arrival ships at the Hong Kong port about every 20 minutes, and then applies AI (artificial intelligence)-based models to predict ship risk, which is represented by the combination of the number of deficiencies and the detention probability of each ship to facilitate the Marine Department to select ships with the highest risks for inspection. Especially, the displayed items as well as their formats on the website, including not only the outputs of the AI models but also the ships' identify information, type, flag, last port of call, agent, current location, etc., are compliant with the preference of the decision makers. This means that the system users do not

---
[*]https://sites.google.com/site/wangshuaian/research/

need to do anything to use the system, and they can acquire the information they are interested in for high-risk ship selection directly from the system. Therefore, the newly developed ship selection system based on data-driven models is in compliance with the users' past working habits and preference, and thus does not pose extra operational burdens to them.

Another issue in this category is regarding data-driven models' applicability: as such models are developed using data collected from certain region and entities, while different regions and entities might be quite different from each other, they do not possess the property of universal applicability. For example, the ship risk prediction models developed in this book all use PSC inspection records at the Hong Kong port, and thus they may be hard to be applied to ports outside of the Tokyo MoU, as different ship selection methods are used, and even to the other ports within the Tokyo MoU, as the expertise and background of the inspectors in different ports might vary a lot. Therefore, if one wishes to apply the ship risk prediction models developed in this book to ports other than the Hong Kong port, new prediction models using the corresponding data sets should be developed, and more advanced methods, such as transfer learning, can also be used.

12.3.4 Target

To promote the acceptance and application of data-driven models, the predicted output, i.e., the target values given by the data-driven models, should comply with shipping domain knowledge. Otherwise, it is highly likely that the users would doubt the accuracy and fairness of the prediction models developed. The following Example 12.5 shows this issue in high-risk ship selection in PSC.

Example 12.5: In many existing studies on ship risk prediction, ship flag, RO, and company performances are considered as input features to the ship risk prediction models, which can be found in References [27–29]. Roughly speaking, these three factors are calculated based on the performance of the ships under their management in the previous three years, and can thus influence their reputation and the inspection priority of their ships in future PSC inspection. Therefore, responsible ship flag, RO, and company put many efforts to maintain their ships to be in satisfactory conditions, and they hope that their hard efforts can gain them good reputation and thus a lower inspection priority of their vessels. Therefore, it is justifiable to say that for two ships with all the other conditions being equal, the ship with better flag, RO, or company performance than the other ship should be predicted to be in a lower risk (i.e., less number of deficiencies or lower detention probability), and thus is less likely to be selected for inspection. If the predicted targets given by a data-driven model conflict with such shipping domain knowledge, the decision makers, as well as the ships' flags, ROs, and companies might hardly trust the predictions. Therefore, developers of the data-driven models should take such shipping domain knowledge into account in model development to guarantee model fairness. Details of this issue will be covered in the later chapters of this part.

To conclude, issues that deserve more attention in developing data-driven models to address practical problems in maritime transport as discussed above are summarized as follows:

1. data: relevance and comprehensiveness;
2. model: model selection and evaluation, black-box nature of ML models;
3. users: putting no extra burden on users, model universal applicability;
4. targets: considering shipping domain knowledge.

It should be noted that there are still a long way to go to develop and apply systems based on data-driven models in the traditional and conservative maritime industry, and the issues covered are only a small part of the whole picture that need more attention and efforts.

References

[1] Talley W.K. 'Maritime transportation research: topics and methodologies'. *Maritime Policy & Management*. 2013;**40**(7):709–25.

[2] Woo S.-H., Pettit S., Beresford A., Kwak D.-W. 'Seaport research: a decadal analysis of trends and themes since the 1980s'. *Transport Reviews*. 2012;**32**(3):351–77.

[3] Woo S.-H., Pettit S.J., Kwak D.-W., Beresford A.K.C. 'Seaport research: a structured literature review on methodological issues since the 1980s'. *Transportation Research Part A*. 2011;**45**(7):667–85.

[4] Woo S.-H., Kang D.-J., Martin S. 'Seaport research: an analysis of research collaboration using social network analysis'. *Transport Reviews*. 2013;**33**(4):460–75.

[5] Shi W., Li K.X. 'Themes and tools of maritime transport research during 2000-2014'. *Maritime Policy & Management*. 2017;**44**(2):151–69.

[6] Yan R., Wang S., Zhen L., Laporte G. 'Emerging approaches applied to maritime transport research: past and future'. *Communications in Transportation Research*. 2021;**1**:100011.

[7] Liu X., He W., Xie J., Chu X. 'Predicting the trajectories of vessels using machine learning'. *5th International Conference on Control, Robotics and Cybernetics (CRC)*; Wuhan, China, IEEE, 2020. pp. 66–70. Available from https://ieeexplore.ieee.org/xpl/mostRecentIssue.jsp?punumber=9253444

[8] Mao S., Tu E., Zhang G., et al. 'An automatic identification system (AIS) database for maritime trajectory prediction and data mining'. *Proceedings of ELM-2016*; Springer, 2018. pp. 241–57.

[9] Virjonen P., Nevalainen P., Pahikkala T., Heikkonen J. 'Ship movement prediction using k-NN method'. *2018 Baltic Geodetic Congress (BGC Geomatics)*; Olsztyn, 2020. pp. 304–09.

[10] Murray B., Perera L.P. 'A dual linear autoencoder approach for vessel trajectory prediction using historical AIS data'. *Ocean Engineering*. 2020;**209**:107478.

[11] Huang Y., Chen L., Chen P., Negenborn R.R., van Gelder P.H.A.J.M. 'SHIP collision avoidance methods: state-of-the-art'. *Safety Science*. 2020;**121**:451–73.

[12] Li W., Zhang C., Ma J., Jia C. 'Long-term vessel motion predication by modeling trajectory patterns with AIS data'. *2019 5th International Conference on Transportation Information and Safety (ICTIS)*; Liverpool, IEEE, 2020. pp. 1389–94.

[13] Gang L., Ma J., Yao J. 'Decision-making of vessel collision avoidance based on support vector regression'. *ICAIIS 2021*; Chongqing, New York, NY, 2021. pp. 1–6. Available from https://dl.acm.org/doi/proceedings/10.1145/3469213

[14] Gao M., Shi G.Y., Liu J. 'Ship encounter azimuth MAP division based on automatic identification system data and support vector classification'. *Ocean Engineering*. 2020;**213**:107636.

[15] Abebe M., Noh Y., Seo C., Kim D., Lee I. ' Developing a SHIP collision risk index estimation model based on dempster-shafer theory'. *Applied Ocean Research*. 2021;**113**:102735.

[16] Wu B., Yan X., Yip T.L., Wang Y. 'A flexible decision-support solution for intervention measures of grounded ships in the Yangtze River'. *Ocean Engineering*. 2017;**141**:237–48.

[17] Wu B., Tang Y., Yan X., Guedes Soares C. 'Bayesian network modelling for safety management of electric vehicles transported in ropax ships'. *Reliability Engineering & System Safety*. 2017;**209**:107466.

[18] Zhang G., Thai V.V., Law A.W.K., Yuen K.F., Loh H.S., Zhou Q. 'Quantitative risk assessment of seafarers' nonfatal injuries due to occupational accidents based on Bayesian network modeling'. *Risk Analysis*. 2020;**40**(1):8–23.

[19] Wang L., Yang Z. 'Bayesian network modelling and analysis of accident severity in waterborne transportation: a case study in China'. *Reliability Engineering & System Safety*. 2018;**180**:277–89.

[20] Fan S., Blanco-Davis E., Yang Z., Zhang J., Yan X. 'Incorporation of human factors into maritime accident analysis using a data-driven Bayesian network'. *Reliability Engineering & System Safety*. 2020;**203**:107070.

[21] Fan S., Yang Z., Blanco-Davis E., Zhang J., Yan X. 'Analysis of maritime transport accidents using Bayesian networks'. *Proceedings of the Institution of Mechanical Engineers, Part O*. 2020;**234**(3):439–54.

[22] Li B., Lu J., Lu H., Li J. 'Predicting maritime accident consequence scenarios for emergency response decisions using optimization-based decision tree approach'. *Maritime Policy & Management*. 2020:1–23.

[23] Lema E., Papaioannou D., Vlachos G.P. 'Investigation of coinciding shipping accident factors with the use of partitional clustering methods'. *PETRA '14*; Rhodes Greece, New York, NY, 2014. pp. 1–4. Available from https://dl.acm.org/doi/proceedings/10.1145/2674396

[24] Zhang Y., Sun X., Chen J., Cheng C. 'Spatial patterns and characteristics of global maritime accidents'. *Reliability Engineering & System Safety*. 2021;**206**:107310.

[25] Yang Z., Yang Z., Yin J. 'Realising advanced risk-based Port state control inspection using data-driven Bayesian networks'. *Transportation Research Part A*. 2018;**110**:38–56.

[26] Wang S., Yan R., Qu X. 'Development of a non-parametric classifier: effective identification, algorithm, and applications in Port state control for maritime transportation'. *Transportation Research Part B*. 2019;**128**:129–57.

[27] Yan R., Wang S., Fagerholt K. 'A semi- " smart predict then optimize " (semi-SPO) method for efficient SHIP inspection'. *Transportation Research Part B*. 2020;**142**:100–25.

[28] Yan R., Wang S., Cao J., Sun D. 'Shipping domain knowledge informed prediction and optimization in Port state control'. *Transportation Research Part B*. 2021;**149**:52–78.

[29] Yan R., Wang S., Peng C. 'An artificial intelligence model considering data imbalance for SHIP selection in Port state control based on detention probabilities'. *Journal of Computational Science*. 2021;**48**:101257.

[30] Xu R.-F., Lu Q., Li W.-J., Li K.X., Zheng H.-S. 'Presented at 2007 International Conference on Machine Learning and Cybernetics'. Hong Kong.

[31] Wu S., Chen X., Shi C., Fu J., Yan Y., Wang S. 'Ship detention prediction via feature selection scheme and support vector machine (SVM)'. *Maritime Policy & Management*. 2022;**49**(1):140–53.

[32] Tsou M.C. 'Big data analysis of port state control ship detention database'. *Journal of Marine Engineering & Technology*. 2019;**18**(3):113–21.

[33] Chung W.-H., Kao S.-L., Chang C.-M., Yuan C.-C. 'Association rule learning to improve deficiency inspection in Port state control'. *Maritime Policy & Management*. 2020;**47**(3):332–51.

[34] Yan R., Zhuge D., Wang S. 'Development of two highly-efficient and innovative inspection schemes for PSC inspection'. *Asia-Pacific Journal of Operational Research*. 2021;**38**(3):2040013.

[35] Pedersen B.P., Larsen J. 'Prediction of full-scale propulsion power using artificial neural networks'. *Proceedings of the 8th International Conference on Computer and IT Applications in the Maritime Industries (COMPIT'09)*; Budapest: Hungary, 2009. pp. 10–12.

[36] Petersen J.P. Mining of SHIP operation data for energy conservation. DTU Informatics; 2012.

[37] Rudzki K., Tarelko W. 'A decision-making system supporting selection of commanded outputs for A ship's propulsion system with A controllable pitch propeller'. *Ocean Engineering*. 2016;**126**:254–64.

[38] Du Y., Meng Q., Wang S., Kuang H. 'Two-phase optimal solutions for ship speed and trim optimization over a voyage using voyage report data'. *Transportation Research Part B*. 2019;**122**:88–114.

[39] Peng Y., Liu H., Li X., Huang J., Wang W. 'Machine learning method for energy consumption prediction of ships in Port considering green ports'. *Journal of Cleaner Production*. 2020;**264**:121564.

[40] Yan R., Wang S., Du Y. 'Development of a two-stage shipfuel consumption prediction and reduction model for a dry bulk ship'. *Transportation Research Part E*. 2020;**138**:101930.

[41] Wang S., Ji B., Zhao J., Liu W., Xu T. 'Predicting ship fuel consumption based on LASSO regression'. *Transportation Research Part D*. 2018;**65**:817–24.

[42] Gkerekos C., Lazakis I., Theotokatos G. 'Machine learning models for predicting ship main engine fuel oil consumption: a comparative study'. *Ocean Engineering*. 2019;**188**:106282.

[43] Soner O., Akyuz E., Celik M. 'Statistical modelling of ship operational performance monitoring problem'. *Journal of Marine Science and Technology*. 2019;**24**(2):543–52.

[44] Uyanık T., Karatuğ Ç., Arslanoğlu Y. 'Machine learning approach to SHIP fuel consumption: a case of container vessel'. *Transportation Research Part D*. 2020;**84**:102389.

[45] Leonov Y., Nikolov V. 'A wavelet and neural network model for the prediction of dry bulk shipping indices'. *Maritime Economics & Logistics*. 2012;**14**(3):319–33.

[46] Şahin B., Gürgen S., Ünver B. 'Forecasting the baltic dry index by using an artificial neural network approach'. *Turkish Journal of Electrical Engineering & Computer Sciences*. 2021;**26**(3):1673–84.

[47] Bae S.H., Lee G., Park K.S. 'A baltic dry index prediction using deep learning models'. *Journal of Korea Trade*. 2021;**25**(4):17–36.

[48] Han Q., Yan B., Ning G., Yu B. 'Forecasting dry bulk freight index with improved SVM. *Math Probl Eng*. 2014;**2014**:1–12.

[49] Bao J., Pan L., Xie Y. 'A new BDI forecasting model based on support vector machine'. Presented at 2016 IEEE Information Technology, Networking, Electronic and Automation Control Conference (ITNEC); Chongqing, China. IEEE,

[50] Cv S., HJv M., Breitner M.H. 'Spot and freight rate futures in the tanker shipping market: short-term forecasting with linear and non-linear methods'. Presented at In: Operations Research Proceedings 2012; Springer,

[51] Santos A.A.P., Junkes L.N., Pires Jr F.C.M. 'Forecasting period charter rates of VLCC tankers through neural networks: a comparison of alternative approaches'. *Maritime Economics & Logistics*. 2014;**16**(1):72–91.

[52] Yang Z., Mehmed E.E. 'Artificial neural networks in freight rate forecasting'. *Maritime Economics & Logistics*. 2019;**21**(3):390–414.

[53] Ubaid A., Hussain F.K., Charles J. 'Machine learning-based regression models for price prediction in the Australian container shipping industry: case study of asia-oceania trade Lane'. *In: International Conference on Advanced Information Networking and Applications*; Springer, 2020. pp. 52–59.

[54] Næss P.A. Investigation of multivariate freight rate prediction using machine learning and AIS data. NTNU; 2018.

[55] Yang D., Wu L., Wang S. 'Can we trust the AIS destination Port information for bulk ships? –implications for shipping policy and practice'. *Transportation Research Part E*. 2021;**149**:102308.

[56] Pallotta G., Vespe M., Bryan K. 'Vessel pattern knowledge discovery from AIS data: a framework for anomaly detection and route prediction'. *Entropy*. 2018;**15**(12):2218–45.

[57] Daranda A. 'Neural network approach to predict marine traffic'. *Baltic Journal of Modern Computing*. 2016;**4**(3):483.

[58] Amariei C., Diac P., Onica E., Roşca V. 'Cell grid architecture for maritime route prediction on AIS data streams'. *DEBS '18*; Hamilton, New Zealand, New York, NY, 2018. pp. 202–04. Available from https://dl.acm.org/doi/proceedings/10.1145/3210284

[59] Bodunov O., Schmidt F., Martin A., Brito A., Fetzer C. 'R-time destination and eta prediction for maritime traffic'. *DEBS '18*; Hamilton, New York, NY, 2018. pp. 198. Available from https://dl.acm.org/doi/proceedings/10.1145/3210284

[60] Nguyen D.D., Le Van C., Ali M.I. 'Vessel destination and arrival time prediction with sequence-to-sequence models over spatial grid'. *DEBS '18*; Hamilton, New York, NY, 2018. pp. 217–20. Available from https://dl.acm.org/doi/proceedings/10.1145/3210284

[61] Ruiz-Aguilar J.J., Moscoso-López J.A., Urda D., González-Enrique J., Turias I. 'A clustering-based hybrid support vector regression model to predict container volume at seaport sanitary facilities'. *Applied Sciences*. 2020;**10**(23):8326.

[62] Wang S., Wang S., Gao S., Yang W. 'Daily SHIP traffic volume statistics and prediction based on automatic identification system data'. *9th International Conference on Intelligent Human-Machine Systems and Cybernetics (IHMSC)*; Hangzhou, IEEE, 2017. pp. 149–54.

[63] Cariou P., Wolff F.C. 'Do Port state control inspections influence flag-and class-hopping phenomena in shipping?'. *Journal of Transport Economics and Policy (JTEP)*. 2011;**45**(2):155–77.

[64] *Annual report*. Tokyo: Tokyo MOU. Available from http://www.tokyo-mou.org/doc/ANN19-f.pdf

Chapter 13
Incorporating shipping domain knowledge into data-driven models

As mentioned in section 12.3.4 of hapter 12, one of the issues regarding the target of applying data-driven models to solve practical problems in maritime transportation is that the predicted target may not comply with shipping domain knowledge. The term "domain knowledge" refers to rules and common senses widely believed by the practitioners. Domain knowledge is based on the practitioners' understanding of the disciplines and activities of the industry, and is gained from longtime experience of the practitioners in this industry as well as their own expertise, professions, specializations, and judgment. For example, in the maritime industry, regarding the activity of ship selection for inspection by PSC (port state control), it is generally believed that given all other conditions being equal, an older ship would have a larger number of deficiencies than a younger ship. Regarding ship fuel consumption prediction, ship sailing speed is the most significant determinant, and it is widely believed that a ship's fuel consumption rate is proportional to its sailing speed to the power of $\alpha = 3$, i.e., $r \propto \beta \times v^{\alpha}$, where r is the hourly or daily fuel consumption, β is a coefficient, and v is the average sailing speed. In practice, α can be higher than 3, especially for large vessels like container ships where it can be 4, 5, or even higher. In the former example, if there is a ship risk prediction model that gives opposite prediction results, i.e., a younger ship has more deficiencies than an older ship under the condition that the other features of these two ships are identical, the prediction results as well as the prediction model are expected to be hardly accepted or used by the port state officers, because they may conclude that the proposed model is inaccurate and unfair based on such prediction. Similarly, in the latter example, if the predicted ship fuel consumption rate does not take such convex and increasing relationship with ship sailing speed, it may not convince the users. Given the condition that practitioners in the conservative while classic maritime industry might be reluctant to replace the current rule-based decision support systems with data-driven ones, it is thus of vital importance to make sure that the prediction results given by the data-driven models to address practical problems comply with the corresponding shipping domain knowledge. Otherwise, the practitioners are more likely to be very skeptical of the models together with their results, and thus are not willing to use them to assist their decision-making.

One way to guarantee that the developed data-driven models constructed from practical data are in compliance with shipping domain knowledge is to explicitly impose the constraints when developing the models. The following two sections in this chapter introduce some initial thoughts on how to incorporate shipping domain knowledge into the development of data-driven models used to deal with a specific problem in maritime transportation. To be specific, section 13.1 discusses how to consider feature monotonicity into a tree-based model developed to predict ship risk, and section 13.2 discusses how to jointly consider feature monotonicity and convexity into ship fuel consumption rate prediction.

13.1 Considering feature monotonicity in ship risk prediction

The content covered in this section is mainly from a paper published by Yan *et al.* in 2021 [1]. That paper aims to assist the Marine Department of Hong Kong to select ships of higher risk among all the foreign visiting ships for inspection by using state-of-the-art ML (machine learning) models with shipping domain knowledge considered. An eXtreme Gradient Boosting (XGBoost) model is first developed to predict the number of deficiencies each foreign visiting ship has. In particular, the XGBoost model takes domain knowledge regarding ship flag performance, RO (recognized organization) performance, and company performance in PSC into account, i.e., given two ships with all the other conditions equal (e.g. with the same age, of the same type, with the same last inspection time, and inspection results), the ship that has worse flag performance, or RO performance, or company performance, should be predicted to have a larger number of deficiencies and thus, should have a high priority/probability to be inspected. The ship risk prediction results are then input to a PSC officer scheduling model to help the maritime authorities to allocate scarce inspection resources such that as much ship noncompliance as possible can be detected. As only the first part of this study is related to ML model development, we only cover that part in this section. Our aim is to discuss how to incorporate domain knowledge into the model construction process of an XGBoost model using more plain words, so as to make readers with a weaker background in data analytics easier to understand.

13.1.1 Introduction of monotonicity in the ship risk prediction problem

The monotonicity property of three features: ship flag performance, RO performance, and company performance, which are also used by the Tokyo MoU to determine ship risk profile (SRP) in the New Inspection Regime (NIR) since January 1, 2014, is considered in this section. The SRP in NIR considers several ship-related parameters including type of ship, age of the ship, ship flag performance, ship RO performance, ship company performance, deficiency conditions, and detention conditions in the previous 36 months [2]. Different weighting points are attached to different states of the parameters, and the final ship risk is determined by the total weighting points. Especially, ship flag performance is established on an annual basis taking into account the performance of the ships under a certain flag state regarding both inspection and detention conditions over the preceding three calendar years. The flag performance is given by a black-gray-white list published by the Tokyo MoU (Memorandum of Understanding) each year in its annual report. That is, ship flag performance from the best to the worse is white, gray, and black. If the ships under a flag state are inspected less than 30 times by all the member states of the Tokyo MoU, the flag's performance is set to "undefined." An RO is an authorized (and recognized) organization by a flag administration to carry out inspections and surveys on its behalf on the ships under its registration. The performance of RO can be high, medium, low, and very low from the best to the worst, which is determined by its ships' inspection and detention history over the preceding 3 calendar years, and is published by the Tokyo MoU in its annual report. Similar to ship flag performance, if the number of inspections of the ships in an RO is less than 60 times over the last 3 years, the performance of this RO is "undefined." The calculation of ship company performance is quite different from the above two indicators. Ship company here refers to a ship's international safety management company. Its performance considers the detention and deficiency conditions of the ships in its fleet, which is updated on a daily basis for a running 36-months period. Unlike ship flag and RO performance, there is no lower limit on the number of inspections to quantify the performance of a company over the last three years. For a company with no inspection in the previous three years, it will be given two weighting points. The company's performance is determined by its detention index and deficiency index, and it also has four states, namely high, medium, low, and very low.

The above three risk factors are considered in the NIR in the following way. The criteria for a ship to be an LRS (low-risk ship) are that its flag is white and is of the IMO (International Maritime Organization) audit, its RO is recognized by at least one member authority of the Tokyo MoU and it should have high performance, and its company should have high performance. In addition, if a ship's flag is on the black list or a ship's RO has low or very low performance, 1 weighting point is attached to the ship, respectively. If a ship's company performance is low or very low, or there is no inspection on the ships belonging to the company in the previous 36 months, 2 weighting points are attached to the ship. It should be noted that these three indicators considered in the NIR are subjective, as they are quantified using criteria established by expert knowledge regarding the division of the states, the ways to generate the states, and the weighting points attached to the states. The common sense or domain knowledge, regarding these three risk factors is that for two ships with the other features identical, the ship that has worse flag/RO/company performance is expected to perform worse in the current PSC inspection, i.e., to have a larger number of deficiencies and/or a higher probability of detention. This domain knowledge should also be followed by data-driven models developed for ship risk prediction for the following reasons. It is obvious that ship flag, RO, and company play an important role in ship management, operation, and maintenance, and hence their performance is taken into account by the widely adopted NIR for ship risk calculation and selection for PSC inspection. In return, their ships' performance in PSC inspections would influence their performance evaluated by the corresponding

MoUs and thus their reputation. Therefore, it is justifiable to conclude that given all other conditions being equal, a ship should be estimated to have worse performance if its flag, RO, or company gets worse. Only by following such a rule can a ship risk prediction model be regarded to be in compliance with shipping domain knowledge, and thus achieve fair prediction. Theoretically, such domain knowledge can be learned from practical inspection records. Nevertheless, as actual data are featured with noise and error, it cannot be guaranteed that such domain knowledge can be preserved in the data-driven models constructed. In other words, if no constraint is imposed in the model training process, such property may not be fully followed, leading to unexpected prediction results.

In Yan *et al.* [1], the problem of ship deficiency number prediction in PSC to assist in high-risk ship selection for PSC inspection is addressed. The dataset contains 1974 initial inspection records at the Hong Kong port, and the input to the ship risk prediction model includes several ship-related features as well as ship historical inspection features. To be more specific, 14 features are considered, namely ship age, GT (gross tonnage), length, depth, beam, type, flag performance, RO performance, company performance, the last inspection date in the Tokyo MoU, the number of deficiencies in the last inspection in the Tokyo MoU, the total number of detentions in all historical PSC inspections, ship flag change times, and whether the ship has a casualty in the last 5 years. Inspection records with ship flag and RO performance undefined are first deleted from the initial dataset, and 1926 records remain in the dataset. Then, the states of ship flag performance are encoded by setting white to 1, gray to 2, and black to 3, and the states of ship RO and company performances are encoded by setting high to 1, medium to 2, low to 3, and very low to 4. Therefore, the domain knowledge regarding ship flag/RO/company performance and the ship deficiency number can be viewed as a monotonic increasing relationship between them: as the value of ship flag/RO/company performance gets large (i.e. from good to bad), the ship deficiency number also gets large.

13.1.2 Integration of monotonic constraint into XGBoost

After data processing, an XGBoost model possessing such monotonicity property is developed for ship deficiency number prediction. XGBoost is an implementation of a gradient boosting machine, whose principle is to construct new weak learners that are maximally correlated with the negative gradient of the loss function associated with the whole ensemble in the current state. We do not introduce how to construct an XGBoost model in detail here, and we refer the readers to Reference [3], which originally proposes the XGBoost model, and Reference [1], which develops an XGBoost model for ship deficiency number prediction in PSC with details given to model construction process, for more information. In this section, we only discuss how to combine feature monotonicity characteristics into the development process of an XGBoost model.

The base learner of an XGBoost takes a tree structure constructed by the CART algorithm, and the learning objective of a base learner is to minimize the balance between model prediction error (i.e. the difference between the predicted and actual numbers of ship deficiencies, where half of the MSE (mean squared error) is often

Incorporating shipping domain knowledge into data-driven models 183

used) and model complexity (evaluated by the number of leaf nodes contained in a decision tree). Recall that as introduced in Chapter 9, a classic DT (decision tree) does not possess the monotonicity property. Therefore, we will introduce in detail how to enforce a monotonic constraint to one feature (which can be ship flag performance, RO performance, or company performance) regarding the number of deficiencies predicted.

We first assume that feature monotonicity is imposed on one feature. As feature monotonicity works in the context that all the other features are equal except for the feature imposed by the monotonicity property, without loss of generality, we assume that the feature imposed by such property is ship flag performance. Suppose that we consider a total of m features for ship deficiency number prediction where the number of deficiencies of ship s is denoted by y_s, and the feature vector is denoted by $\mathbf{x} = (x^1, ..., x^{m'}, ..., x^m)$. We further denote ship flag performance by $\bar{m}$ where the monotonic increasing constraint is imposed. This means that for two ships s_1 and s_2, the monotonicity property of ship flag performance works when $x_{s_1}^{m'} = x_{s_2}^{m'}, m' = 1, ..., \bar{m} - 1, \bar{m} + 1, ..., m$, and given $x_{s_1}^{\bar{m}} < x_{s_2}^{\bar{m}}$, we should have the predicted numbers of ship deficiencies $\hat{y}_{s_1} \leq \hat{y}_{s_2}$. As the property of feature monotonicity works in examples with all features being equal except for the feature enforced by the monotonicity property, the tree can be simplified by only containing the splits using the monotonic feature, as using all other features (which are identical to all the examples considered) for node splitting will always lead these examples to the same tree nodes and thus have the same output.

The structure of a simplified tree in an XGBoost model which only takes the monotonic feature into account is shown in Figure 13.1. It should be emphasized

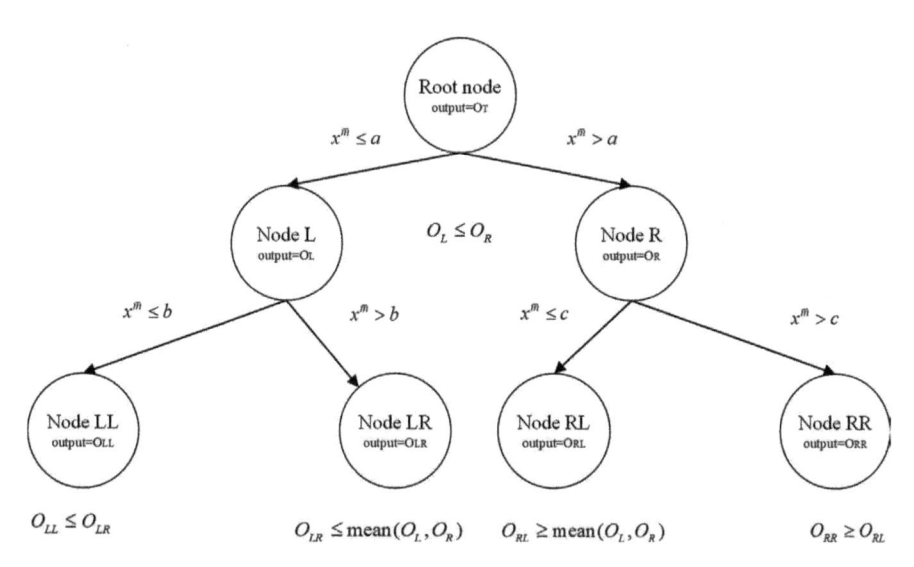

Figure 13.1 An illustration of imposing monotonicity constraint to a tree in an XGBoost model

that only the values of the feature imposed by the monotonically increasing constraint will be used for node splitting, as this constraint works in the context that all the other features are identical for the related ships. Similar to a traditional CART regression tree, the tree starts to split from the root node at the top which contains the whole training set, and the output of this node is denoted by O_T. Suppose further that a is used as the threshold of feature $x^{\bar{m}}$ to split the root node into the left child node L and right child node R, i.e., for examples with feature value no larger than a, they are split to the left child node whose output is O_L, and the other examples are split to the right child node whose output is O_R. To guarantee the monotonically increasing constraint put on feature $x^{\bar{m}}$, constraint $O_L \leq O_R$ is imposed to the outputs of the child nodes of the root node. If this constraint cannot be satisfied when using a as the threshold for splitting, it will not be used to split the root node. Furthermore, if no candidate value of feature $x^{\bar{m}}$ can satisfy the constraint, the node will not be split. The above two rules apply for all nodes in a tree. Therefore, it can be guaranteed that if the root node can be split, we have $O_L \leq O_R$. Similarly, constraint $O_{LL} \leq O_{LR}$ is imposed to node L and $O_{RR} \geq O_{RL}$ is imposed to node R, such that the monotonically increasing constraint can be guaranteed for nodes L and R, and we can expect that node L and node R are either not split, or have $O_{LL} \leq O_{LR}$ and $O_{RR} \geq O_{RL}$ in the child nodes, respectively. Furthermore, as the output of node LR should be no more than the output of node R, an upper bound is enforced to the output of node LR, i.e., O_{LR}, as $O_{LR} \leq \text{mean}(O_L, O_R) = \frac{O_L + O_R}{2} \leq O_R$. Similarly, a lower bound is enforced to the output of node RL, O_{RL}, as $O_{RL} \geq \text{mean}(O_L, O_R) = \frac{O_L + O_R}{2} \geq O_L$. Therefore, the outputs of the nodes in this layer satisfy that $O_{LL} \leq O_{LR} \leq \text{mean}(O_L, O_R) \leq O_{RL} \leq O_{RR}$. As the tree is split in a recursive manner, the monotonicity increasing property of the whole tree and thus on feature $x^{\bar{m}}$ can be guaranteed. As the gradient boosting machine is in an additive manner, the monotonicity property of the whole XGBoost model can be guaranteed, and so are the final predicted ship deficiency numbers.

An XGBoost model imposed by monotonic constraint is denoted by monotonic XGBoost by Yan et al. [1]. Yan et al. further compare the monotonic XGBoost model with other popular and state-of-the-art regression models, including standard XGBoost, monotonic LightGBM, DT, RF (random forest), GBRT (gradient boosting decision tree), LASSO (least absolute shrinkage and selection operator) regression, ridge regression, and SVR (support vector regression), and conclude that the prediction performance of the monotonic XGBoost is the best among all the models for comparison regarding both MSE and MAE. This complies with the comment that model prediction performance can be improved if reasonable monotonic constraints on certain features are imposed, showing that such constrained models can generalize better [4–6].

13.2 Integration of convex and monotonic constraints into ANN (artifical neural network)

This section aims to address the ship fuel consumption rate prediction problem while taking shipping domain knowledge regarding ship sailing speed and fuel consumption into account. As the shipping industry is mainly powered by heavy fuel oil, a large environmental footprint has been created by shipping activities due to the emissions of greenhouse gases (GHGs) and air pollutants. According to the report of the fourth IMO GHG emission, shipping was responsible for 2.76% of global anthropogenic GHG emissions in 2012, and the proportion increased to 2.89% in 2018 [7]. In recent years, the sustainability of shipping has become a public concern, and various emission control measures and regulations have been imposed on ships involving both national and international transportation activities to reduce the adverse impacts on the environment. Meanwhile, the cyclical downturn of the world economy and high bunker prices make it necessary and urgent for ship owners and managers to operate their ships in a more cost-effective way while still accomplishing the global trade amount. In ship's daily operations, bunker fuel costs account for a large proportion of vessel operating costs. For large vessels such as container ships, when the bunker fuel price is about 500 USD per ton, the bunker cost can contribute to about 75% of the operating cost [8, 9]. Under these conditions, shipping companies are constantly making hard efforts to optimize ship energy efficiency, i.e., to use as little fuel as possible to accomplish the required transport tasks.

Various factors can influence ship energy efficiency due to the complexity of vessel engine systems and the surrounding sea and weather conditions in a voyage. According to Yan *et al.* [10], common features that can influence ship fuel consumption rates can be divided into four categories: ship mechanical features, ship operational features, ship maintenance features, and sea and weather conditions. Especially, ship mechanical data include ship dimension features (e.g. length, beam, and gross tonnage) and power system features (e.g. engine parameters and designed speed). Ship operational features include ship voyage and sailing behavior information like sailing speed over ground, sailing speed through water, type of fuel used, fuel density, and temperature, etc., as well as ship mechanical conditions while operating, such as the conditions of propeller pitch, rudder angle, main engine load and working hours, hull and propeller fouling conditions, and wetted surface area. Ship maintenance features mainly include ship dry docking records. Sea and weather condition data mainly refer to the sea states and weather information along the voyage, where the sea states include sea depth, sea water temperature and density, wave, swell, and current conditions, etc., while weather information includes the direction and value of wind, air density, and temperature, etc.

Among all the influencing factors on ship fuel consumption rates, ship sailing speed is widely believed to be the most significant determinant of ship fuel consumption. In many existing studies, a ship's fuel consumption rate at sea is usually treated as proportional to its sailing speed to a power of α, where α is empirically shown to be from 1.452 to 4.8 as summarized in Reference [10]. This shows that ship hourly fuel consumption takes a monotonically increasing and convex relationship with ship sailing speed. In other words, the fuel consumption rate is higher when a ship's speed is higher when all other things being equal, and the ship fuel consumption rate increases faster as the sailing speed increases, as the ship fuel consumption rate is approximately proportional to speed to a power larger than 1. The domain knowledge regarding the relationship between ship sailing speed and ship

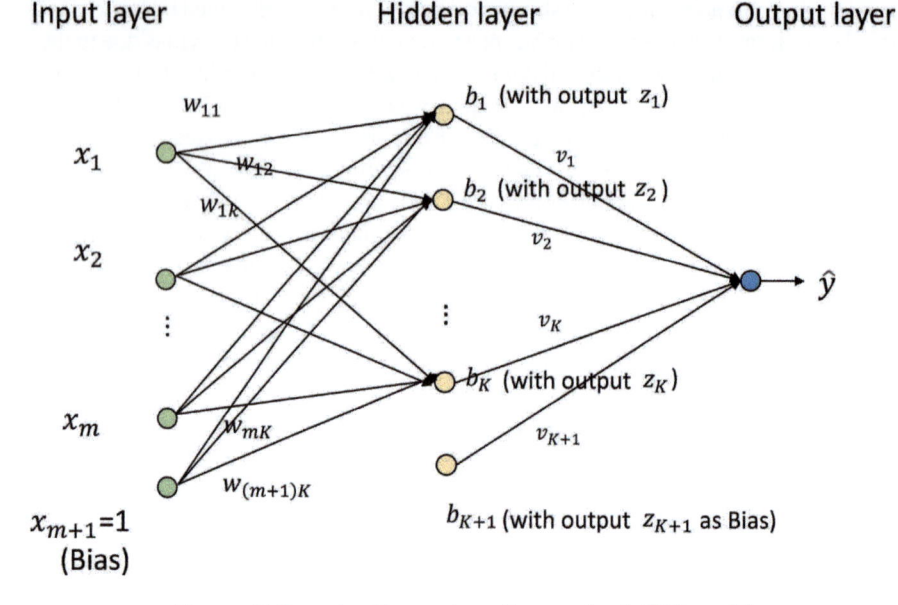

Figure 13.2   An illustration of a standard ANN model

fuel consumption rate should be taken into account when developing data-driven models for ship fuel consumption prediction. This section discusses the idea of incorporating such domain knowledge into the development of ANN models.

Traditional ANNs are introduced in chapter 8 of this book. Suppose that we have a traditional ANN model shown in Figure 13.2 on hand, where the input layer with $m + 1$ neurons is used to receive the input of m features, which can be the average sailing speed, sailing condition, draft, air density, sea water temperature, and wave and wind conditions, and the $(m + 1)$th neuron is the bias term. The hidden layer has $K + 1$ neurons, where the $(K + 1)$th neuron is the bias. The output layer gives the predicted ship fuel consumption rate and is denoted by $\hat{y}$. Without sacrificing generality, we denote by node x_1 in the input layer ship sailing speed, which is imposed by both convex and monotonically increasing constraints, while the other nodes are free from constraints imposed. To preserve such domain knowledge, we introduce a special type of neuron in the hidden layer named convex and monotonic neuron, which has convex and monotonically increasing activation functions and non-negative weights connected to them. These weights include those from the input layer and those to the output layer. In addition, it is also required that node x_1 representing the feature of ship sailing speed in the input layer only connects to convex and monotonic neurons in the hidden layer. The ANN model imposed by the above features based on Figure 13.2 is denoted by ANN-DK and is shown in Figure 13.3.

Figure 13.3 shows that node x_1 representing ship sailing speed is connected with nodes b_1 to b_{K_1} in the hidden layer marked in orange that has monotonically increasing and convex activation functions using non-negative weights. These nodes are also connected with the output layer using non-negative weights. It is also noted that node x_1 is only connected to the aforementioned nodes with monotonically increasing and convex activation functions while there is no connection between node x_1 and the other nodes without constraints imposed in the hidden layer. Moreover, to guarantee the non-negativity of the weights, if the weight values updated in each round of training is positive or zero, there is no further action; otherwise, their values are changed to zero. For the other nodes in the input layer, i.e., nodes x_2 to x_{m+1} without constraints imposed, they are connected to nodes b_1 to b_{K_1} as well as the other normal nodes in the hidden layer marked in yellow. The following Theorems 13.1 and 13.2 guarantee that the ANN-DK model is monotonically increasing and convex in x_1, and thus the predicted output is in compliance with shipping domain knowledge.

Theorem 13.1 The following relation holds, which guarantees the monotonicity property of x_1 to the predicted fuel consumption rates:

$$\frac{\partial \hat{y}}{\partial x_1} \geq 0. \tag{13.1}$$

Proof:

$$\begin{aligned} \frac{\partial \hat{y}}{\partial x_1} &= \sum_{k=1}^{K} \frac{\partial \hat{y}}{z_k} \frac{z_k}{t_k} \frac{t_k}{x_1} \\ &= \sum_{k=1}^{K} v_k f'(t_k) w_{1k} \\ &= \sum_{k=1}^{K_1} v_k f'(t_k) w_{1k} + \sum_{k=K_1+1}^{K} v_k f'(t_k) w_{1k} \\ &= \sum_{k=1}^{K_1} v_k f'(t_k) w_{1k} \geq 0, \end{aligned} \tag{13.2}$$

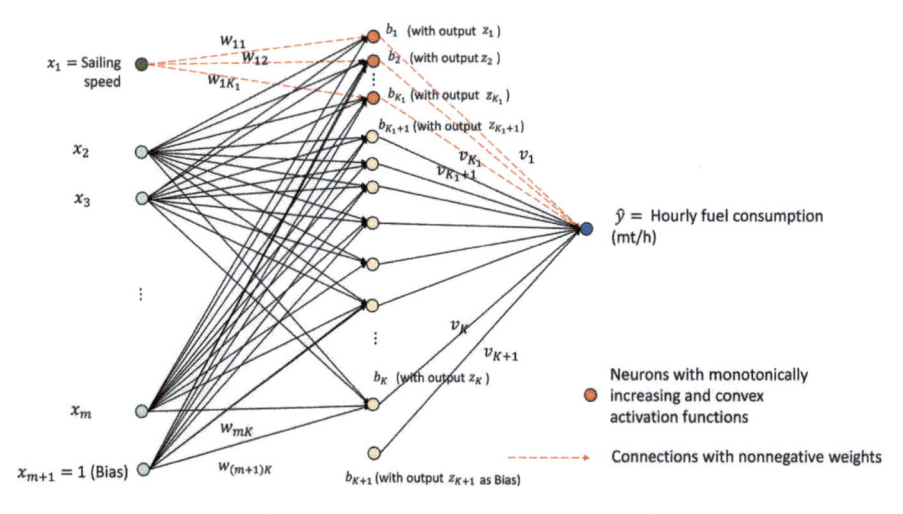

Figure 13.3 An illustration of a domain knowledge informed ANN model

where t_k is the weighted sum of the inputs to node b_k in the hidden layer. As the activation functions are monotonically increasing in the input to nodes b_1 to b_{K_1} in the hidden layer, $f'(t_k)$ is positive. Furthermore, considering that weights v_k and $w_{1k}, k = 1, ..., K_1$ are non-negative, it can be guaranteed that $\sum_{k=1}^{K_1} v_k f'(t_k) w_{1k} \geq 0$. Meanwhile, as there is no connection between node x_1 and nodes b_{K_1+1} to b_K, we have $\sum_{k=K_1+1}^{K} v_k f'(t_k) w_{1k} = 0$. Therefore, $\frac{\partial \hat{y}}{\partial x_1}$ is non-negative, showing that x_1 is monotonically increasing in the predicted fuel consumption rate by the ANN-DK model.

Theorem 13.2 The following relation holds, which guarantees the convexity of the predicted fuel consumption rate in x_1:

$$\frac{\partial^2 \hat{y}}{\partial x_1^2} \geq 0. \tag{13.3}$$

Proof:

$$\begin{aligned}
\frac{\partial^2 \hat{y}}{\partial x_1^2} &= \frac{\partial \sum_{k=1}^{K} v_k f'(t_k) w_{1k}}{\partial x_1} \\
&= \sum_{k=1}^{K} v_k w_{1k} f''(t_k) \frac{\partial t_k}{\partial x_1} \\
&= \sum_{k=1}^{K_1} v_k w_{1k} f''(t_k) \frac{\partial t_k}{\partial x_1} + \sum_{k=K_1+1}^{K} v_k w_{1k} f''(t_k) \frac{\partial t_k}{\partial x_1} \\
&= \sum_{k=1}^{K_1} v_k w_{1k}^2 f''(t_k) \frac{\partial t_k}{\partial x_1} \geq 0.
\end{aligned} \tag{13.4}$$

The above inequality holds because the activation functions of nodes b_1 to b_{K_1} are convex, there are no connections between node x_1 and nodes b_{K_1+1} to b_K in the hidden layer, and weight w_{1k} is non-negative.

Therefore, it can be safely concluded that as long as Theorems 13.1 and 13.2 can be satisfied in an ANN model developed for ship fuel consumption prediction, the convex and monotonically increasing relationship between ship sailing speed and hourly fuel consumption rate can be preserved. Furthermore, such constraints can be imposed on more features. For example, when the direction of wind/swell is the same as the direction of a ship's heading, a larger wind force/swell would generally decrease the hourly fuel consumption rate. To consider such a monotonic relationship, another new type of neuron with a monotonic activation function (which is not necessarily convex) can be introduced in the hidden layer, which is also connected by non-negative weights with the corresponding neurons imposed by the monotonic constraint in the input layer and the neurons in the output layer. ANNs with the above properties can also be applied to address other problems where similar domain knowledge should be preserved both within and outside of the area of maritime transportation.

References

[1] Yan R., Wang S., Cao J., Sun D. 'Shipping domain knowledge informed prediction and optimization in port state control'. *Transportation Research Part B*. 2021;**149**:52–78.

[2] *Information sheet on the new inspection regime (NIR)* [online]. Tokyo: Tokyo MoU. 2022. Available from http://www.tokyo-mou.org/doc/NIR-information%20sheet-r.pdf

[3] Chen T., Guestrin C. 'XGBoost: a scalable tree boosting system'. *Proceedings of the 22nd ACM SIGKDD International Conference on Knowledge Discovery and Data Mining*; 2016. pp. 785–94.

[4] Sill J. 'Monotonic networks' in *Advances in Neural Information Processing Systems*; 1997.

[5] Daniels H., Velikova M. 'Monotone and partially monotone neural networks'. *IEEE Transactions on Neural Networks*. 2010;**21**(6):906–17.

[6] Pei S., Hu Q., Chen C. 'Multivariate decision trees with monotonicity constraints'. *Knowledge-Based Systems*. 2016;**112**:14–25.

[7] *Fourth IMO GHG study 2020 – final report* [online]. London: IMO Publications. 2020 Aug 25. Available from https://www.imo.org/en/OurWork/Environment/Pages/Fourth-IMO-Greenhouse-Gas-Study-2020.aspx

[8] Ronen D. 'The effect of oil price on containership speed and fleet size'. *Journal of the Operational Research Society*. 2011;**62**(1):211–16.

[9] Wang S., Meng Q., Liu Z. 'Bunker consumption optimization methods in shipping: a critical review and extensions'. *Transportation Research Part E*. 2013;**53**:49–62.

[10] Yan R., Wang S., Psaraftis H.N. 'Data analytics for fuel consumption management in maritime transportation: status and perspectives'. *Transportation Research Part E*. 2021;**155**:102489.

Chapter 14

Explanation of black-box ML models in maritime transport

In addition to model fairness, another serious drawback of using ML (machine learning) models to address practical problems in maritime transport is related to the black-box property of most ML models: despite their success in many real-world applications to not only the maritime industry but also other industries including defense, medicine, finance, and law, thanks to their high prediction performance, they are opaque in terms of explainability, which makes users and even developers difficult to understand, trust, and manage such powerful AI (artificial intelligence) applications. Especially, when decisions derived from black-box systems based on ML models affect human's life, safety, and the environment, there is a more urgent need for explaining and understanding how such decisions are furnished by AI methods. Especially, in the relatively traditional and conservative maritime industry, decision-makers and stakeholders are more likely to be reticent to adopt decision support tools powered by new technologies such as AI which they can hardly interpret, control, and thus trust. This chapter aims to first introduce the necessity of explaining black-box ML models in the maritime industry, especially in the case of PSC (port state control). Then, popular methods to achieve black-box model explanation are introduced.

14.1 Necessity of black-box ML model explanation in the maritime industry

14.1.1 What is the explanation for ML models

Before discussing the details of providing explanations for black-box ML models, we first distinguish two similar yet different terms: interpretability and explanation. We have to say that there is no unified definition of either term in the existing literature within the context of ML or AI, and they may have different meanings in different scenarios. We borrow the definition of interpretability given by Doshi-Velez and Kim [1], which says that

Definition 14.1: *Interpret means to explain or to present the understandable terms.*

In the domain of ML, the authors further define interpretability as the ability to explain or present in understandable terms to a human. Then, according to Arrieta *et al.* (2020) [2], for ML models, interpretability is "a passive characteristic of a model referring to the level at which a given model makes sense for a human observer." The authors further explain that this characteristic can be interpreted as model transparency. This shows that model interpretability is an inherent property of an ML model, indicating that the ML model itself processes the ability to make the developers and users understand its reasoning process and the predictions generated. Typical ML models falling in this category include rule-based learning and reasoning, linear regression (LR), decision tree (DT), and kNN (k-nearest neighbor). Details of these ML models will be covered in the next sections. In contrast, explainability is "an active characteristic of a model, denoting any action or procedure taken by a model with the intent of clarifying or detailing its internal functions." Therefore, model explainability can be understood as explaining the ML model of black-box nature developed using other external techniques, where such techniques can be local (i.e. for the prediction of a single example) or global (i.e. for the overall performance of the whole model), and model-specific (which are limited to specific model classes) or model-agnostic (which can be used to any ML models). A detailed introduction of the related methods will be given in the next sections.

We say that most of the ML models are of black-box nature because we do not know how the features fed into the model are processed and used to give the final output. According to Doshi-Velez and Kim [1], the black-box nature of ML models is not a big deal when (1) there are no significant consequences for unacceptable results given by an ML model; or (2) the problem is sufficiently well studied and validated in practice, and thus, the users trust the decisions recommended by the system even if it is imperfect, i.e., the prediction results are not always of 100% accuracy. Otherwise, an ML model explanation should be given to address the *incompleteness* in the problem formalization, which means that getting to know what if the prediction is not enough for some problems, and it is also expected that why such prediction is made should be explained [3]. Unfortunately, neither condition is held in the maritime industry. The main reason is that this industry involves several heterogeneous and conservative stakeholders and decision-makers, who make strategic, tactical, and operational decisions heavily dependent on long-term experience instead of emerging data-driven models, especially ML models. Consequently, recommendations generated by ML models in black-box nature without convincing explainability provided can be seldom acceptable to them, even if in many cases the recommendations given by these models can be much more efficient and reasonable than transparent but naive rules or expert systems.

One example in maritime transportation is ship selection decisions in PSC made on each workday at port authorities: although there are various high-performing data-driven models developed for ship risk prediction or high-risk ship recommendation in existing studies, which have also been empirically shown to be more efficient than the rule-based SRP (ship risk profile) ship selection method adopted by most MoUs (Memorandum of Understandings), these data-driven models are seldom adopted by port authorities at the moment. Meanwhile, for ship risk prediction problem in PSC, the first condition mentioned above is not satisfied, because ship selection decision is vital for both port authorities and ship operators. For port authorities, the number of foreign visiting ships is large each day, while the available inspection resources can be scarce. Moreover, it can be seen from the annual reports of the MoUs that only some of the inspected ships are with deficiency/deficiencies identified, while only a small proportion of them are actually detained, i.e., with serious deficiency/deficiencies found. Therefore, only a small proportion of all the foreign visiting ships can be and should be inspected. It is thus crucial to allocate the limited inspection resources to the ships with the highest risks (i.e. with the worst performance) in PSC inspection to improve the efficiency of inspection. From the perspective of ship operators, too frequent and unnecessary inspections would delay the shipping schedule, and thus, reduce the efficiency of the fast turnover of the shipping logistics system and cause financial loss. Besides, if too few substandard ships are inspected by the port states, ship operators may lack the motivation to intensively maintain their ships to be in good condition. On the contrary, if too many qualified ships are inspected by a certain port, ship operators may turn to other destinations with looser inspection strategies, and such behavior is referred to as "port shop," which may affect overall vessel quality and ports' reputation. Besides, the second condition mentioned above is also not satisfied in ship selection for inspection by PSC, because although this topic is widely studied in existing literature where several data-driven models have been developed to assist port authorities in high-risk ship selection, the proposed models are not applied by the ports, and thus, the results have not yet been fully validated. Consequently, there is still a long way to go to make the users trust the prediction results given by the data-driven models in PSC.

In addition to fulfilling the needs of model explanation according to Doshi-Velez and Kim [1], providing explanations to black-box models developed to address practical problems in maritime transport can bring the following advantages:

1. Trust: conservative stakeholders and decision-makers in the traditional maritime industry are reluctant to accept and use recommendations given by data-driven models based on empirical data as these models are not as tractable and controllable as the expert knowledge systems currently adopted. Therefore, to gain their trust in the newly proposed data-driven models and thus, promote the use of these models in practical decision-making, explanations are needed to help them understand and verify the data-driven models' internal schemes, working processes, and strengths as well as weaknesses.
2. Transferability: since explanation can shed light on understanding and implementation of the black-box models developed, it can help the users to know how well the data-driven prediction models generalize in different scenarios and thus, can promote the use of the data-driven models.
3. Fairness: as explanations reveal how a black-box data-driven model works and why a prediction is given, model fairness can be verified to ensure that the predicted results do not discriminate against particular stakeholders. Especially, as decisions in the maritime industry would influence a variety of stakeholders, such as ship owners, operators, management companies, port authorities, and

shipping service providers, model fairness is a key to guarantee that the predictions are reasonable and thus, verify the recommendations given by the black-box models to be compliant with ethical standards and common beliefs.
4. Extensibility: explanations given for black-box data-driven models can help the developers to further adjust the settings and hyperparameters of the ML models more effectively, so as to improve their prediction accuracy. They can also help the model users generate general rules and extract new knowledge from the massive data for future decision-making.

14.1.2 Propose and evaluate explanations for black-box ML models

Then, in order to provide a proper explanation of a black-box model, what elements should be considered when explaining? Basically, two questions need to be answered when explaining black-box ML models: how and why [4]. The "how" question refers to how the black-box model works, i.e., its internal working mechanism, and the "why" question refers to why a certain output is predicted for a certain example. However, it is also pointed out in existing literature that different explanations are expected to be given to distinct behaviors, distinct problems, and distinct users in Vilone and Longo [4]. In other words, black-box model explanations should consider the following four elements: the user, i.e., the receiver of explanations; the goals, i.e., what questions need to be answered by the explanations; the contents, what information is expected to be obtained from the explanations; and the language, i.e., what terminology should be used when presenting the explanations. These four elements are correlated with each other. Details and relations between them are summarized in Table 14.1.

After presenting an explanation to a black-box model considering the above four elements, a natural question then is which explanation is good? In other words, what are the criteria to evaluate the quality of explanations given to a black-box model. Actually, it is still not that clear what is the most effective way to structure explanations according to Vilone and Longo [4], and how to select and evaluate model explanation methods for a particular problem and audience as mentioned in Murdoch *et al.* [5]. This indicates that there might be a lack of unified ways to evaluate the explanations given to a specific problem, as problems themselves can vary a lot from each other. The evaluation process is quite different in model explanation and model construction using supervised ML techniques as introduced in the earlier chapters of this book, where there are unified and widely recognized metrics to quantitatively evaluate model prediction performance. For model explanation evaluation, we introduce a framework called PDR, which is short for predictive accuracy, descriptive accuracy, and relevancy as proposed in Murdoch *et al.* [5] for selecting and evaluating the most suitable explanations given to a black-box prediction model for a specific problem. Details of the three points are provided as follows:

- Predictive accuracy: predictive accuracy tends to evaluate the quality of the black-box prediction model itself, as only explanations given to accurate

Table 14.1   Four elements of black-box model explanation

User	Goals	Contents	Language
Researchers and developers	Understand how the entire data-driven model works to verify its fairness, transferability, and extensibility, and to improve its overall performance.	Comprehensive and technical explanations.	Can be sophisticated and full of terminologies to maintain professionalism, and might be hard to understand by people who are not the researchers or developers of the model.
Model users	Make sure that the current model can meet the requirements and purposes according to their expectations.	Reasoning process of all components of the model and its application scenarios.	The terminologies used should be within the context of model practical use and avoid too professional terms.
People affected by the model but not direct users	Get to know the rationale behind the black-box model without being familiar with all of the model's components.	The general model working mechanism and decision process.	Plain words without professional terminologies.

prediction models are meaningful. This requires that the prediction models to be explained should learn the underlying relationships in the data well and thus, generalize well to unknown examples. In general, model predictive accuracy is evaluated by the accuracy of the test set using proper metrics. Under the context of predictive accuracy evaluation for a model explanation, one should pay attention to the data used to check the predictive accuracy: the test set data must resemble the interest of the model users. For example, to evaluate a high-risk ship selection model developed for the Hong Kong port using inspection records there, it is not reasonable to evaluate the prediction model's performance using the inspection records at a port other than the Hong Kong port, say the Port of Singapore. Moreover, sometimes using average prediction accuracy might not be enough, as it is also expected that model performance should be stable when there are data and model perturbations within a reasonable range. This is because if the model has dramatic changes when there are slight changes in the data and model, the explanations generated from the model might not be trustworthy.

- Descriptive accuracy: when explaining the behavior of a black-box prediction model, a critical point is that the explanations should capture the relationships between features and output learned by the prediction model precisely. Otherwise, the explanations are meaningless. Nevertheless, it should be acknowledged that accurately capturing such relations can be hard, especially for complex black-box models such as deep neural networks. Therefore, there is a trade-off between model accuracy and model explanation in practical use.
- Relevancy: in addition to high accuracy, explanations should be relevant considering a particular audience for a chosen domain problem. As shown in Table 14.1, different audiences, e.g., researchers and developers, model users, and people affected by the model but not direct users, have different expectations of the explanations provided to them. Taking the prediction model of ship risk for PSC inspection as an example, the first class of users, who are developers and maintenance staff of the ship selection system, might be interested in the explanations of the inner working mechanism, i.e., how the model captures the relationship between the features and the target, what are the intermediate outputs within the model, and the rationale behind the predictions, so as to evaluate and improve model performance. The second class of users, who are usually officials in port authorities, might focus more on the ship risk prediction process, especially why a ship is predicted to be of certain risk and why a ship is predicted to be of higher risk than the other ship. Moreover, they may also be interested in prediction fairness as discussed in chapter 13. The third class of users, who are usually ship operators, managers, and owners, might be interested in the performance of their ships predicted by the ship risk prediction model so as to reduce the frequency of inspection of their ships and improve their ships' performance as well as model prediction fairness similar to the second class of users.

One point that needs to be added is that there is a trade-off between predictive accuracy and descriptive accuracy: usually, a more complex black-box prediction model has higher predictive accuracy, but this means that its descriptive accuracy might be low, as such a model might be hard to analyze. In contrast, a simple white-box prediction model can have a high descriptive accuracy as its behavior is easier to be captured. However, its predictive accuracy might not be high enough.

14.2 Popular methods for black-box ML model explanation

14.2.1 Forms and types of explanations

Explanations of the predicted results of an ML model can have different forms, where the main forms are summarized as follows:

1. Model internal: this type of explanation is usually given by interpretable models intrinsically, such as the weights in LR models and the tree structure as well as the node splitting criteria in DTs.

2. Feature importance: this method aims to show how important a feature is to the prediction of the target, i.e., how much does the feature contribute to the prediction of the target. It can be expressed as the importance value of a single feature, or in a more complex form such as the pairwise feature interaction strengths.
3. Partial dependence: this method aims to show the marginal contribution of the values of a feature on the predicted output, which is often called partial dependence plots. As marginal contribution needs to be calculated, all the other features other than the feature to be explored need to remain unchanged, and different values of the feature to be explored should be traversed.
4. Data point: this method aims to use existent or newly created examples to show the decision process of an ML model. One typical method on a single example level is called counterfactual explanation, which aims to find similar examples of the example to be explained by changing some of its features and observing the changes in the output, so as to understand how decisions are made by the prediction model.
5. Surrogate model: this method aims to find a surrogate model which is inherently interpretable, and can mimic the behavior of the black-box ML model to be explained.

There are several criteria that can be followed to classify explanations given to black-box ML prediction models from different perspectives. One criterion is based on model property as discussed above: an ML model is intrinsically explainable or external explanation methods should be applied to explain the ML model. The first class of explanation is called intrinsic explanation and the second class is called post hoc explanation. Intrinsically explainable models are interpretable ML models with simple structures and are self-explainable. A typical example is LR which takes an additive linear form, and the coefficients of the terms (i.e. features) can be viewed as their weighting points (or the approximate importance scores if feature values are normalized) while the final predicted target of a particular example is given by the weighted sum of its feature values. Another typical example is DT, especially shallow DTs, in which the reasoning process is shown by the (feature and value) pairs selected for node splitting, and the predicted output of a particular example is determined by the average/majority outputs of the examples in the node where this example falls. In addition, feature importance can be generated from a DT: a feature is regarded to be more important than the other if it can reduce node impurity to a larger extent. It should be noted that the explanations given by an intrinsic model can be from relatively limited aspects, and such an explanation sacrifices the accuracy of the prediction model. In contrast, post hoc explanation refers to the process of developing an ML model of black-box nature first and then using external explanation methods to explain the model. If necessary, it can also be applied to explain intrinsic models from wider perspectives.

Another two criteria consider the application scope of the explanation method. One is that if the explanation is given to a single example, it is local. Alternatively, if the explanation is given to the entire model, it is global. The other is that if the explanation method can only be used to a specific model, e.g., artificial neural networks or

DTs, the explanation method is model-specific. On the contrary, if the explanation method can be applied to all types of ML models, it is model-agnostic.

In the following sections, we first use DT as an example to show how to gain explanations from an intrinsic explanation model. Then, we introduce a post hoc, local, and model-agnostic explanation method to explain black-box ML models called Shapley additive explanations (SHAP).

14.2.2 Introduction of intrinsic explanation model using DT as an example

We use a regression tree for ship deficiency number prediction as an example to show how to generate explanations from a DT. The dataset used to construct the regression DT is the same as that used in Example 1.2 in Chapter 9. To reduce the complexity of the tree for ease of illustration, we set hyperparameters *max_depth* to 3 and *min_samples_leaf* to 1. The core code to generate a visualized DT using the *scikit-learn* API [6] and the *PyDotPlus* API [7] is as follows:

```
dot_data = tree.export_graphviz(final_model, # The model to
    be explained
feature_names=feature, # Specify the feature names shown in
    the figure
class_names=target, # Specify the target name shown in the
    figure
filled=True, # Fill the nodes by colors
rounded=True, # Round the digits shown in the figure
)
graph = pydotplus.graph_from_dot_data(dot_data)
```

The visualized DT is presented in Figure 14.1. The node at the top of the tree shown in Figure 14.1 is the root node, which contains all the 1,500 ships, i.e., the entire training set, used for tree training. The first line in the node shows the feature and its value selected to split the node, i.e., feature gross tonnage (GT, whose unit is 100 cubic feet) is selected to split the root node with a threshold value of 6,451.5. This means that ships whose GT is no more than the threshold are split to their left child node, where the process is shown by the "True" arrow under the root node, meaning that splitting criteria "GT <= 6,451.5" is satisfied in these ships. Other ships, i.e., ships whose GT is more than 6,451.5 the unit of GT has been added to the place where feature GT is mentioned, are split to the right child node, as shown by the "False" arrow under the root node. The second line in the root node shows the current MSE (mean squared error), i.e., impurity, of the node. The MSE of the root node is 27.732. The third line shows the number of ships contained by the node. The fourth line is the predicted ship deficiency number of this node, which is the average deficiency number of all the ships contained in the node. This value is 4.309 in the root node, indicating that the average number of deficiencies of the ships in the entire training set is 4.309.

Explanation of black-box ML models in maritime transport 199

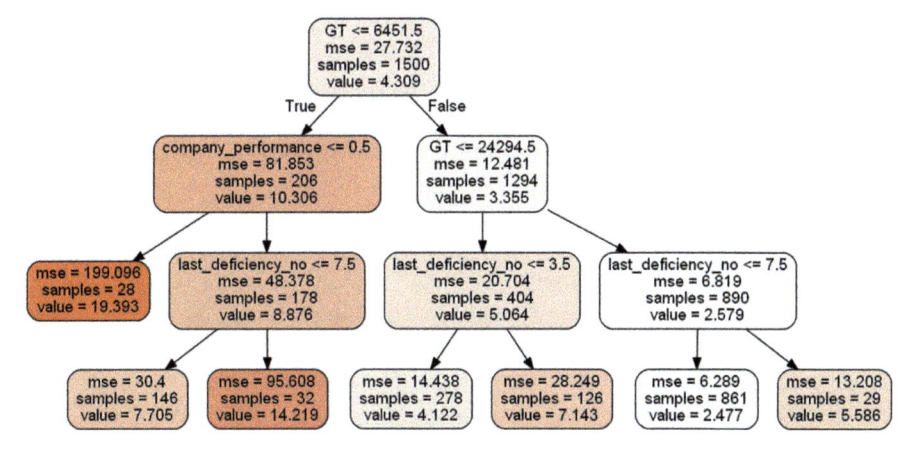

Figure 14.1 Visualization of a regression DT for ship deficiency number prediction

The tree is split in a recursive manner until the stopping criteria of *max_depth* as 3 is met. Therefore, we can see that the depth of the tree is 3. The total number of ships contained in all leaf nodes is equal to 1,500. It is also interesting to find that although splittings in DTs aim to reduce impurity in subsequent nodes, which is represented by MSE in this example, the MSE may not necessarily be reduced after splitting a node. An example is the split of the root node to its left child node, where the MSE increases from 27.732 to 81.853. This indicates that the left child node might contain examples with extreme output values (i.e. very large ship deficiency numbers) that have a big influence on the MSE of the node. This phenomenon is also shown by the outputs of leaf nodes: for leaf nodes with the predicted ship deficiency number close to the output of the root nodes, i.e., nodes 3, 5, and 6, their MSE values are less than or even much less than the MSE of the root node. The outputs of the other nodes are large, which is mainly because they contain ships with very large deficiency number. Consequently, their MSE values are higher than the MSE of the root node. This shows that extreme values are indeed contained in the training set, and they can have a very large impact on a DT's performance.

The above analysis shows the intrinsic explainable property of DT models: the inner working mechanism of a DT, i.e., how the DT is split from the root node to leaf nodes is shown by the (feature and value) pair; the examples contained, output and error can be obtained for each node; and if there comes a new ship, how to classify the ship to a leaf node and decide its predicted deficiency number are all clear. Model fairness can be verified based on these explanations. For example, from the root node, it can be seen that a smaller ship tends to have a larger deficiency number in PSC inspection, as it may not be managed that well compared to larger ships. Moreover, from the second node on the right in tree depth 2, it can be seen that if a ship has a larger deficiency number in the last initial inspection, it is predicted to have a larger deficiency number by the DT for the current PSC inspection, which complies with expert knowledge. Based on the splitting criteria and the MSE of each node, feature importance can be generated. As only three features are used for node splitting in the tree in Figure 14.1, i.e., GT, company_performance, and last_deficiency_no, only these features have importance scores, which are 6.8689, 1.7838, and 1.4505, respectively, showing that ship GT is the most important feature determining ship deficiency number, followed by ship company performance and the number of deficiencies in the last PSC initial inspection. For the detailed methods to calculate feature importance score in regression DT, readers are referred to Reference 8 for more information. In addition, decision rules can be extracted from the DT in Figure 14.1. For example, it can be concluded that if a ship has GT no more than 6,451.5 and has unknown company performance, its deficiency number is predicted to be 19.393, which is relatively high and deserves much attention from the port authority. Another rule is that if a ship has a GT larger than 6,451.5 while less than 24,294.5 and its last deficiency number is no larger than 3.5, its predicted deficiency number is 4.122.

14.2.3 SHAP method

SHAP is proposed by Lundberg and Lee in 2017 [9], which is theoretically based on Shapley value in game theory. SHAP is used to explain the output of the prediction of an individual example (i.e. local) of an arbitrary machine learning model (i.e. model-agnostic) after it is constructed (i.e. post hoc). A base value, which is the average output of the examples in the training set, is first calculated, and the contribution of each feature value of the example to the predicted target is calculated. Then, the final output of the example is the sum of the base value and the contributions of its feature values where such contribution is represented by marginal contribution, which can be viewed to be similar to a linear model taking a sum format. SHAP is motivated by coalition game theory, where the "game" refers to the prediction task of an example, the "players" are the features included in the prediction model, and the "gain" is the difference between the final predicted target and the base value, i.e., the sum of the contributions of the feature values. Given an example x_i with m features, the contribution of feature m', $m' = 1, ..., m$, e.g., x_i is denoted by $\phi_{m'}^i$. Furthermore, given that the base value is $\bar{y}$, which is the average output in the training set, the predicted output of the example x_i denoted by $f(x_i)$ explained by SHAP can be represented by the following additive linear function form:

$$f(\mathbf{x}_i) = \bar{y} + \sum_{m'=1}^{m} \phi_{m'}^i. \tag{14.1}$$

The key point in SHAP is how to calculate the marginal contribution of the value of each feature for the example to be explained, i.e., $\phi_{m'}^i$, e.g., i's feature m'. The basic idea is to calculate the difference when considering and ignoring this feature, which is thus the marginal contribution of the feature. Several efficient algorithms are proposed to calculate the SHAP values of feature values efficiently. Readers are referred to Lundberg and Lee [9] and Lundberg et al. [10] for more information, and we do not cover the detailed algorithms in this chapter. Using SHAP to explain the prediction of black-box ML models has been implemented by the SHAP API

Table 14.2   Feature values and the corresponding SHAP values of the example ship

Feature name	Value	SHAP value
GT	131,332	−1.584329
age	5	−0.317698
flag_performance	1	−0.038422
RO_performance	1	−0.059324
company_performance	2	−0.053150
last_inspection_time	10	−0.171819
last_deficiency_no	5	0.355471
last_inspection_state	0	−0.038518
type_bulk_carrier	0	−0.058035
type_container_ship	1	0.007297
type_general_cargo/multipurpose	0	−0.173660
type_other	0	0.005028
type_passenger_ship	0	−0.002885
type_tanker	0	0.047903

developed by Lundberg [11]. Following is an example to use the SHAP method to explain the ship deficiency number prediction model based on random forest (RF) in Example 9.4 of Chapter 9.

We first explain the predicted number of deficiencies of ships in the test set. The following code is used to obtain the matrix of SHAP values on the test set based on the RF model constructed by the training set and the base value in the training set:

```
import shap # import the corresponding library
explainer_test = shap.TreeExplainer(final_model) # Use SHAP
    method developed for tree models to improve the
    computational efficiency
shap_values_test = explainer_test.shap_values(X_test) # Get
    the matrix of SHAP values on the test set
shap.initjs()
print('The base value is: ',explainer_test.expected_value) #
    Output the model's base value: 4.320877
```

Then, we explain the prediction of a random ship in the test set. Feature values and their corresponding SHAP values of the selected example ship are shown in Table 14.2.

The sum of SHAP values of all features is −2.082140. Given the base value as 4.320877, the final output of the deficiency number of this ship is 2.238737. The real deficiency number of this ship is 3, and thus, the absolute prediction error of this ship is 0.761263. Main contributors to the final predicted deficiency number can also be shown in a visual manner in Figure 14.2.

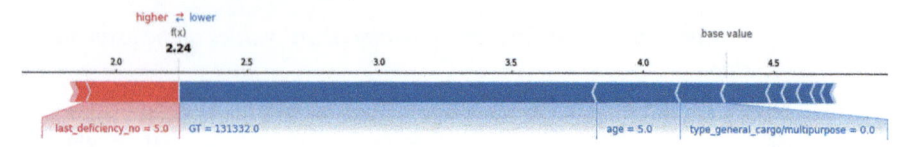

202 *Machine learning and data analytics for maritime studies*

Figure 14.2 Visualization of the main contributors to the example ship's predicted deficiency number

As shown in Figure 14.2, the base value is 4.32 and the predicted value is 2.24, where both are shown directly on the axis. The main features of this ship that decrease the predicted deficiency number are shown in blue, which are having GT at 131,332 (being a relatively large ship), of ship age 5 (being a young ship), and not being ship type general_cargo/multipurpose (not of a ship type associated with a larger number of deficiencies). Meanwhile, the major feature that increases the predicted ship deficiency number is having 5 deficiencies in the last initial inspection within the Tokyo MoU, which is larger than the average deficiency number over the training set at 4.32. The figure further shows that the overall effects of the values of all the features lead to the ship having 2.08 less deficiencies than the base value, and thus, the final predicted deficiency number of the ship is 4.32 − 2.08 = 2.24 in the current PSC inspection. Then, the decision process of this ship, i.e., why it is predicted to have 2.24 deficiencies and what are the main contributors to the final target, is clearly presented to the users.

We then use SHAP to generate more insights from the training set. We first use a beeswarm plot with the *y*-axis representing each feature and the *x*-axis representing the features' SHAP values with each dot representing a single ship in the training set as shown in Figure 14.3. Feature values from low to high are shown by gradient colors as shown on the right side of the figure. When multiple dots land at the same *x* position, they pile up to show the density. The core code to generate this figure is given as follows:

```
import matplotlib.pyplot as plt
my_dpi=216
plt.figure(figsize=(1000/my_dpi, 1000/my_dpi), dpi=my_dpi)
shap.summary_plot(shap_values, X_train,
    max_display=X_train.shape[1], show=False)
```

We then explore more on the feature values on the predicted deficiency number.

- GT: for feature values shown by blue dots, i.e., for relatively small feature values, many of their SHAP values are larger than 0, indicating that being a smaller ship is more likely to have the worse condition and thus, increases the number of deficiencies. It is also noted that when the feature values are shown in blue, they can also have negative SHAP values which would reduce a ship's predicted deficiency number. This shows that the conditions of small ships might

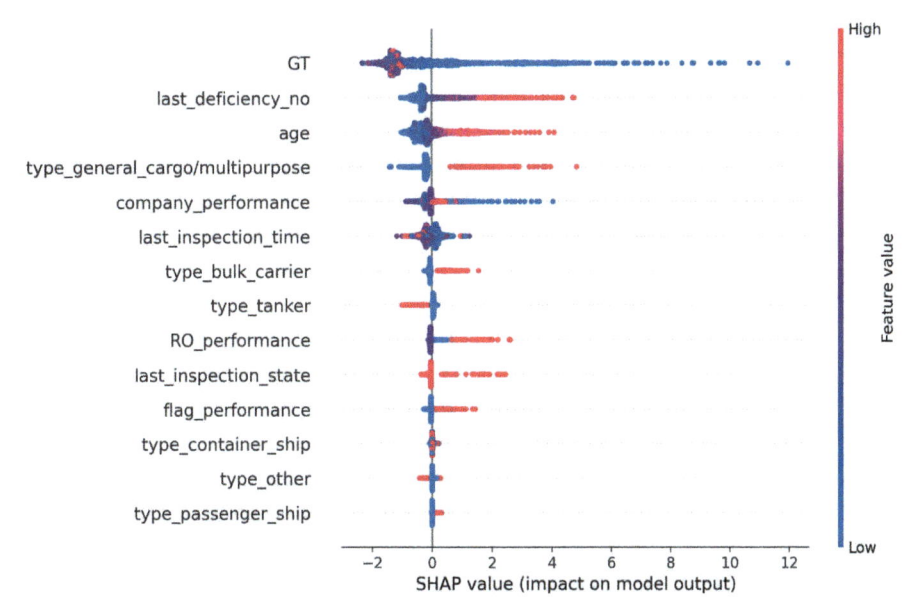

Figure 14.3   Visualization of local explanation summary on the training set using SHAP

be varied. In contrast, when the feature values are shown in red which refers to large ships, their SHAP values are always less than −1, showing that large ships are generally in good condition.

- Last_deficiency_number: having a larger number of deficiencies in the last initial PSC inspection in the Tokyo MoU would no doubt increase the predicted number of deficiencies in the current PSC inspection, and these ships are shown by the red dots with positive SHAP values, as this indicates that the ship is in a relatively bad condition. Bad performance in the last PSC inspection can increase the final prediction by up to 6. Meanwhile, it is also interesting to find that even if a ship performs relatively well in the last PSC inspection, it does not necessarily mean that the ship could perform well in the current inspection, and these ships are shown by the blue dots with positive SHAP values. This is mainly because ships' conditions can vary from time to time, especially when the last inspection is long ago. In addition, different port states may have different inspection strategies and different degrees of strictness. In general, it can be seen that having less deficiency number in the last PSC inspection can reduce the number of deficiencies identified in the current PSC inspection.
- Age: the general condition of this feature is that older ships tend to have positive and slightly negative SHAP values while younger ships tend to have negative SHAP values. This is intuitive and easy to understand: the manufacturing technology of older ships is not as exquisite as that of younger ships, and older ships may also have more wear and tear as well as damage after a long period of sailing.

- Company_performance: recall that the values of ship company performance are encoded in this way: unknown->0, high->1, medium->2, low->3, and very low->4. Therefore, when the values of company performance are small as shown by blue dots, they are usually associated with nearly zero or negative SHAP values shown in a cluster, while some of them are also attached with positive SHAP values shown by sporadic dots on the right-hand side. When the values of this feature are large as shown by red dots, the associated SHAP values can be either positive or negative. This shows that generally, a ship's worse performance company would increase its predicted deficiencies and vice versa. Nevertheless, it is not absolute as there are many ships within a company's fleet and their conditions can vary a lot.
- RO_performance: recall that the values of ship RO (recognized organization) performance are encoded in this way: unknown->0, high->1, and medium->2. As a larger value of RO_performance indicates that the RO has worse performance, and thus, it is with more positive SHAP values as shown by the red dots on the right-hand side. When RO_performance has a medium value, i.e., 2, the associated SHAP value is near 0. When an RO's performance is unknown, it is also slightly positive.
- Flag_performance: recall that the values of ship flag performance are encoded in this way: unknown->0, white->1, gray->2, and black->3. The SHAP values associated with ship flag performance show a clear tendency: when the value is high as shown in red dots, i.e., when the flag is on the gray or black list, the associated SHAP values are usually positive. Otherwise, when the flag is on the white list or when it is unknown, the associated SHAP values are near 0 or slightly negative.
- Last_inspection_time: last_inspection_time refers to the period in the month between the current and the last PSC inspections within the Tokyo MoU. The basic trends of a ship's last inspection time and its deficiency number in current PSC inspection are that usually, a shorter last inspection time indicates a larger number of deficiencies. This is mainly caused by the ship inspection priority according to the SRP ship selection method: the length of the inspection time window becomes longer from HRS (high-risk ship) to SRS (standard-risk ship) and LRS (low-risk ship). Therefore, ships in worse conditions would be more frequently inspected, and thus, their last inspection time would be shorter.
- Last_inspection_state: recall that the values of ships' last inspection states are encoded in this way: none->-1, no->0, yes->1. Most of the ships are not detained in the last PSC inspection (1,376 out of 1,500), while 49 are detained and 75 have no previous PSC inspection. Therefore, large values of this feature are scattered on the right-hand side associated with positive SHAP values, while smaller values might be bunched near 0 with heavy coverage. Therefore, most of the dots for this feature is in red while seldom of them are in blue.
- Ship type: this feature includes six sub-features: type_bulk_carrier, type_container_ship, type_general_cargo/multipurpose, type_passenger_ship, type_tanker, and type_other. Especially, the risk tendency of ships belonging to type_bulk_carrier, type_general_cargo/multipurpose, type_passenger_ship,

and type_tanker is obvious: bulk carrier, general cargo carrier/ship, multipurpose ship, and passenger ships are associated with large SHAP values, while the tanker is associated with small SHAP values. In contrast, a container ship is generally associated with non-negative SHAP values, while a ship type other than a container ship can be attached with slightly negative, zero, and slightly positive SHAP values. Finally, for other ship types, as they contain various types of ships, including but not limited to gas carrier, heavy load ship, special purpose ship, livestock carrier, and tug, they could either increase the predicted ship deficiency number (i.e. associated with positive SHAP values) or decrease the predicted ship deficiency number (i.e. associated with negative SHAP values).

Based on the analysis of the beeswarm plot representing the relationship between ship feature values and their corresponding SHAP values, how different values of the features contribute to the final predicted ship deficiency number is clear, and it can also be verified that the predictions given by the RF model are generally in compliance with shipping domain knowledge. The overall importance of each feature on the whole training set can be further generated from the beeswarm plot shown in Figure 14.3 by calculating the mean *absolute* SHAP values of the feature among all the examples in the training set. This means that even if SHAP values of a certain feature can be positive or negative, their effects would not be offset. In this way, the local SHAP method can be extended to a global one as feature importance is at the global level. It can be anticipated that the larger a feature's mean absolute SHAP, the greater influence the feature has on the final prediction as it can change the predicted target more. Feature importance shown in the bar chart is presented in Figure 14.4.

It is shown in Figure 14.4 that ship GT is the major determinant of ship deficiency number, followed by the last deficiency number and ship age. Intuitively, a ship's last inspection state, i.e., whether the ship is detained in the last PSC initial inspection, should have a large influence on the predicted deficiency number in the current PSC inspection. However, as ships with detention only constitute a very small proportion of all the ships in the training set (49 detentions among all the 1,500 ships), its importance is reduced, and it only ranks 10th among all the features regarding their importance. For ship management parties, ship company performance has the largest impact, followed by RO performance and flag performance. As ship type is divided into six sub-types, it can be hard to evaluate the overall influence of ship type. Nevertheless, Figure 14.4 suggests that general cargo carrier/ship and multipurpose ship, bulk carrier, and tanker can have a large influence on the predicted ship deficiency number.

Based on the above illustration and analysis, it can be seen that the SHAP method can be used to explain the predictions given to individual ships by calculating marginal contributions of the values of each feature for the example to be explained in a local manner, to show the overall influence of feature values on the prediction target over the whole data set in a local manner, and to generate feature importance in a global manner. Based on the explanations, the model reasoning process and thus,

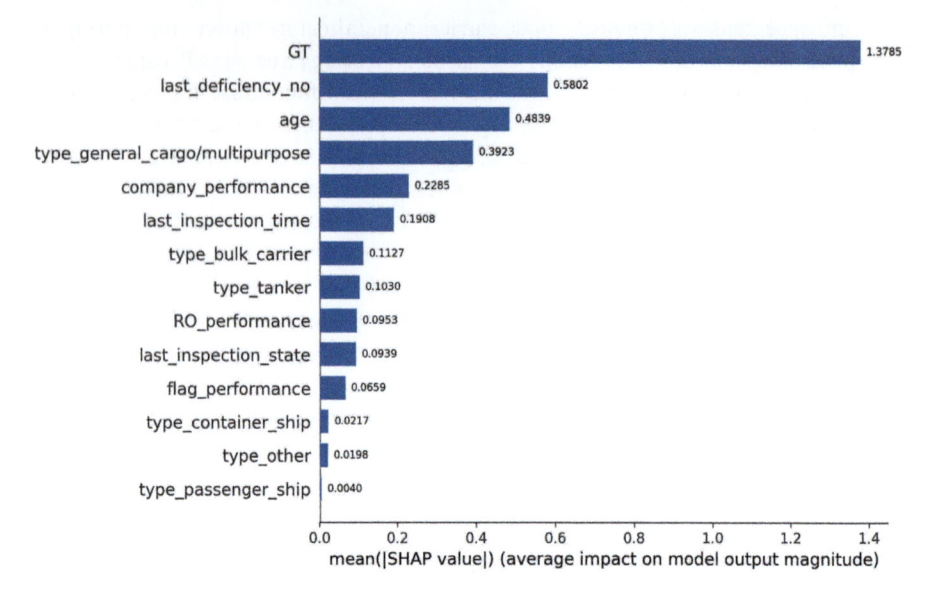

Figure 14.4   Visualization of global feature importance using SHAP

fairness can be verified, so as to convince the users regarding the rationale behind the black-box model to improve model acceptance and practical applicability.

References

[1] Doshi-Velez F., Kim B. 'Towards a rigorous science of interpretable machine learning'. *ArXiv:170208608*. 2017.
[2] Barredo Arrieta A., Díaz-Rodríguez N., Del Ser J, *et al*. 'Explainable artificial intelligence (XAI): concepts, taxonomies, opportunities and challenges toward responsible AI'. *Information Fusion*. 2020;**58**:82–115.
[3] Molnar C. Interpretable machine learning. Lulu.com; 2020.
[4] Vilone G., Longo L. 'Notions of explainability and evaluation approaches for explainable artificial intelligence'. *Information Fusion*. 2021;**76**:89–106.
[5] Murdoch W.J., Singh C., Kumbier K., Abbasi-Asl R., Yu B. 'Definitions, methods, and applications in interpretable machine learning'. *Proceedings of the National Academy of Sciences of the United States of America*. 2019;**116**(44):22071–80.
[6] Pedregosa F., Varoquaux G., Gramfort A., *et al*. 'Scikit-learn: machine learning in python'. *Journal of Machine Learning Research*. 2011;**12**:2825–30.
[7] *PyDotPlus pydotplus homepage* [online] [PyDotPlus Developers]. 2022 Jul 15. Available from https://pydotplus.readthedocs.io/reference.html
[8] *Sklearn.tree.decisiontreeregressor* [online]. *scikit-learn developers*. 2022 Jul 19. Available from https://scikit-learn.org/stable/modules/generated/sklearn.tree.DecisionTreeRegressor.html

[9] Lundberg S.M., Lee S.I. 'A unified approach to interpreting model predictions'. *Advances in Neural Information Processing Systems*. 2017;**30**.
[10] Lundberg S.M., Erion G.G., Lee S.I. 'Consistent individualized feature Attribution for tree ensembles'. *ArXiv:180203888*. 2018.
[11] *Shap API reference* [online] [Scott Lundberg Revision]. 2022 Aug 19. Available from https://shap-lrjball.readthedocs.io/en/latest/api.html

Chapter 15
Linear optimization

We will discuss linear optimization in this chapter, as linear optimization is important in itself and is the foundation of more advanced optimization techniques.

15.1 Basics

Example 15.1. *Consider a service with the port rotation below:*

Melbourne (M, 1) → Sydney (S, 2) → Brisbane (B, 3) → Melbourne (M, 1)

Ships with a capacity of 1000 (TEUs) are deployed to provide a weekly frequency. The container shipment demand is: Melbourne to Sydney q^{MS} = 800, Melbourne to Brisbane q^{MB} = 700. The profit of transporting 1 TEU from Melbourne to Sydney is $200, and from Melbourne to Brisbane is $300. Develop an optimization model to evaluate the maximum profit ($/week) the company can make. □

To formulate the above problem as a mathematical model, first, we need decision variables, or variables for short. Decision variables are generally what we can control. Some models have auxiliary decision variables simply for the purpose of ease of modeling. In this example, the demand q^{MS} = 800 is beyond our control and hence is a parameter for the model (a parameter is a given/known constant). Here, we can control how many containers to transport from Melbourne to Sydney, denoted by y^{MS}, and how many from Melbourne to Brisbane, denoted by y^{MB}.

Second, we need to define the objective function. The objective function is an expression of the target associated with the decision variables. Here, we choose the profit as the objective function:

$$200y^{MS} + 300y^{MB}$$

We seek to find values for y^{MS} and y^{MB} that could maximize the objective function.

Third, we need to define the constraints because the decision variables cannot take all real values. For example, y^{MB} cannot be 1 billion because of limited demand and limited ship capacity; y^{MB} cannot be −4, either.

Solution. Let y^{MS} and y^{MB} be the decision variables representing the volumes of containers transported from Melbourne to Sydney and Melbourne to Brisbane, respectively. The model is:

max $200y^{MS} + 300y^{MB}$ (maximize the total profit)

subject to

$y^{MS} + y^{MB} \le 1000$

(ship capacity constraint on the leg from M to S)

$y^{MB} \le 1000$

(cannot carry more containers than the ship capacity on the leg from S to B)

$y^{Ms} \le 800$

(cannot carry more containers for the OD pair (M S) than the demand)

$y^{MB} \le 700$

(demand constraint for the OD pair (M, B))

$y^{MS} \ge 0$

(cannot carry a negative number of containers for the OD pair (M, S))

$y^{MB} \ge 0$

(nonnegativity constraint on y^{MB}).

We can see from the above solution that, in an optimization model we need to (1) define the decision variables; (2) define the objective function and explain it; and (3) define the constraints and explain each constraint. We can use any symbol to represent a decision variable; however, we often use symbols that are easy to remember. If the meaning of the objective function or a constraint is straightforward, it is acceptable not to explain it. In the above solution, "max" means we want to maximize the objective function; we use "min" if we want to minimize the objective function (e.g., when the objective function represents cost). The model for Example 1 has two variables and six constraints.

Example 15.2. *Reconsider the service in Example 15.1. Ships with a capacity of 1500 (TEUs) are deployed to provide a weekly frequency. The container shipment demand is: Melbourne to Sydney $q^{MS} = 800$, Melbourne to Brisbane $q^{MB} = 700$, and Sydney to Brisbane $q^{SB} = 900$. The profit of transporting 1 TEU from Melbourne to Sydney is $200, from Melbourne to Brisbane is $300, and from Sydney to Brisbane is $100. Develop an optimization model to evaluate the maximum profit ($/week) the company can make.*

Solution. Let y^{MS}, y^{MB}, and y^{SB} be the decision variables representing the volumes of containers transported from M to S, M to B, and S to B, respectively. The model is:

max $200y^{MS} + 300y^{MB} + 100y^{SB}$

subject to

$y^{MS} + y^{MB} \leq 1500$

(ship capacity constraint on the leg from M to S)

$y^{MB} + y^{SB} \leq 1500$

(cannot carry more containers than the ship capacity on the leg from S to B)

$y^{Ms} \leq 800$

$y^{MB} \leq 700$

$y^{SB} \leq 900$

$y^{MS} \geq 0$

$y^{MB} \geq 0$

$y^{SB} \geq 0.$

Example 15.3. *Reconsider Example 15.1. Among the 800 TEUs to transport from Melbourne to Sydney, 200 TEUs are from a major customer and must be transported (otherwise the customer may be lost in the future). Develop an optimization model to evaluate the maximum profit ($/week) the company can make.*

Solution. The answer is the model in the solution to Example 1 plus the following constraint: $y^{MS} \geq 200$.

Example 15.4. *Reconsider Example 15.1. Suppose that if the company rejects 1 TEU from Melbourne to Sydney, it loses $30; and if it rejects 1 TEU from Melbourne to Brisbane, it loses $20 (e.g., loss of good will). Formulate a linear optimization model to help the company make the decision of how to transport containers.*

Solution. The answer is the model in the solution to Example 15.1, except that the objective function is changed to

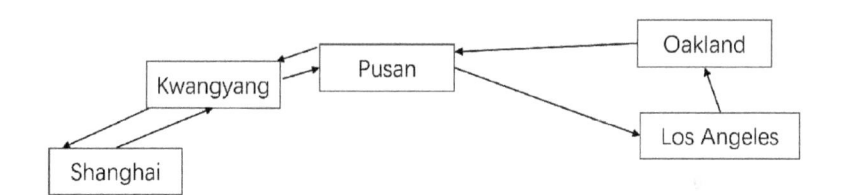

From/To	Los Angeles (Thu)	Oakland (Mon)
Pusan (Sun)	11	15
Kwangyang (Fri)	13	17
Shanghai (Thu)	14	18

From/To	Pusan (Sat)	Kwangyang (Mon)	Shanghai (Wed)
Oakland (Tue)	18	20	22
Los Angeles (Sun)	20	22	24

Figure 15.1   CC1 service by OOCL

212 Machine learning and data analytics for maritime studies

$$\max 200y^{MS} + 300y^{MB} - 30(800 - y^{MS}) - 20(700 - y^{MB}).$$

Example 15.5. *Consider the CC1 service in* Figure 15.1 *with the port rotation*

Shanghai (S, 1) → Kwangyang (K, 2) → Pusan (P, 3) → Los Angeles (L, 4) → Oakland (O, 5) → Pusan (P, 6) → Kwangyang (K, 7) → Shanghai (S, 1)

Ships with a capacity of 8000 (TEUs) are deployed to provide a weekly frequency. The container shipment demand is: Shanghai to Los Angeles q^{SL} = 4500, Kwangyang to Los Angeles q^{KL} = 1000, Pusan to Los Angeles q^{PL} = 1500, Oakland to Shanghai q^{OS} = 3700, and Shanghai to Pusan q^{SP} = 1900. The profit of transporting 1 TEU from Shanghai to Los Angeles is $1800, Kwangyang to Los Angeles is $1900, Pusan to Los Angeles is $1600, Oakland to Shanghai is $900, and Shanghai to Pusan is $500. Develop an optimization model to evaluate the maximum profit ($/week) the company can make.

Solution. Let y^{SL}, y^{KL}, y^{PL}, y^{OS}, and y^{SP} be the decision variables representing the volumes of containers transported from S to L, K to L, P to L, O to S, and S to P, respectively. The model is:

$\max 1800y^{SL} + 1900y^{KL} + 1600y^{PL} + 900y^{OS} + 500y^{SP}$

subject to

$y^{SL} + y^{SP} \leq 8000$

(ship capacity constraint on the leg from S to K)

$y^{SL} + y^{KL} + y^{SP} \leq 8000$

(ship capacity constraint on the leg from K to P)

$y^{SL} + y^{KL} + y^{PL} \leq 8000$

(ship capacity constraint on the leg from P to L)

$y^{OS} \leq 8000$

(ship capacity constraint on the leg from O to P)

$y^{OS} \leq 8000$

(ship capacity constraint on the leg from P to K)

$y^{OS} \leq 8000$

(ship capacity constraint on the leg from K to S)

$0 \leq y^{SL} \leq 4500$

$0 \leq y^{KL} \leq 1000$

$0 \leq y^{PL} \leq 1500$

$0 \leq y^{OS} \leq 3700$

$0 \leq y^{SP} \leq 1900.$

15.2 Classification of linear optimization models according to solutions

The examples in section 15.1 are all linear optimization models. A linear optimization model has a linear objective function to be maximized or minimized, and a set of linear constraints. The following functions are linear:

$$5x + 6y - 3.4z$$
$$100x_1 - 0.001x_2 + 1002.$$

The following functions are nonlinear:

$$5x + 6y^2 - 3.4z$$
$$100x_1 - 0.001 \sin x_2 + 1002$$
$$3xy.$$

The following constraints are linear:

$$5x + 6y - 3.4z \leq 23$$
$$100x_1 - 0.001x_2 + 1002 \geq 100$$
$$100x_1 - 0.001x_2 + 1002 = 100.$$

It should be noted that we do not usually consider "<" or ">" constraints in linear optimization, e.g., $x_1 + x_2 < 5$.

A feasible solution to a linear optimization model is a vector of values of the decision variables that satisfies all the constraints. For example, in the model for Example 1, $(y^{MS}, y^{MB}) = (0, 1)$ is a feasible solution. The set of all feasible solutions is called the feasible set. We can calculate the objective function value for each feasible solution. For example, the objective function value of the solution $(y^{MS}, y^{MB}) = (0, 1)$ is 300. An optimal solution is the "best" feasible solution, which has the largest objective function value for a maximization model, and the smallest objective function value for a minimization model. The objective function value of an optimal solution is called the optimal objective function value.

Example 15.6. *Write down three distinct feasible solutions to the model in the solution to Example 1, and calculate their objective function values.*

Solution. There are many feasible solutions. For example, the objective function value of the solution $(y^{MS}, y^{MB}) = (0, 1)$ is 300; the objective function value of the solution $(y^{MS}, y^{MB}) = (0, 0)$ is 0; the objective function value of the solution $(y^{MS}, y^{MB}) = (10, 10)$ is 5000; the objective function value of the solution $(y^{MS}, y^{MB}) = (100, 100)$ is 50,000.

A linear optimization model can be classified into three categories according to solutions: infeasible, unbounded, and having an optimal solution. A linear optimization model may have no feasible solution, i.e., the model is infeasible, e.g.:

$$\max \ x + y$$

subject to

$$x+y \geq 1$$
$$x \leq 0$$
$$y \leq 0.$$

A linear optimization model is infeasible if and only if its feasible set is empty, i.e., no solution satisfies all the constraints simultaneously. Whether a linear optimization model is infeasible or not has nothing to do with its objective function. A linear optimization model either has 0, or 1, or an infinite number of feasible solutions. A linear optimization model is feasible if it has at least one feasible solution.

A linear optimization model may be unbounded, i.e., for a maximization model, the objective function value can be infinitely large and for a minimization model, the objective function value can be infinitely small, e.g.:

$$\max x+y$$

subject to

$$x+y \geq 1$$
$$x \geq 0$$
$$y \leq 0.$$

Note that if a linear optimization model is unbounded, its feasible set must be unbounded. If the feasible set of a linear optimization model with decision variables $x_1, x_2, \cdots, x_n$ is bounded, i.e., there exists a positive number M such that any feasible solution satisfies $-M \leq x_1 \leq M, -M \leq x_2 \leq M, \cdots, -M \leq x_n \leq M$, then the model will not be unbounded. If the feasible set is unbounded, the model may be bounded (e.g., minimizing x subject to $x \geq 0$) or unbounded (e.g., minimizing $-x$ subject to $x \geq 0$). For most practical problems, we do not worry about whether they unbounded, because in reality the absolute values of the decision variables cannot be infinitely large.

If a linear optimization model is feasible and not unbounded, then it has an optimal solution, e.g.:

$$\max x+y$$

subject to

$$x+y \geq 1$$
$$x \leq 1$$
$$y \leq 1.$$

We often add the superscript "$*$" to a decision variable to represent its value in an optimal solution. For example, we often use (x^*, y^*) to represent the optimal values of the decision variables (x, y).

If a linear optimization model has an optimal solution, the optimal solution may not be unique, e.g.:

$$\max x+y$$

subject to

$$x + y \leq 1$$
$$x \leq 1$$
$$y \leq 1.$$

If a linear optimization model has more than one optimal solution, then it has an infinite number of optimal solutions. In most cases, we are only interested in obtaining one of them.

If an optimization model has an optimal solution, then there is no loss of generality to say "suppose that it is a minimization model with a positive objective function." For example, if we aim to maximize $3x - 4y$, then we can minimize $-(3x - 4y) + z$ subject to $z = 10^{10}$ (i.e., we let the new decision variable z equal a very large positive number).

Example 15.7. *Suppose that all linear optimization models in this question have optimal solutions. (1) Consider a minimization linear optimization model with several constraints, and one of them is $3x - 4y \leq 5$. If the constraint is changed to $3x - 4y \leq 6$, how will the feasible set change? How will the optimal objective function value change? (2) Consider a linear optimization model with several constraints and minimizing $3x - 4y$. If the objective function is changed to $6x - 8y$, how will the feasible set change? How will the optimal solution change? How will the optimal objective function value change?*

Solution. (1) The feasible set will be larger or not change. The optimal objective function value will be smaller or not change. (2) The feasible set will not change. The optimal solution will not change. The optimal objective function value will be twice as large as before.

Example 15.8. *Reconsider the model in the solution to Example 1. (1) If the ship capacity increases, how will the feasible set change? How will the optimal objective function value change? (2) If the demand from Melbourne to Brisbane decreases, how will the feasible set change? How will the optimal objective function value change? (iii) If the profit of transporting 1 TEU from Melbourne to Sydney is $400 instead of $200, and from Melbourne to Brisbane is $600 instead of $300, how will the feasible set change? How will the optimal solution change? How will the optimal objective function value change?*

Solution. (1) The feasible set will be larger or not change. The optimal objective function value will be larger or not change. (2) The feasible set will be smaller or not change. The optimal objective function value will be smaller or not change. (3) The feasible set will not change. The optimal solution will not change. The optimal objective function value will be twice as large as before.

15.3 Equivalence between different formulations

It is helpful to understand that some forms of a linear optimization model are equivalent. When we say transform Model A to Model B, we mean that the Model B is equivalent to Model A in the sense that gives an optimal solution to Model B, we can easily derive an optimal solution to Model A.

Regarding linear optimization, first, as we have seen, a maximization objective can be transformed to a minimization objective, and vice versa. For example, maximizing $5x_1 - 6x_2$ is equivalent to minimizing $-(5x_1 - 6x_2)$. Second, "$\leq$" and "$\geq$" constraints can be transformed to each other. For example, $5x_1 - 6x_2 \leq 10$ is equivalent to $-(5x_1 - 6x_2) \geq -10$. Third, "$=$" constraints can be transformed to "$\leq$" and "$\geq$" constraints. For example, $5x_1 - 6x_2 = 10$ is equivalent to the combination of $5x_1 - 6x_2 \leq 10$ and $5x_1 - 6x_2 \geq 10$.

Example 15.9. *Transform the following linear optimization model to one with a minimization objective and "$\leq$" constraints.*

$$\max 5x + 6y$$

subject to

$$3x - 4y \geq 34$$
$$45x - 98y = 9$$
$$x \geq 1$$
$$y \leq 0.$$

Solution

$$\min -5x - 6y$$

subject to

$$-3x + 4y \leq -34$$
$$45x - 98y \leq 9$$
$$-(45x - 98y) \leq -9$$
$$-x \leq -1$$
$$y \leq 0.$$

Example 15.10. *Transform the model in Example 9 to one with a maximization objective and "$\geq$" constraints.*

Solution

$$\max 5x + 6y$$

subject to

$$3x - 4y \geq 34$$
$$45x - 98y \geq 9$$
$$-(45x - 98y) \geq -9$$
$$x \geq 1$$
$$-y \geq 0.$$

Fourth, "$\leq$" and "$\geq$" constraints can be transformed to a combination of "$=$" constraints and nonnegativity constraints on the decision variables (i.e., constraints imposing that the decision variables are nonnegative). For example, $5x_1 - 6x_2 \leq 10$ is equivalent to the combination of $y = 10 - (5x_1 - 6x_2)$ and $y \geq 0$.

Fifth, we can transform a model to one in which all decision variables are nonnegative. For example, if we have a constraint $x \leq 0$, we can let $u = -x$ and replace x by $-u$ in the model. If we have a constraint $x \geq 3$, we can let $u = x - 3$ and replace x by $u + 3$ in the model. If we have a constraint $x \leq 3$, we can let $u = 3 - x$ and replace x by $3 - u$ in the model. If the model does not specify whether x is nonnegative or nonpositive, we can define $u_1 \geq 0, u_2 \geq 0$ and replace x by $u_1 - u_2$ in the model.

Example 15.11. *Transform the model in Example 5.9 to one with only "=" constraints and nonnegative decision variables.*

Solution. As $x \geq 1$ and $y \leq 0$, we let $s = x - 1$ and $t = -y$. Hence, $x = s + 1, s \geq 0$ and $y = -t, t \geq 0$. The model is

$$\max 5(s+1) + 6(-t) \text{ (i.e., } 5s - 6t + 5)$$

subject to

$$3(s+1) - 4(-t) \geq 34 \text{ (i.e., } 3s + 4t \geq 31)$$
$$45(s+1) - 98(-t) = 9 \text{ (i.e., } 45s + 98t = -36)$$
$$s \geq 0$$
$$t \geq 0.$$

Letting $r = (3s + 4t) - 31$, the model is equivalent to

$$\max 5s - 6t \text{ (note that the constant 5 does not affect the model)}$$

subject to

$$3s + 4t - r = 31$$
$$45s + 98t = -36$$
$$r \geq 0$$
$$s \geq 0$$
$$t \geq 0.$$

Sixth, we can transform the objective function to one that is equal to a decision variable. For example, maximizing $5x_1 - 6x_2$ is equivalent to maximizing u, subject to $u = 5x_1 - 6x_2$. It is also equivalent to maximizing u, subject to $u \leq 5x_1 - 6x_2$. In plain words, this property means that the objective function can somehow be considered as a constraint; it also means that if we can solve an optimization model with complex constraints, we can also solve an optimization model with complex constraints and a complex objective function.

Finally, in some problems there is no objective function, for instance, when we are only interested in finding a feasible solution. In this case, we can aim to "min 0."

Two general forms of linear optimization models are frequently used in theoretical analysis: the canonical form that maximizes $c^T x$ subject to $Ax \leq b$ and $x \geq 0$, and the standard form that maximizes $c^T x$ subject to $Ax = b$ and $x \geq 0$, where x is a column vector representing n decision variables, A is an $m \times n$ matrix, c is a column vector with n rows, b is a column vector with m rows, and A, b, and c are all parameters.

Example 15.12. *Transform the model in Example 15.9 to the canonical form.*
Solution. Letting $x = u - v$, $u \geq 0$, $v \geq 0$, and $y = -w$, $w \geq 0$, the model is transformed to

$$\max 5(u - v) + 6(-w)$$

subject to

$$-3(u - v) + 4(-w) \leq -34$$
$$45(u - v) - 98(-w) \leq 9$$
$$-45(u - v) + 98(-w) \leq -9$$
$$-(u - v) \leq -1$$
$$u \geq 0$$
$$v \geq 0$$
$$w \geq 0.$$

Hence, in the canonical form, $x = [\ u \quad v \quad w\ ]^T$, $A = \begin{bmatrix} -3 & 3 & -4 \\ 45 & -45 & 98 \\ -45 & 45 & -98 \\ -1 & 1 & 0 \end{bmatrix}$,

$b = [\ -34 \quad 9 \quad -9 \quad -1\ ]^T$, and $c = [\ 5 \quad -5 \quad -6\ ]^T$, with $x = u - v$ and $y = -w$.

15.4 Graphical method for models with two variables

Linear optimization models with two variables can be solved intuitively using graphs. In practical applications, hardly any problem has only two variables. However, learning the graphical method is helpful for appreciating the properties of linear optimization models.

In the graphical method, we first draw constraints that define an upper or lower bound for a decision variable (e.g., $x \geq 0, y \leq 100$), then draw the other constraints, and finally draw a line or a series of parallel lines that represent the objective function.

Example 15.13. *Consider the model below:*

$$\max x_1 + x_2$$

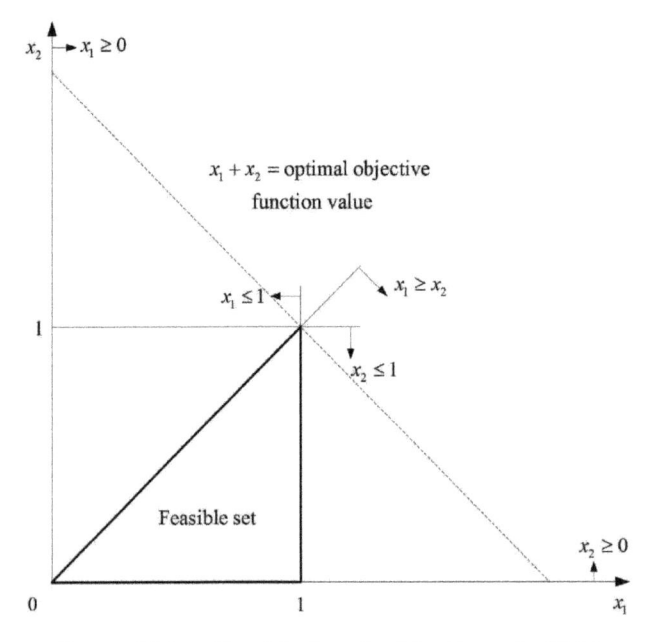

Figure 15.2   Graphical method for Example 15.13

subject to

$$x_1 \geq x_2$$
$$x_1 \leq 1$$
$$x_2 \leq 1$$
$$x_1 \geq 0$$
$$x_2 \geq 0.$$

Use the graphical method to find the optimal solution.

Solution. See Figure 15.2. The optimal solution corresponds to the intersection of the lines $x_1 = x_2$ and $x_1 = 1$. Therefore, the optimal solution is $x_1^* = 1, x_2^* = 1$. The optimal objective function value is 2.

Example 15.14. *Consider the model below:*

$$\min x_2 - x_1$$

subject to

$$x_1 \geq x_2$$
$$x_1 \leq 1$$
$$x_2 \leq 1$$
$$x_1 \geq 0$$
$$x_2 \geq 0.$$

Use the graphical method to find the optimal solution.

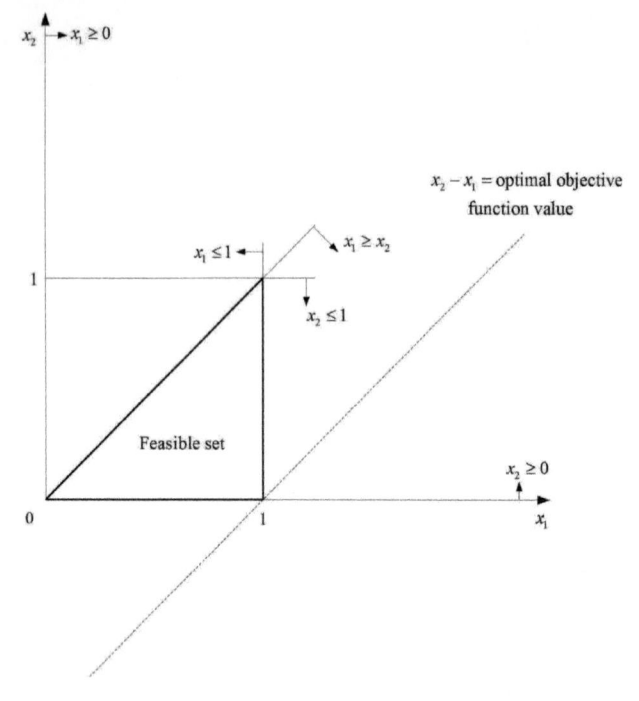

Figure 15.3 Graphical method for Example 15.14

Solution. See Figure 15.3. The optimal solution corresponds to the intersection of the lines $x_1 = 1$ and $x_2 = 0$. Therefore, the optimal solution is $x_1^* = 1, x_2^* = 0$. The optimal objective function value is -1.

Example 15.15. *Consider the model below:*

$$\max 2x_1 + x_2$$

subject to

$$x_1 + x_2 \geq 1$$
$$x_1 \leq 2$$
$$x_2 \leq 1$$
$$x_1 \geq 0$$
$$x_2 \geq 0.$$

Use the graphical method to find the optimal solution.
Solution. See Figure 15.4. The optimal solution corresponds to the intersection of the lines $x_1 = 2$ and $x_2 = 1$. Therefore, the optimal solution is $x_1^* = 2, x_2^* = 1$. The optimal objective function value is 5.

Example 15.16 *Consider the model below:*

$$\max 2x_1 + x_2$$

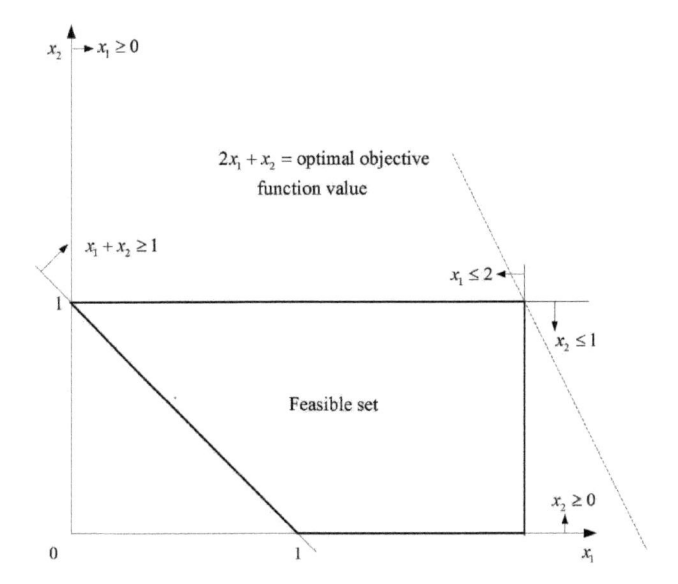

Figure 15.4   Graphical method for Example 15.15

subject to

$$x_1 + x_2 \geq 1$$
$$x_1 \leq 3$$
$$x_2 \leq 1$$
$$x_1 \geq 0$$
$$x_2 \geq 0.$$

Use the graphical method to find the optimal solution.

Solution. See Figure 15.5. The optimal solution corresponds to the intersection of the lines $x_1 = 3$ and $x_2 = 1$. Therefore, the optimal solution is $x_1^* = 3, x_2^* = 1$. The optimal objective function value is 7.

Example 15.17. *Consider the model below:*

$$\max 2x_1 + x_2$$

subject to

$$x_1 \leq 1$$
$$x_2 \leq 1$$
$$x_1 + x_2 = 1$$
$$x_1 \geq 0$$
$$x_2 \geq 0.$$

Use the graphical method to find the optimal solution.

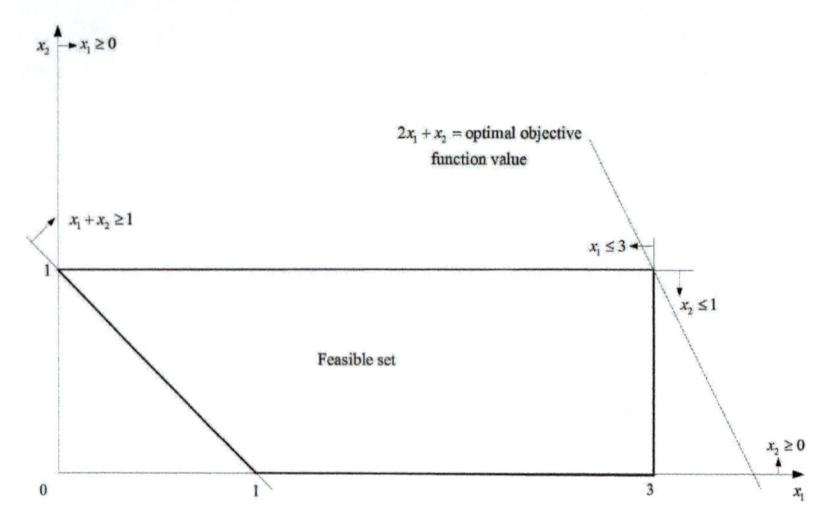

222 *Machine learning and data analytics for maritime studies*

Figure 15.5   Graphical method for Example 15.16

Solution. See Figure 15.6. Note that the feasible set is a line segment. The optimal solution corresponds to the intersection of the lines $x_1 + x_2 = 1$ and $x_1 = 1$. Therefore, the optimal solution is $x_1^* = 1, x_2^* = 0$. The optimal objective function value is 2.

Example 15.18. *Consider the model below:*

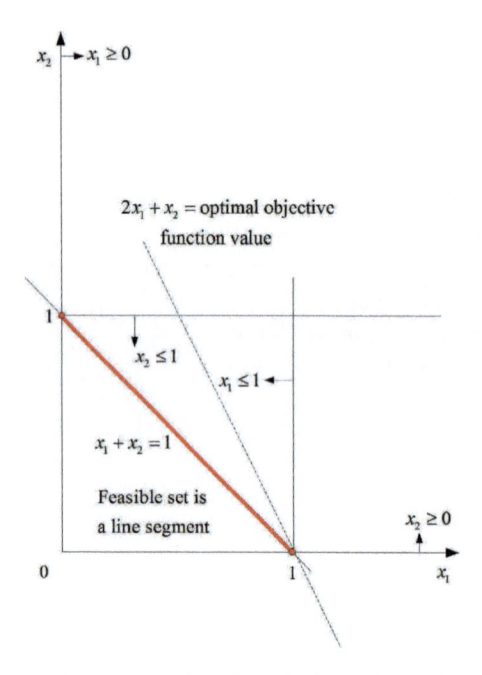

Figure 15.6   Graphical method for Example 15.17

Linear optimization 223

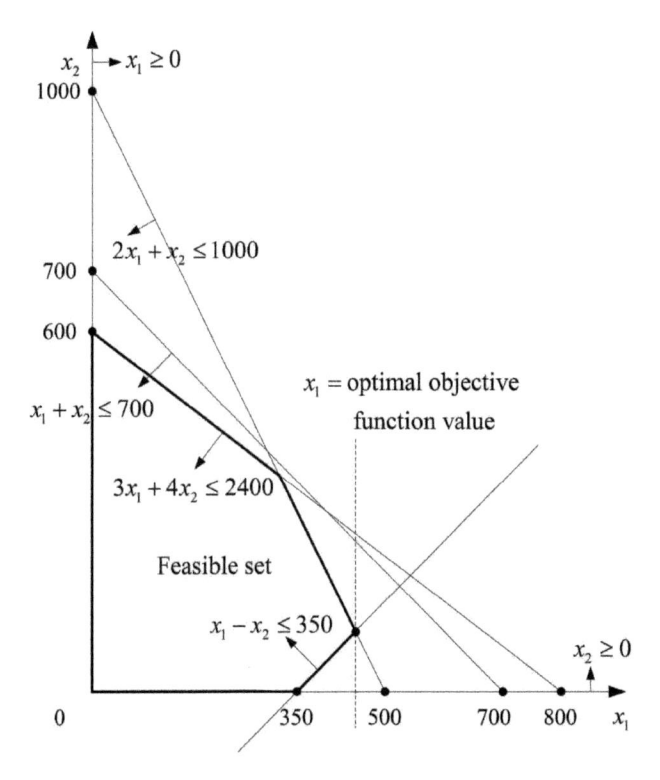

Figure 15.7   Graphical method for Example 15.18

$$\max x_1$$

subject to

$$2x_1 + x_2 \leq 1000$$
$$3x_1 + 4x_2 \leq 2400$$
$$x_1 + x_2 \leq 700$$
$$x_1 - x_2 \leq 350$$
$$x_1 \geq 0$$
$$x_2 \geq 0.$$

Use the graphical method to find the optimal solution.

Solution. See Figure 15.7. The optimal solution corresponds to the intersection of the lines $2x_1 + x_2 = 1000$ and $x_1 - x_2 = 350$. Therefore, the optimal solution is $x_1^* = 450, x_2^* = 100$. The optimal objective function value is 450.

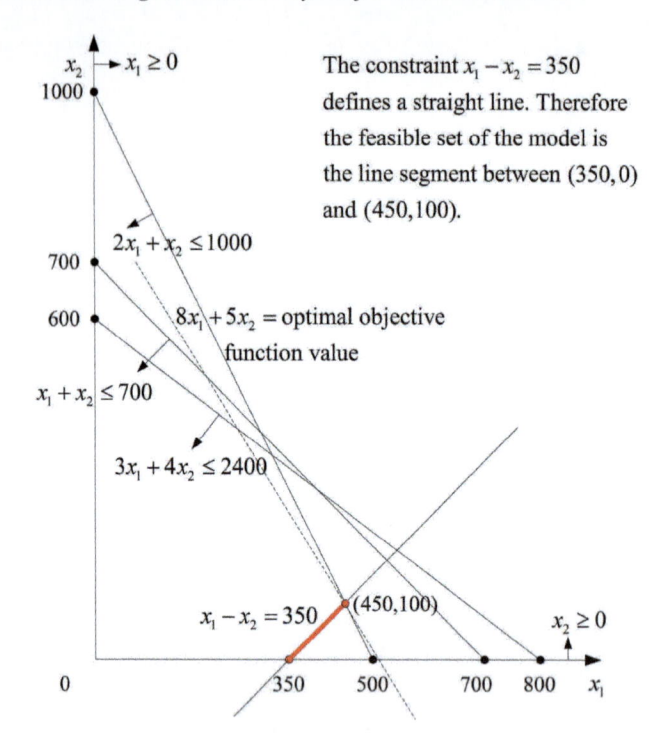

Figure 15.8 Graphical method for Example 15.19

Example 15.19. *Consider the model below:*

max $8x_1 + 5x_2$

subject to

$$2x_1 + x_2 \leq 1000$$
$$3x_1 + 4x_2 \leq 2400$$
$$x_1 + x_2 \leq 700$$
$$x_1 - x_2 = 350$$
$$x_1 \geq 0$$
$$x_2 \geq 0.$$

Use the graphical method to find the optimal solution.
Solution. See Figure 15.8. Note that the feasible set is a line segment. The optimal solution corresponds to the intersection of the lines $2x_1 + x_2 = 1000$ and $x_1 - x_2 = 350$. Therefore, the optimal solution is $x_1^* = 450, x_2^* = 100$. The optimal objective function value is 4100.

We can see from the above examples that if a linear optimization model has an optimal solution, then there exists an optimal solution that is at a "corner point" of the feasible set[*].

The graphical method can also be used to identify whether a linear optimization model is infeasible, unbounded, or has an infinite number of optimal solutions.

15.5 Using software to solve linear optimization models

Linear optimization models can be solved by simplex method, ellipsoid method, or interior point method.

When we say "solve a linear optimization model," we mean finding an optimal solution to it, or proving that it is infeasible or unbounded. If we know an optimal solution, it is very easy to calculate the optimal objective function value. Knowing the optimal objective function value without knowing an optimal solution is of little value in most cases.

There are many free and commercial solvers for linear optimization models, e.g., Excel, MATLAB, lpSolve, GUROBI, and CPLEX. A linear optimization solver is a software package that can do the following things for a linear optimization model: (1) check whether it is infeasible; (2) check whether it is unbounded; and (3) return an optimal solution if there is one. Note that if a linear optimization model has more than one optimal solution, a solver generally returns only one optimal solution that is at a corner point.

Generally speaking, linear optimization models are "easy" to solve: models with 100,000 variables and 100,000 constraints can be solved in 1 minute[†]. Since free solvers can also solve linear optimization models very efficiently, hardly any commercial solver is solely dedicated to linear optimization models. In fact, commercial solvers are mainly aimed to solve the much more difficult mixed-integer linear optimization models.

We can learn how to use Excel to solve a few simple linear optimization models to appreciate the use of solvers. Since the solvers are updated frequently, we do not discuss the detailed steps. Readers can easily Google "Excel linear optimization" to learn more about how to configure Excel to solve linear optimization models. Note that to use Excel, we need the "Solver Add-in" function.

Example 15.20. *Use a solver (e.g., Excel) to solve the following linear optimization model:*

$$\max 200x + 300y$$

---

[*]There are some linear optimization models with no corner points, e.g., maximizing $0x + 2y$ subject to $0 \leq y \leq 1$. We generally do not need to worry about them in practical applications.
[†]The exact time depends on the parameters of the model, the solver, and the computer.

subject to
$$x + y \le 1000$$
$$y \le 1000$$
$$x \le 800$$
$$y \le 700$$
$$x \ge 0$$
$$y \ge 0.$$

Solution. The optimal solution is $x^* = 300, y^* = 700$.

Example 15.21. *Use a solver (e.g., Excel) to solve the following linear optimization model:*

$$\max 200x + 300y$$

subject to
$$x + y \ge 1000$$
$$y \ge 1000$$
$$x \ge 800$$
$$y \ge 700$$
$$x \ge 0$$
$$y \ge 0.$$

Solution. If Excel is used, then the results are "The Objective Cell values do not converge. Solver can make the Objective Cell as large (or as small when minimizing) as it wants," which indicates that the model is unbounded.

Example 15.22. *Use a solver (e.g., Excel) to solve the following linear optimization model:*

$$\max 200x + 300y$$
subject to
$$x + y \le 1000$$
$$y \ge 1000$$
$$x \ge 800$$
$$y \ge 700$$
$$x \ge 0$$
$$y \ge 0.$$

Solution. If Excel is used, the results are "Solver could not find a feasible solution. Solver cannot find a point for which all Constraints are satisfied," which indicates that the model is infeasible.

Example 15.23. *Use a solver (e.g., Excel) to solve the following linear optimization model:*

$$\max x + y$$

subject to

$$x + y \leq 1$$
$$x \geq 0$$
$$y \geq 0.$$

Solution. Although there are an infinite number of optimal solutions, Excel only provides one optimal solution, e.g., $x^* = 1, y^* = 0$.

15.6 An in-depth understanding of linear optimization

The following questions are on linear optimization models. For each question, you should either answer that such a linear optimization model does not exist (and you do not need to provide the reason), or give an example of such a linear optimization model.

Example 15.24. *Give a model that is both infeasible and unbounded.*
Solution. Such a model does not exist.
Example 15.25. *Give a model whose optimal objective function value is 1, and after removing one constraint, the optimal objective function value is 0.*
Solution. Minimizing x subject to $x \geq 1, x \geq 0$. After removing the constraint $x \geq 1$, the optimal objective function value is 0.
Example 15.26.
1. *A model is infeasible, and after removing one constraint, it is feasible.*
2. *A model is infeasible, and after removing one constraint, it is unbounded.*
3. *A model is infeasible, and after removing one constraint, it has an optimal solution.*
4. *A model is infeasible, and after removing one constraint, it has an infinite number of optimal solutions.*
5. *A model has an optimal solution, and after adding one constraint, it is unbounded.*
6. *A model has an optimal solution, and after adding one constraint, it is infeasible.*
7. *A model has exactly one optimal solution, and after adding one constraint, it has an infinite number of optimal solutions.*
8. *A model has an infinite number of optimal solutions, and after adding one constraint, it has exactly one optimal solution.*
9. *A model is infeasible, and after changing its objective function, it has an optimal solution.*
10. *A model has an optimal solution, and after changing its objective function, it is infeasible.*
11. *A model has an optimal solution, and after changing its objective function, it is unbounded.*
12. *A model is unbounded, and after changing its objective function, it has exactly one optimal solution.*

13. *A model is unbounded, and after changing its objective function, it has an infinite number of optimal solutions.*

   Solution.

   1. Maximize x subject to $x \geq 1$ and $x \leq 0$. Remove the first constraint.
   2. Minimize x subject to $x \geq 1$ and $x \leq 0$. Remove the first constraint.
   3. Maximize x subject to $x \geq 1$ and $x \leq 0$. Remove the first constraint.
   4. Maximize 0 subject to $x \geq 1$ and $x \leq 0$. Remove the first constraint.
   5. Not possible.
   6. Minimize x subject to $x \geq 1$. Add $x \leq 0$.
   7. Maximize x subject to $x \geq 0$, $y \geq 0$, and $x + y \leq 1$. Add $x \leq 0.5$.
   8. Maximize x subject to $x \geq 0$, $y \geq 0$, $x + y \leq 1$, and $x \leq 0.5$. Add $x + y \leq 0.5$.
   9. Not possible.
   10. Not possible.
   11. Minimize x subject to $x \geq 0$. Change the objective function to $\max x$.
   12. Maximize x subject to $x \geq 0$. Change the objective function to $\min x$.
   13. Maximize $x + y$ subject to $x \geq 0$, $y \geq 0$, and $x + y \geq 1$. Change the objective function to $\min x + y$.

15.7 Useful applications of linear optimization solvers

Although a linear optimization solver only provides one optimal solution if there are an infinite number of optimal solutions, we can take advantage of the solver in a number of smart ways to address more problems.

Example 15.27. *Given a linear optimization model with a non-empty and bounded feasible set defined by $Ax \leq b$ and $x \geq 0$, where the vector of decision variables x has n elements and A, b, and c are parameters of appropriate dimensions, how to use a linear optimization solver to check whether it has only one feasible solution or an infinite number of feasible solutions.*

Solution. The number of feasible solutions is independent of the objective function. We solve the following two models: maximizing x_1 subject to $Ax \leq b$ and $x \geq 0$, and minimizing x_1 subject to $Ax \leq b$ and $x \geq 0$. Here both models have an optimal solution. If their optimal objective function values are different, then there are an infinite number of feasible solutions to the original model. Otherwise we check models that maximize and minimize $x_2, x_3, \cdots, x_n$. If all of the optimal objective function values of the n maximization models are the same as the corresponding values of the minimization models, there is only one feasible solution.

Example 15.28. *Given a linear optimization model that is feasible and not unbounded: maximizing $c^T x$ subject to $Ax \leq b$ and $x \geq 0$, where the vector of decision variables x has n elements and A, b and c are parameters of appropriate dimensions, how to use a linear optimization solver to check whether it has only one optimal solution or an infinite number of optimal solutions.*

Solution. Obtain an optimal solution, denoted by x^*. Now the problem becomes how to check whether there is one or an infinite number of feasible solutions to the set of constraints $Ax \leq b$, $x \geq 0$, and $c^T x = c^T x^*$, which is Example 15.27.

Example 15.29. *Given a set of constraints $Ax \leq b$ and $x \geq 0$, where the vector of decision variables $x = [\ x \ \ y\ ]^T$ and A, b, and c are parameters of appropriate dimensions. We want to find the optimal values of the decision variables satisfying two conditions. Condition 1: the expression $3x - 4y$ is maximized. Condition 2: among all the feasible solutions satisfying Condition 1, we want to find the one that minimizes $2x + 3y$. How to use a linear optimization solver to address this problem*[‡]
Solution. The first approach is to use the big-M method. Our first priority is to maximize $3x - 4y$, and our second priority is to maximize $-2x - 3y$. Therefore, we can solve the model maximizing $10^8(3x - 4y) + (-2x - 3y)$ subject to $Ax \leq b$ and $x \geq 0$. Of course, the solution may not be accurate.

The second approach involves sequential optimization. We first solve the model maximizing $3x - 4y$ subject to $Ax \leq b$ and $x \geq 0$ to obtain the optimal solution $(\hat{x}, \hat{y})$. We then solve the model minimizing $2x + 3y$ subject to $Ax \leq b$, $x \geq 0$, and $3x - 4y = 3\hat{x} - 4\hat{y}$. Its optimal solution (x^*, y^*) is what we want.

Example 15.30. *A linear optimization model has decision variables $x_1, x_2, \cdots, x_n$ and the following constraints:*

$a_{11}x_1 + a_{12}x_2 + ... + a_{1n}xn \geq b_1$

$a_{21}x_1 + a_{22}x_2 + ... + a_{2n}xn \geq b_2$

$\vdots$

$a_{m1}x_1 + a_{m2}x_2 + ... + a_{mn}xn \geq b_m$

$0 \leq x_1 \leq 1$

$0 \leq x_2 \leq 1$

$\vdots$

$0 \leq x_n \leq 1$

where a_{ij}, b_i are all constants, $i = 1, 2, \cdots, m$, $j = 1, 2, \cdots, n$. We do not know its objective function yet. How to use a linear optimization solver to check whether the first constraint $a_{11}x_1 + a_{12}x_2 + \cdots + a_{1n}x_n \geq b_1$ is redundant (i.e., removing the constraint does not affect the model)?
Solution. We actually need to check whether the feasible set defined by all the constraints is the same as the feasible set defined by the constraints excluding the first one. To this end, we solve the model minimizing $a_{11}x_1 + a_{12}x_2 + \cdots + a_{1n}x_n$ subject to the second to the last constraints. If the optimal objective function value is not smaller than b_1, which means, any point that satisfies the second to the last constraints automatically satisfies the first one, then the first constraint is redundant; otherwise it is not redundant.

[‡]For example, in traffic control, Condition 1 might mean maximizing the survival rates in road accidents, and Condition 2 might mean minimizing the travel delay due to road accidents. The objective in Condition 1 is much more important than that in Condition 2.

Example 15.31. *Suppose that Ω_1 is a set of vectors $(x_1, x_2, \cdots, x_n)$ defined by the following $m + 2n$ constraints:*

$$a_{11}x_1 + a_{12}x_2 + \ldots + a_{1n}x_n \geq b_1$$
$$a_{21}x_1 + a_{22}x_2 + \ldots + a_{2n}x_n \geq b_2$$
$$\vdots$$
$$a_{m1}x_1 + a_{m2}x_2 + \ldots + a_{mn}x_n \geq b_m$$
$$0 \leq x_1 \leq 1$$
$$0 \leq x_2 \leq 1$$
$$\vdots$$
$$0 \leq x_n \leq 1$$

Ω_2 is a set of vectors $(x_1, x_2, \cdots, x_n)$ defined by the following $m + 2n$ constraints:

$$c_{11}x_1 + c_{12}x_2 + \cdots + c_{1n}x_n \geq d_1$$
$$c_{21}x_1 + c_{22}x_2 + \cdots + c_{2n}x_n \geq d_2$$
$$\vdots$$
$$c_{m1}x_1 + c_{m2}x_2 + \cdots + c_{mn}x_n \geq d_m$$
$$0 \leq x_1 \leq 1$$
$$0 \leq x_2 \leq 1$$
$$\vdots$$
$$0 \leq x_n \leq 1$$

where a_{ij}, b_i, c_{ij}, d_i are all constants, $i = 1, 2, \cdots, m$, $j = 1, 2, \cdots, n$. How to use a linear optimization solver to check whether the two sets Ω_1 and Ω_2 are the same?

Solution. We first need to check whether the two sets are empty (minimizing 0 subject to the constraints that define each set). If $\Omega_1 = \emptyset$ and $\Omega_2 = \emptyset$, they are the same. If exactly one of them is empty, they are different.

If $\Omega_1 \neq \emptyset$ and $\Omega_2 \neq \emptyset$, we refer to Example 15.30: if each constraint that defines Ω_1 is redundant for Ω_2 (i.e., any point in Ω_2 automatically satisfies all constraints in Ω_1), and each constraint that defines Ω_2 is redundant for Ω_1, then the two sets are the same; otherwise they are different.

Chapter 16
Advanced linear optimization

Basic linear optimization models have been introduced in Chapter 15. In this chapter, more advanced linear optimization models will be covered.

16.1 Network flow optimization

Many practical problems are associated with the optimization of flow in a network, such as transportation, telecommunication, and power transmission networks. A network has nodes and arcs (arcs can also be called links). For example, a city logistics network has intersections (nodes) and roads (arcs). Arcs are directional, i.e., cargoes/passengers can only be transported from the tail node to the head node of an arc. Therefore, a two-way street is usually considered as two arcs.

Example 16.1: Walmart has three warehouses (W1, W2, and W3) that store the same type of product and five supermarkets (S1–S5) that need the products in a city. The number of products available at each warehouse, the number of products needed at each supermarket, and the transportation cost per unit product from each warehouse to each supermarket are shown below. Develop a linear optimization model to help Walmart make the decision of how to transport the products.

Warehouse	Number of products available
W1	100
W2	200
W3	50

Supermarket	Number of products needed
S1	80
S2	90
S3	70
S4	60
S5	50

Unit cost ($)	S1	S2	S3	S4	S5
W1	1	2	4	3	6
W2	5	2	4	4	4
W3	5	1	1	3	2

Solution. Let f_{ij} be the decision variables representing the number of products transported from warehouse $i = 1, 2, 3$ to supermarket $j = 1, 2, 3, 4, 5$. The model is as follows:

$$\min f_{11} + 2f_{12} + 4f_{13} + 3f_{14} + 6f_{15} + 5f_{21} + 2f_{22} +$$
$$4f_{23} + 4f_{24} + 4f_{25} + 5f_{31} + f_{32} + f_{33} + 3f_{34} + 2f_{35}$$

subject to

$$f_{11} + f_{12} + f_{13} + f_{14} + f_{15} \leq 100$$
$$f_{21} + f_{22} + f_{23} + f_{24} + f_{25} \leq 200$$
$$f_{31} + f_{32} + f_{33} + f_{34} + f_{35} \leq 50$$
$$f_{11} + f_{21} + f_{31} = 80$$
$$f_{12} + f_{22} + f_{32} = 90$$
$$f_{13} + f_{23} + f_{33} = 70$$
$$f_{14} + f_{24} + f_{34} = 60$$
$$f_{15} + f_{25} + f_{35} = 50$$
$$f_{ij} \geq 0, i = 1, 2, 3, j = 1, 2, 3, 4, 5.$$

Sometimes, we may use symbols to simplify the notation. We can define sets $I = \{1, 2, 3\}, J = \{1, 2, 3, 4, 5\}$, represent by p_i the number of products available in warehouse $i \in I$, denote by q_j the number of products needed by supermarket $j \in J$, and let c_{ij} be the cost of transporting one product from warehouse $i \in I$ to supermarket $j \in J$. Note that p_i, q_j, and c_{ij} are all known parameters. Let f_{ij} be the decision variables representing the number of products transported from warehouse $i \in I$ to supermarket $j \in J$. The model is as follows:

$$\min \sum_{i \in I} \sum_{j \in J} c_{ij} f_{ij}$$

subject to

$$\sum_{j \in J} f_{ij} \leq p_i, i \in I$$
$$\sum_{i \in I} f_{ij} = q_j, j \in J$$
$$f_{ij} \geq 0, i \in I, j \in J.$$

It can be seen that the above model is very compact. Moreover, it is very general: we might save the two sets I and J and parameters p_i, q_j, and c_{ij} in files, and program the model using a linear optimization solver. Next time when Walmart plans the transportation, it only needs to change the input files. The computer codes for linear

Advanced linear optimization 233

optimization, which might be too complex for logistics managers to understand, do not need to be changed.

Example 16.2: A small furniture company has three stores (S1–S3) and two warehouses (W1 and W2). Stores 1, 2, and 3 need 12, 13, and 14 desks, respectively. Warehouses 1 and 2 have 20 and 25 desks, respectively. The transportation costs per desk from each warehouse to each store are shown below. Assuming that the number of desks can be a fraction, develop a linear optimization model to help the company make the decision of how to transport the desks.

Unit cost ($)	S1	S2	S3
W1	1	2	4
W2	5	2	4

Solution. Let f_{ij} be the decision variables representing the number of desks transported from warehouse $i = 1, 2$ to store $j = 1, 2, 3$. The model is as follows:

$$\min f_{11} + 2f_{12} + 4f_{13} + 5f_{21} + 2f_{22} + 4f_{23}$$

subject to

$$f_{11} + f_{12} + f_{13} \leq 20$$
$$f_{21} + f_{22} + f_{23} \leq 25$$
$$f_{11} + f_{21} = 12$$
$$f_{12} + f_{22} = 13$$
$$f_{13} + f_{23} = 14$$
$$f_{ij} \geq 0, i = 1, 2, j = 1, 2, 3.$$

Example 16.3: This is a maximum flow problem. We have a crude oil pipeline network shown in Figure 16.1. The arrows are the pipelines: crude oil can only flow in the direction of the arrows, and the numbers on the arrows are the capacities (maximum flow rates of crude oil) of the pipelines (1 000 tons/h). Node A is an oil field and node E is a refinery factory. Develop a linear optimization model to find the maximum flow of crude oil per hour from A to E.

In network flow problems, we can use the amount of cargo flow on each arc as the decision variables. Such a formulation often needs the flow conservation equations: if a node does not export or import cargoes, then the total cargo inflow must

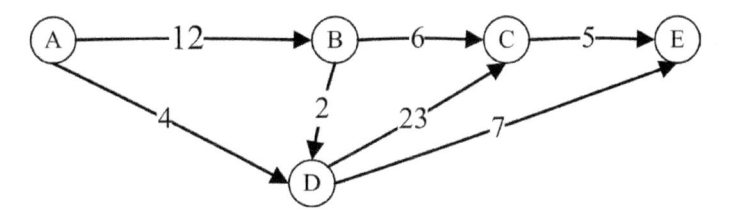

Figure 16.1 *A crude oil pipeline network*

be equal to the total outflow; if the node exports cargoes, then the difference between the total outflow and the total inflow equals the number of exported cargoes; and if the node imports cargoes, then the difference between the total inflow and the total outflow equals the number of imported cargoes. We first present a link flow formulation for the maximum flow problem:

Solution. Let $\mathcal{N} := \{A, B, C, D, E\}$ be the set of nodes and $\mathcal{A}$ be the set of arcs:

$$\mathcal{A} := \{(A, B), (A, D), (B, C), (B, D), (C, E), (D, C), (D, E)\}.$$

Let f_{ij} be the decision variables representing the flow on arcs $(i,j) \in \mathcal{A}$. The model is as follows:

$$\max f_{AB} + f_{AD} \text{ (maximize the total net outflow from node A)}$$

subject to

$$f_{AB} = f_{BC} + f_{BD}$$
(flow conservation at node B)

$$f_{BC} + f_{DC} = f_{CE}$$
(flow conservation at node C)

$$f_{AD} + f_{BD} = f_{DC} + f_{DE}$$
(flow conservation at node D)

$$f_{AB} \leq 12, f_{AD} \leq 4, \quad f_{BC} \leq 6, f_{BD} \leq 2$$
$$f_{CE} \leq 5, f_{DC} \leq 23, \quad f_{DE} \leq 7$$
$$f_{ij} \geq 0, \forall (i,j) \in \mathcal{A}.$$

Note that in the above link flow formulation, the flow conservation equations at nodes B, C, and D ensure that the total inflow to node E, $f_{CE} + f_{DE}$, equals the total outflow from node A, $f_{AB} + f_{AD}$.

Example 16.4: In Example 16.3, develop a link flow linear optimization model to find the maximum flow of crude oil per hour from node B to node E.

Solution. Let $\mathcal{N} := \{A, B, C, D, E\}$ be the set of nodes and $\mathcal{A}$ be the set of arcs:

$$\mathcal{A} := \{(A, B), (A, D), (B, C), (B, D), (C, E), (D, C), (D, E)\}.$$

Let f_{ij} be the decision variables representing the flow on arcs $(i,j) \in \mathcal{A}$. The model is as follows:

$$\max f_{BC} + f_{BD} - f_{AB} \text{(maximize the total net outflow from node B)}$$

subject to

$$f_{AB} + f_{AD} = 0 \text{ (flow conservation at node A)}$$
$$f_{BC} + f_{DC} = f_{CE} \text{ (flow conservation at node C)}$$
$$f_{AD} + f_{BD} = f_{DC} + f_{DE} \text{ (flow conservation at node D)}$$
$$f_{AB} \leq 12, f_{AD} \leq 4, f_{BC} \leq 6, f_{BD} \leq 2$$
$$f_{CE} \leq 5, f_{DC} \leq 23, f_{DE} \leq 7$$
$$f_{ij} \geq 0, \forall (i,j) \in \mathcal{A}.$$

Note that as we seek the maximum flow *from B to E*, it does not make sense to have flow to B. Therefore, it is correct to say that $f_{AB} = 0$ and hence f_{AB} can be removed from the model.

We can also use path flows as the decision variables in the maximum flow problem. To this end, we first need to enumerate all paths from the origin node to the destination node.

Example 16.5: In Example 16.3, develop a path flow linear optimization model to find the maximum flow of crude oil per hour from node *A* to node *E*.

Solution. From node *A* to node *E*, we have the following paths:

Path 1: $A \to B \to C \to E$

Path 2: $A \to B \to D \to E$

Path 3: $A \to B \to D \to C \to E$

Path 4: $A \to D \to E$

Path 5: $A \to D \to C \to E$.

Let f_i be the decision variables representing the flow on path $i \in \{1, 2, 3, 4, 5\}$. The path flow formulation is as follows:

$$\max \sum_{i=1}^{5} f_i \text{(maximize the total flow from node A to node E)}$$

subject to

$$f_1 + f_2 + f_3 \leq 12 \text{ (capacity on arc (A, B))}$$
$$f_4 + f_5 \leq 4 \text{ (capacity on arc (A, D))}$$
$$f_1 \leq 6 \text{ (capacity on arc (B, C))}$$
$$f_2 + f_3 \leq 2 \text{ (capacity on arc (B, D))}$$
$$f_1 + f_3 + f_5 \leq 5 \text{ (capacity on arc (C, E))}$$
$$f_3 + f_5 \leq 23 \text{ (capacity on arc (D , C))}$$
$$f_2 + f_4 \leq 7 \text{ (capacity on arc (D, E))}$$
$$f_i \geq 0, i = 1, 2, 3, 4, 5.$$

It should be noted that in many real problems, the sizes of the networks are very large, i.e., the networks contain a large number of nodes and links. Enumerating all of the paths from one node to another is impossible.

Example 16.6: In Example 16.3, develop a path flow linear optimization model to find the maximum flow of crude oil per hour from node B to node E.

Solution. From node B to node E, we have the following paths:

Path 1 : $B \to C \to E$

Path 2 : $B \to D \to E$

Path 3 : $B \to D \to C \to E$.

Let f_i be the decision variable representing the flow on path. The model is as follows:

$$\max \sum_{i=1}^{3} f_i$$

subject to

$$f_1 \le 6 \text{ (capacity on arc (B, C))}$$
$$f_2 + f_3 \le 2 \text{ (capacity on arc (B, D))}$$
$$f_1 + f_3 \le 5 \text{ (capacity on arc (C, E))}$$
$$f_3 \le 23 \text{ (capacity on arc (D, C))}$$
$$f_2 \le 7 \text{ (capacity on arc (D, E))}$$
$$f_i \ge 0, i = 1, 2, 3.$$

Example 16.7: Similar to Example 16.3, develop a link flow linear optimization model to find the maximum flow of crude oil per hour from node A to node E in the network shown in Figure 16.2.

Solution. Let $\mathcal{N} := \{A, B, C, D, E\}$ be the set of nodes and $\mathcal{A}$ be the set of arcs:

$$\mathcal{A} := \{(A, B), (A, D), (B, C), (B, D), (C, A), (C, E), (D, C), (D, E), (E, C)\}.$$

Let f_{ij} be the decision variables representing the flow on arcs $(i, j) \in \mathcal{A}$. The model is as follows:

$$\max f_{AB} + f_{AD} - f_{CA}$$

subject to

$$f_{AB} = f_{BC} + f_{BD} \text{ (flow conservation at node B)}$$
$$f_{BC} + f_{DC} + f_{EC} = f_{CA} + f_{CE} \text{ (flow conservation at node C)}$$
$$f_{AD} + f_{BD} = f_{DC} + f_{DE} \text{ (flow conservation at node D)}$$
$$f_{AB} \le 12, f_{AD} \le 4, \quad f_{BC} \le 6, f_{BD} \le 2$$
$$f_{CA} \le 1, f_{CE} \le 5, \quad f_{DC} \le 23, f_{DE} \le 7, f_{EC} \le 3$$
$$f_{ij} \ge 0, \forall (i,j) \in \mathcal{A}.$$

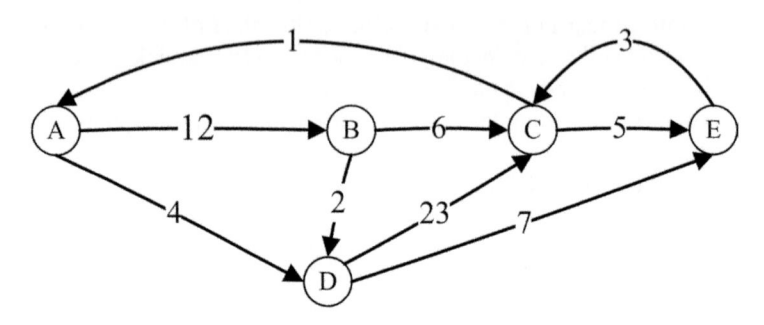

Figure 16.2   A crude oil pipeline network

Advanced linear optimization 237

Example 16.8: This is a minimum cost flow problem. Consider a multimodal transportation problem. Multimodal transportation means transportation with more than one transportation mode, e.g., a combination of trains and trucks. Figure 16.3 shows a multimodal transportation network including rail services and truck services. There are four roads for trucks and three railway lines. The maximum number of cargoes that can be transported on a railway line and on a road is 10 and 5 units, respectively. The cargoes can only be transported in the directions of the arrows. The numbers on the arrows represent the length of the railway lines and roads. The transportation cost per unit distance is 3 for trucks and 2 for trains. Node A represents the port of Los Angeles and node E is Chicago. Develop a linear optimization model to find the minimum cost for transporting 12 units of cargoes from A to E.

We can see that both the maximum flow problem and the minimum cost flow problem involve the determination of flow on each arc. Hence, their decision variables can be the same.

Solution. Let $\mathcal{N} := \{A, B, C, D, E\}$ be the set of nodes and $\mathcal{A}$ be the set of arcs:

$$\mathcal{A} := \{(A, B), (A, D), (B, C), (B, D), (C, E), (D, C), (D, E)\}.$$

Let f_{ij} be the decision variables representing the flow on arcs $(i, j) \in \mathcal{A}$. The model is as follows:

$$\min 2 \times (6f_{AB} + 5f_{CE} + 8f_{DC}) + 3 \times (4f_{AD} + 6f_{BC} + 2f_{BD} + 7f_{DE})$$

subject to

$$f_{AB} + f_{AD} = 12 \text{ (flow conservation at node A)}$$
$$f_{AB} = f_{BC} + f_{BD} \text{ (flow conservation at node B)}$$
$$f_{BC} + f_{DC} = f_{CE} \text{ (flow conservation at node C)}$$
$$f_{AD} + f_{BD} = f_{DC} + f_{DE} \text{ (flow conservation at node D)}$$
$$f_{CE} + f_{DE} = 12 \text{ (flow conservation at node E)}$$

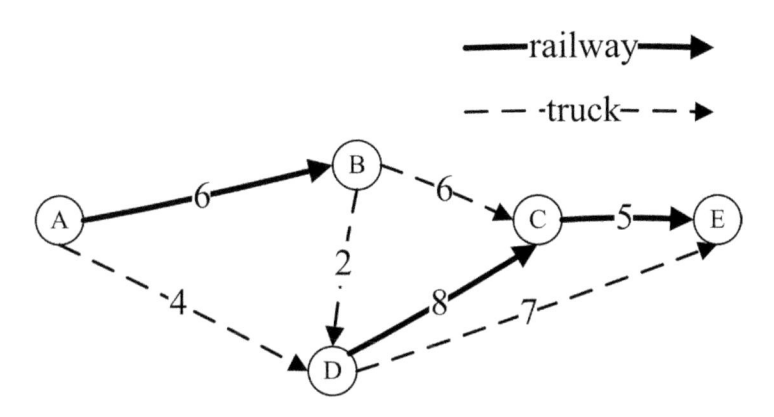

Figure 16.3   A multimodal transportation network

$$f_{AB} \leq 10, f_{CE} \leq 10, \quad f_{DC} \leq 10$$
$$f_{AD} \leq 5, f_{BC} \leq 5, \quad f_{BD} \leq 5, f_{DE} \leq 5$$
$$f_{ij} \geq 0, \forall (i,j) \in \mathcal{A}.$$

Example 16.9: In Example 16.8, develop a linear optimization model to find the minimum cost for transporting 12 units of cargoes from node B to node E.

Solution. Let $\mathcal{N} := \{A, B, C, D, E\}$ be the set of nodes and $\mathcal{A}$ be the set of arcs:

$$\mathcal{A} := \{(A,B), (A,D), (B,C), (B,D), (C,E), (D,C), (D,E)\}.$$

Let f_{ij} be the decision variables representing the flow on arcs $(i,j) \in \mathcal{A}$. The model is as follows:

$$\min 2 \times (6f_{AB} + 5f_{CE} + 8f_{DC}) + 3 \times (4f_{AD} + 6f_{BC} + 2f_{BD} + 7f_{DE})$$

subject to

$$f_{AB} + f_{AD} = 0 \text{ (flow conservation at node A)}$$
$$f_{BC} + f_{BD} - f_{AB} = 12 \text{(flow conservation at node B)}$$
$$f_{BC} + f_{DC} = f_{CE} \text{ (flow conservation at node C)}$$
$$f_{AD} + f_{BD} = f_{DC} + f_{DE} \text{ (flow conservation at node D)}$$
$$f_{CE} + f_{DE} = 12 \text{(flow conservation at node E)}$$
$$f_{AB} \leq 10, f_{CE} \leq 10, \quad f_{DC} \leq 10,$$
$$f_{AD} \leq 5, f_{BC} \leq 5, \quad f_{BD} \leq 5, f_{DE} \leq 5$$
$$f_{ij} \geq 0, \forall (i,j) \in \mathcal{A}.$$

Example 16.10: In Example 16.8, the price of a unit of cargo in Los Angeles is 100, and 115 in Chicago (the price has the same unit as the transportation cost). A company buys cargoes in Los Angeles, transports them to Chicago, and sells them to make a profit. Develop a linear optimization model to find the optimal number of cargoes to purchase in Los Angeles that maximizes the profit.

Solution. Let $\mathcal{N} := \{A, B, C, D, E\}$ be the set of nodes and $\mathcal{A}$ be the set of arcs:

$$\mathcal{A} := \{(A,B), (A,D), (B,C), (B,D), (C,E), (D,C), (D,E)\}.$$

Let f_{ij} be the decision variables representing the flow on arcs $(i,j) \in \mathcal{A}$. Denote by x the units of cargoes to purchase in Los Angeles. The model is as follows:

$$\max(115 - 100)x - \left[2 \times (6f_{AB} + 5f_{CE} + 8f_{DC}) + 3 \times (4f_{AD} + 6f_{BC} + 2f_{BD} + 7f_{DE})\right]$$

subject to

$$f_{AB} + f_{AD} = x \text{ (flow conservation at node A)}$$
$$f_{AB} = f_{BC} + f_{BD} \text{ (flow conservation at node B)}$$
$$f_{BC} + f_{DC} = f_{CE} \text{ (flow conservation at node C)}$$
$$f_{AD} + f_{BD} = f_{DC} + f_{DE} \text{ (flow conservation at node D)}$$

$$f_{CE} + f_{DE} = x \text{ (flow conservation at node E)}$$
$$f_{AB} \leq 10, f_{CE} \leq 10, \quad f_{DC} \leq 10$$
$$f_{AD} \leq 5, f_{BC} \leq 5, \quad f_{BD} \leq 5, f_{DE} \leq 5$$
$$x \geq 0$$
$$f_{ij} \geq 0, \forall (i,j) \in \mathcal{A}.$$

Example 16.11: This is a shortest path problem. Figure 16.4 shows a road transportation network. The numbers on the arrows represent the travel time of each road. Develop a linear optimization model to calculate the shortest travel time from node A to node E. You are not allowed to enumerate all paths from A to E.

Solution. To address the problem, consider transporting one unit of cargo from A to E. The capacity of each road is infinity, and the unit transportation cost on a road is equal to its travel time. Then, in the optimal solution to the corresponding minimum cost flow problem, the cargo will travel on the shortest path. Let $\mathcal{N} := \{A, B, C, D, E, F, G\}$ be the set of nodes and $\mathcal{A}$ be the set of arcs:

$$\mathcal{A} := \{(A, B), (A, D), (A, F), (B, C), (B, D), (B, F), (B, G),$$
$$(C, E), (C, G), (D, C), (D, E), (F, G), (G, E)\}.$$

Let f_{ij} be the decision variables representing the flow on arcs $(i,j) \in \mathcal{A}$. The model is as follows:

$$\min 12 f_{AB} + 4 f_{AD} + 3 f_{AF} + 6 f_{BC} + 2 f_{BD} + f_{BF} + 11 f_{BG} +$$
$$5 f_{CE} + 10 f_{CG} + 23 f_{DC} + 7 f_{DE} + 8 f_{FG} + 9 f_{GE}$$

subject to

$$f_{AB} + f_{AD} + f_{AF} = 1 \text{ (flow conservation at A)}$$
$$f_{AB} = f_{BC} + f_{BD} + f_{BF} + f_{BG} \text{ (flow conservation at B)}$$
$$f_{BC} + f_{DC} = f_{CE} + f_{CG} \text{ (flow conservation at C)}$$

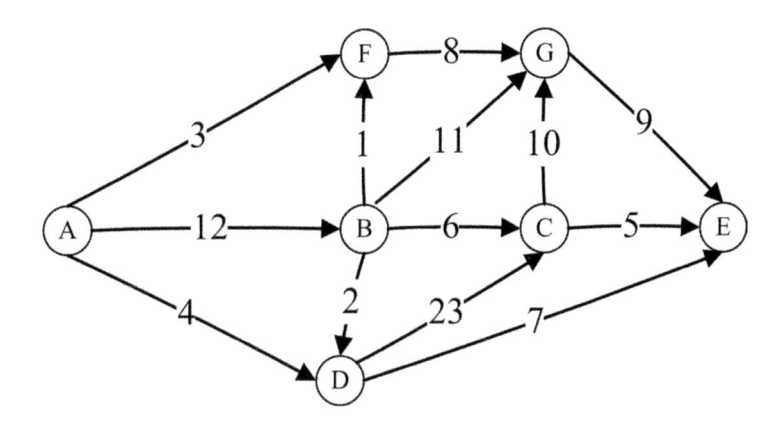

Figure 16.4   A road transportation network

$f_{AD} + f_{BD} = f_{DC} + f_{DE}$ (flow conservation at D)
$f_{CE} + f_{DE} + f_{GE} = 1$ (flow conservation at E)
$f_{AF} + f_{BF} = f_{FG}$ (flow conservation at F)
$f_{BG} + f_{CG} + f_{FG} = f_{GE}$ (flow conservation at G)
$f_{ij} \geq 0, \forall (i,j) \in \mathcal{A}.$

The arcs with positive flows form the shortest path.

It is possible that if there is more than one shortest path in Example 16.11, a minimum cost flow solution may contain several paths. However, all of them are the shortest paths, and it is not difficult to identify one of them. See the example in Figure 16.5.

In the minimum cost flow problem in Example 16.8, we have only one type of cargo (commodity). If we have several commodities that are differentiated by their origins and destinations, we then have the multi-commodity flow problem. The models are similar to that of the minimum cost flow problem, except that we need to differentiate the commodities.

Example 16.12: Consider Example 16.8. Node A is the port of Los Angeles, node D is Atlanta, and node E is Chicago. A total of 12 units of clothes are to be transported from A to E and 2 units of machines are to be transported from D to E. Develop a linear optimization model to find the minimum cost of transporting all the cargoes.

Solution. Let $\mathcal{N} := \{A, B, C, D, E\}$ be the set of nodes and $\mathcal{A}$ be the set of arcs:

$$\mathcal{A} := \{(A,B), (A,D), (B,C), (B,D), (C,E), (D,C), (D,E)\}.$$

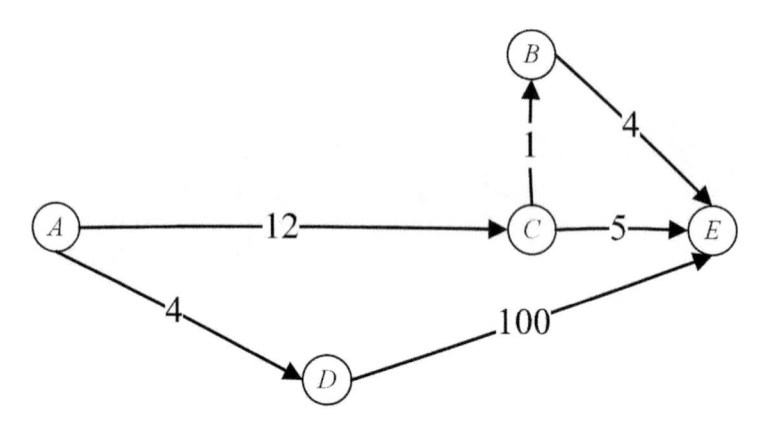

Figure 16.5 A road transportation network with more than one shortest path from A to E

Advanced linear optimization 241

Let f_{ij}^{AE} be the decision variables representing the flow of commodities with origin A and destination E (i.e., clothes) on arcs $(i,j) \in \mathcal{A}$. Let f_{ij}^{DE} be the decision variables representing the flow of commodities with origin D and destination E (i.e., machines) on arcs $(i,j) \in \mathcal{A}$. The model is as follows:

$$\min 2 \times [6(f_{AB}^{AE} + f_{AB}^{DE}) + 5(f_{CE}^{AE} + f_{CE}^{DE}) + 8(f_{DC}^{AE} + f_{DC}^{DE})] +$$
$$3 \times [4(f_{AD}^{AE} + f_{AD}^{DE}) + 6(f_{BC}^{AE} + f_{BC}^{DE}) + 2(f_{BD}^{AE} + f_{BD}^{DE}) + 7(f_{DE}^{AE} + f_{DE}^{DE})]$$

subject to

Flow conservation equations:
$$f_{AB}^{AE} + f_{AD}^{AE} = 12$$
$$f_{AB}^{AE} = f_{BC}^{AE} + f_{BD}^{AE}$$
$$f_{BC}^{AE} + f_{DC}^{AE} = f_{CE}^{AE}$$
$$f_{AD}^{AE} + f_{BD}^{AE} = f_{DC}^{AE} + f_{DE}^{AE}$$
$$f_{CE}^{AE} + f_{DE}^{AE} = 12$$
$$f_{AB}^{DE} + f_{AD}^{DE} = 0$$
$$f_{AB}^{DE} = f_{BC}^{DE} + f_{BD}^{DE}$$
$$f_{BC}^{DE} + f_{DC}^{DE} = f_{CE}^{DE}$$
$$(f_{DC}^{DE} + f_{DE}^{DE}) - (f_{AD}^{DE} + f_{BD}^{DE}) = 2$$
$$f_{CE}^{DE} + f_{DE}^{DE} = 2$$

Capacity constraints:
$$f_{AB}^{AE} + f_{AB}^{DE} \le 10, f_{CE}^{AE} + f_{CE}^{DE} \le 10, f_{DC}^{AE} + f_{DC}^{DE} \le 10$$
$$f_{AD}^{AE} + f_{AD}^{DE} \le 5, f_{BC}^{AE} + f_{BC}^{DE} \le 5, f_{BD}^{AE} + f_{BD}^{DE} \le 5, f_{DE}^{AE} + f_{DE}^{DE} \le 5$$

Nonnegativity constraints:
$$f_{ij}^{AE} \ge 0, f_{ij}^{DE} \ge 0, \forall (i,j) \in \mathcal{A}.$$

Example 16.13: Figure 16.6 shows a liner container shipping network. The circles represent ports. There are three routes r_1, r_2, and r_3. The capacity of a ship deployed on route r_1 is E_1 (TEUs). E_2 and E_3 have similar meanings. The volumes of containers between different OD pairs are shown in the figure. For example, is the demand (TEUs/week) from port 1 to port 2. The profit for transporting one TEU from port 1 to port 2 is ($/TEU) and have similar meanings. Assume that container handling costs are 0. Develop a linear optimization model for finding the maximum profit that can be gained from transporting containers.

It should be noted that container liner shipping networks are usually very sparse. For example, if we assume that there is at most one arc from one node to another, then a network with 100 nodes has at most 100 ×99 = 9 900links. If the number of

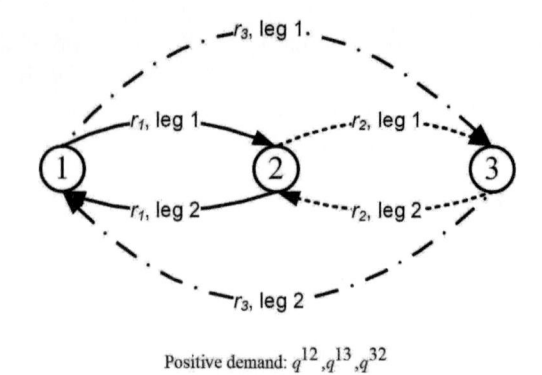

Figure 16.6 A container liner shipping network with direct delivery and transshipment

links is much smaller than this maximum number, then we say that the network is sparse. A sparse network does not have many paths from one node to another, and hence we can enumerate all paths[*].

Solution. The set of OD pairs is $\mathcal{W} = \{(1,2),(1,3),(3,2)\}$. Let $\mathcal{H}^{od}$ be the set of itineraries (paths) for OD pair $(o,d) \in \mathcal{W}$. To simplify the notation, we use $<r, i>$ to represent leg i of route r. Then $\mathcal{H}^{12}$ consists of the following:

$$h_1 :<1,1>$$
$$h_2 :<3,1> \to <2,2>$$

$\mathcal{H}^{13}$ consists of:

$$h_3 :<3,1>$$
$$h_4 :<1,1> \to <2,1>$$

$\mathcal{H}^{32}$ consists of:

$$h_5 :<2,2>$$
$$h_6 :<3,2> \to <1,1>.$$

Define $\mathcal{H} := \cup_{(o,d) \in \mathcal{W}} \mathcal{H}^{od}$. Let y_h be the decision variables representing the flow on itinerary $h \in \mathcal{H}$ (TEUs/week). The model is as follows:

$$\max \sum_{(o,d) \in \mathcal{W}} g^{od} \sum_{h \in \mathcal{H}^{od}} y_h \text{ (maximize the total profit)}$$

subject to

$$\sum_{h \in \mathcal{H}^{od}} y_h \leq q^{od}, (o,d) \in \mathcal{W} \text{ (limited demand)}$$
$$y_1 + y_4 + y_6 \leq E_1 \text{ (capacity on leg } <1,1>)$$

---

[*]Another advantage of using path flow formulations over link flow formulations is that it is easy for the former to incorporate business considerations (cf Reference 1).

$$y_4 \leq E_2 \text{ (capacity on leg} <2,1>)$$
$$y_2 + y_5 \leq E_2 \text{ (capacity on leg} <2,2>)$$
$$y_2 + y_3 \leq E_3 \text{ (capacity on leg} <3,1>)$$
$$y_6 \leq E_3 \text{ (capacity on leg} <3,2>)$$
$$y_h \geq 0, h \in \mathcal{H}.$$

Example 16.14: Using the notation similar to Example 16.13, develop a linear optimization model to find the maximum profit that can be gained from transporting containers in Figure 16.7.

Solution. The set of OD pairs is $\mathcal{W} = \{(1,2), (1,3), (2,3)\}$. Let $\mathcal{H}^{od}$ be the set of itineraries for OD pair $(o,d) \in \mathcal{W}$. Then $\mathcal{H}^{12}$ consists of the following:

$h_1 :< 1, 1 >$
$h_2 :< 1, 4 >$

$\mathcal{H}^{13}$ consists of:

$h_3 :< 1, 1 > \to < 1, 2 >$
$h_4 :< 1, 4 > \to < 1, 2 >$

$\mathcal{H}^{23}$ consists of the following:

$h_5 :< 1, 2 >$.

Define $\mathcal{H} := \cup_{(o,d) \in \mathcal{W}} \mathcal{H}^{od}$. Let y_h be the decision variables representing the flow on itinerary $h \in \mathcal{H}$. The model is as follows:

$$\max \sum_{(o,d) \in \mathcal{W}} g^{od} \sum_{h \in \mathcal{H}^{od}} y_h$$

subject to

$$\sum_{h \in \mathcal{H}^{od}} y_h \leq q^{od}, (o,d) \in \mathcal{W}$$
$$y_1 + y_3 \leq E_1 \text{ (capacity on leg} <1,1>)$$
$$y_3 + y_4 + y_5 \leq E_1 \text{ (capacity on leg} <1,2>)$$
$$y_2 + y_4 \leq E_1 \text{ (capacity on leg} <1,4>)$$
$$y_h \geq 0, h \in \mathcal{H}.$$

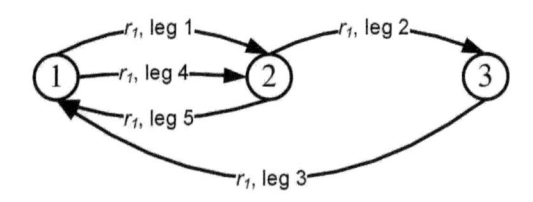

Positive demand: q^{12}, q^{13}, q^{23}

Figure 16.7 A route with two identical legs

Example 16.15. Similar to Example 16.13, develop a linear optimization model to find the maximum profit that can be gained from transporting containers in Figure 16.8.

Solution. The set of OD pairs is $\mathcal{W} = \{(1,4),(1,6),(2,5),(2,6),(3,4),(3,5),(3,6)\}$. Let $\mathcal{H}^{od}$ be the set of itineraries for OD pair $(o,d) \in \mathcal{W}$. Then $\mathcal{H}^{14}$ consists of the following:

$h_1 :< 1, 1 > \to < 3, 1 >$

$\mathcal{H}^{16}$ consists of the following:

$h_2 :< 1, 1 > \to < 3, 1 > \to < 5, 1 >$

$\mathcal{H}^{25}$ consists of the following:

$h_3 :< 2, 1 > \to < 3, 1 > \to < 4, 1 >$

$\mathcal{H}^{26}$ consists of the following:

$h_4 :< 2, 1 > \to < 3, 1 > \to < 5, 1 >$

$\mathcal{H}^{34}$ consists of the following:

$h_5 :< 3, 1 >$

$\mathcal{H}^{35}$ consists of the following:

$h_6 :< 3, 1 > \to < 4, 1 >$

$\mathcal{H}^{36}$ consists of the following:

$h_7 :< 3, 1 > \to < 5, 1 >.$

Define $\mathcal{H} := \cup_{(o,d) \in \mathcal{W}} \mathcal{H}^{od}$. Let y_h be the decision variable representing the flow on itinerary $h \in \mathcal{H}$. The model is as follows:

$$\max \sum_{(o,d) \in \mathcal{W}} g^{od} \sum_{h \in \mathcal{H}^{od}} y_h$$

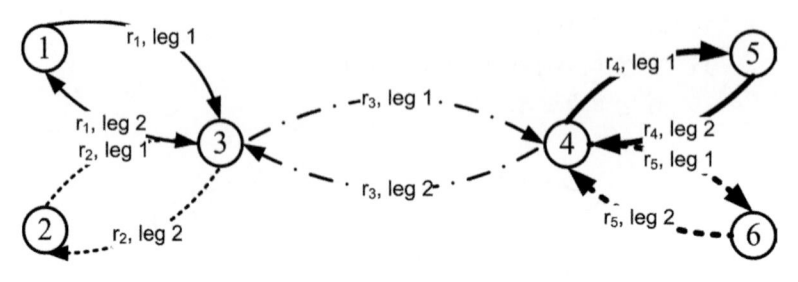

Positive demand: $q^{14}, q^{16}, q^{25}, q^{26}, q^{34}, q^{35}, q^{36}$

Figure 16.8   A hub-and-spoke liner shipping network

subject to

$$\sum_{h \in \mathcal{H}^{od}} y_h \leq q_{od}, (o, d) \in \mathcal{W}$$
$$y_1 + y_2 \leq E_1 \text{(capacity on leg} < 1, 1 >)$$
$$y_3 + y_4 \leq E_2 \text{(capacity on leg} < 2, 1 >)$$
$$y_1 + y_2 + y_3 + y_4 + y_5 + y_6 + y_7 \leq E_3 \text{(capacity on leg} < 3, 1 >)$$
$$y_3 + y_6 \leq E_4 \text{(capacity on leg} < 4, 1 >)$$
$$y_2 + y_4 + y_7 \leq E_5 \text{(capacity on leg} < 5, 1 >)$$
$$y_h \leq 0, h \in \mathcal{H}.$$

We encapsulate this sub-section by stating the problems discussed in mathematical languages. We are given a network denoted by $g = (\mathcal{N}, \mathcal{A})$, where $\mathcal{N}$ is the set of nodes and $\mathcal{A} \subset \mathcal{N} \times \mathcal{N}$ is the set of arcs. (i) In the maximum flow problem, each arc $(i,j) \in \mathcal{A}$ has a capacity denoted by $s_{i,j} \geq 0$, and the objective is to find the maximum flow from node $n^{source} \in \mathcal{N}$ to node $n^{sink} \in \mathcal{N}/\{n^{source}\}$. (ii) In the shortest path problem, each arc $(i,j) \in \mathcal{A}$ has a unit cost denoted by $c_{ij} \geq 0$, and the objective is to find the minimum cost path from node $n^{source} \in \mathcal{N}$ to node $n^{sink} \in \mathcal{N}/\{n^{source}\}$. (iii) In the minimum cost flow problem, each arc $(i,j) \in \mathcal{A}$ has a unit cost denoted by $c_{ij} \geq 0$ and a capacity denoted by $s_{ij} \geq 0$, and the objective is to find the minimum total cost for transporting units of cargoes from node $n^{source} \in \mathcal{N}$ to node $n^{sink} \in \mathcal{N}/\{n^{source}\}$. (iv) In the multi-commodity flow problem[†], each arc $(i,j) \in \mathcal{A}$ has a unit cost denoted by $c_{ij} \geq 0$ and a capacity denoted by $s_{ij} \geq 0$. There are a set of OD pairs denoted by $\mathcal{W} \subset \mathcal{N} \times \mathcal{N}$. Cargoes between different OD pairs are different. Between OD pair $(o, d) \in \mathcal{W}$, q^{od} units of cargoes must be transported. The objective is to find the minimum total cost for transporting all cargoes between all the OD pairs in $\mathcal{W}$ [‡].

16.2 Dummy nodes and links

We have studied the maximum flow problem, the shortest path problem, the minimum cost flow problem, and the multi-commodity flow problem. Some new problems look different but can be transformed into these known problems after some modifications. As a result, we can use approaches that we are familiar with to model the new problems.

[†] More linear optimization formulations for the multi-commodity flow problem can be found in Reference 2.
[‡] In fact, the maximum flow problem and the shortest path problem are special cases of the minimum cost flow problem; the minimum cost flow problem is a special case of the multi-commodity flow problem.

Example 16.16: Reconsider Example 16.1. How do you add dummy[§] nodes and links to transform the problem to a minimum cost flow problem?

Solution. See Figure 16.9. The nodes W1–W3 represent the three warehouses, and S1–S5 represent the five supermarkets. The links from a warehouse to a supermarket represent the transportation of products. The unit cost of such a link is equal to the transportation cost, and the capacity is infinite. We further add a dummy source node that is connected to each warehouse. A dummy arc from the dummy source node to a warehouse has no cost and a capacity that is equal to the number of products available in the warehouse. A dummy sink node is added and is connected to each supermarket. A dummy arc from a supermarket to the dummy sink node has no cost and a capacity that is equal to the number of products required by the supermarket. The problem now becomes: how to transport $80 + 90 + 70 + 60 + 50 = 350$ products from the dummy source node to the dummy sink node at the lowest total cost.

Example 16.17: Reconsider Example 16.12. How do you add dummy nodes and links to transform the problem into a minimum cost flow problem?

Solution. See Figure 16.10. The transportation network is not changed: the unit cost and capacity of rail and truck links are the same as their physical meanings. A dummy source node is added and is connected to nodes A and D by two dummy arcs. The dummy arc from the dummy source node to node A has no cost and a capacity of 12; the dummy arc from the dummy source node to node D has no cost and a capacity of 2. A dummy sink node is added and is connected by a dummy arc from node E. The dummy arc has no cost and infinite capacity. The problem now becomes: how to

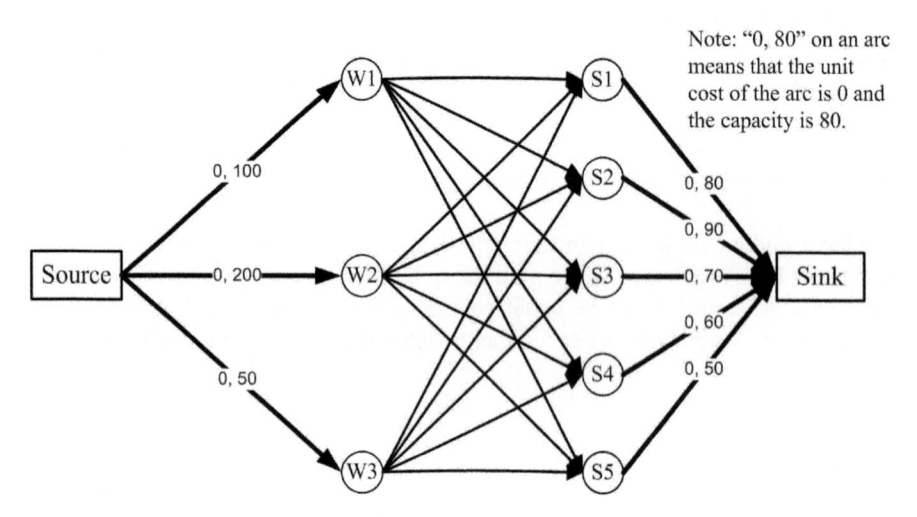

Figure 16.9   A transportation problem with dummy nodes and links

---

[§]The word "dummy" means something that does not exist or is imaginary. For example, we can almost always add a dummy arc to a network while imposing that its capacity is 0.

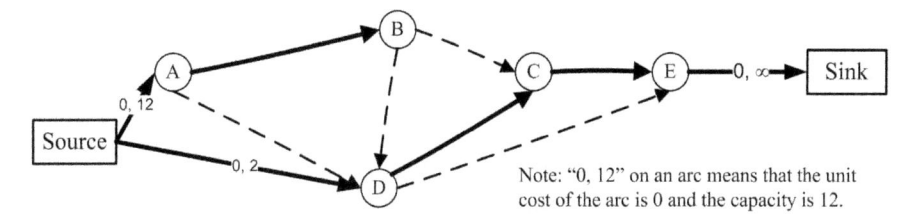

Figure 16.10   A multi-commodity flow problem with only one destination

transport 12 + 2 = 14 units of cargoes from the dummy source node to the dummy sink node at minimum cost.

It should be noted that this example is a special multi-commodity flow problem in that all commodities have the same destination. Therefore, we can transform it into a minimum cost flow problem. If all commodities have the same origin, we can also transform the problem into a minimum cost flow problem. However, a general multi-commodity flow problem cannot be transformed into a minimum cost flow problem.

Because of trade imbalance, for example, China exports more cargoes to the United States than the imported cargoes from the United States, some ports (e.g., ports in the United States) have surplus empty containers, and some ports (e.g., ports in China) are in shortage of empty containers (these ports are called deficit ports). Hence, empty containers have to be repositioned from surplus ports to the deficit ports. Unlike laden containers, all empty containers can be considered identical.

Example 16.18: This is an empty container repositioning problem. Reconsider Example 16.13. Suppose that the remaining capacity of leg $<r,i>$ after carrying laden containers is $\tilde{E}_{ri}$. The remaining capacity is used for repositioning empty containers. The number of surplus empty containers at port $p \in \{1,2,3\}$ is denoted by $\tilde{q}_p$. Here $\tilde{q}_1 < 0$, $\tilde{q}_2 > 0$, and $\tilde{q}_3 > 0$, which means port 1 is a deficit port and ports 2 and 3 are surplus ports. Moreover, $\sum_{p=1}^{3} \tilde{q}_p = 0$. We need to check whether the network has sufficient capacity to reposition all the empty containers from ports 2 and 3 to port 1. How to transform the problem into a maximum flow problem?

Solution. Since all empty containers are identical, we can add a dummy source node, connected to each surplus port with an arc whose capacity equals the volume of the surplus empty containers at the port, and a dummy sink node, connected from each deficit port with an arc whose capacity equals the volume of the deficit empty containers at the port, as shown in Figure 16.11. Now the problem is transformed into a maximum flow problem: all empty containers can be repositioned if and only if the maximum flow from the source to the sink equals $\tilde{q}_2 + \tilde{q}_3$.

Example 16.19: Reconsider Example 16.18. Suppose that $\tilde{q}_1 < 0$, $\tilde{q}_2 > 0$, $\tilde{q}_3 < 0$, and $\sum_{p=1}^{3} \tilde{q}_p = 0$. How do you add dummy nodes and links to transform the empty container repositioning problem into a maximum flow problem?

Solution. See Figure 16.12. Now the problem is transformed into a maximum flow problem: all empty containers can be repositioned if and only if the maximum flow from the source to the sink equals $\tilde{q}_2$.

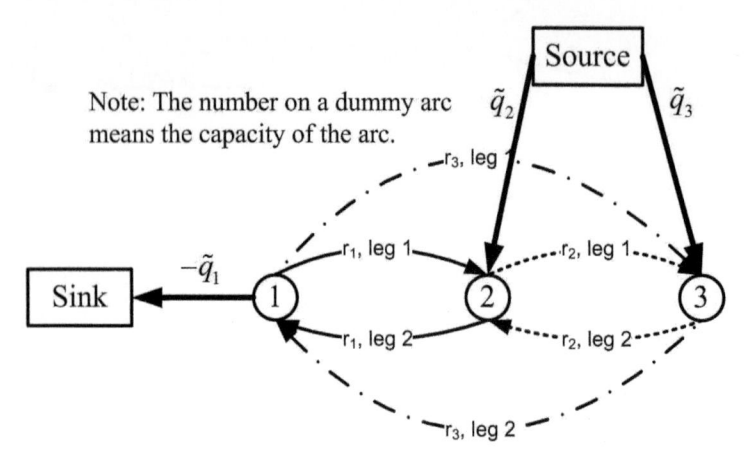

248 Machine learning and data analytics for maritime studies

Figure 16.11 A liner shipping network for empty container repositioning with dummy nodes and links

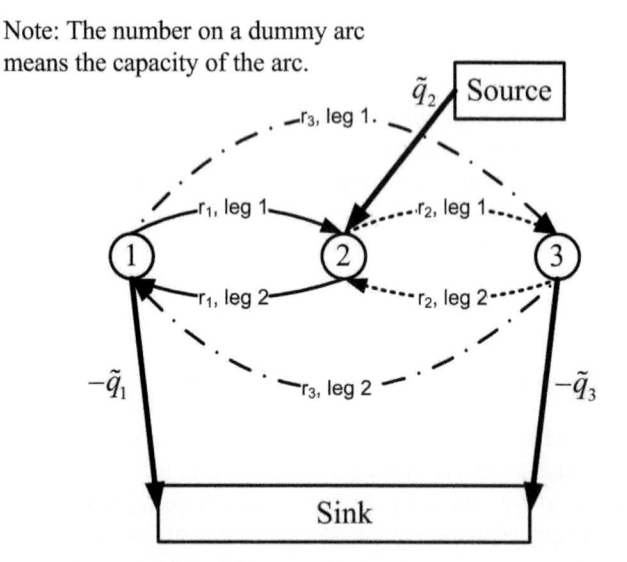

Figure 16.12 A liner shipping network for empty container repositioning with dummy nodes and links

16.3 Using linear formulations for nonlinear problems

Some seemingly nonlinear problems can be transformed into linear problems. For example, the constraint $|x + 2y| \leq 3$ is equivalent to the combination of $x + 2y \leq 3$ and $x + 2y \geq -3$; the constraint $\max\{x - 2y, 2x + z, x + y - z\} \leq z + 1$ is equivalent to the combination of $x - 2y \leq z + 1$, $2x + z \leq z + 1$, and $x + y - z \leq z + 1$.¶ Minimizing the objective function $x + 2y + \max\{x - y, -2x + 3y\}$ is equivalent to minimizing $x + 2y + u$, subject to $u \geq x - y$ and $u \geq -2x + 3y$. Transforming a nonlinear constraint or a nonlinear model to a linear one is called linearization of the constraint or model.

Example 16.20: We have a set of vectors Ω defined by all of the vectors (x, y, z) satisfying:

$$x + 2y \leq 6$$
$$x - y - z \geq 4$$
$$3x + 2z \leq 5$$
$$0 \leq x \leq 5$$
$$0 \leq y \leq 5$$
$$0 \leq z \leq 5.$$

Formulate a linear optimization model to find the vector $(x, y, z) \in \Omega$ with the smallest L_1 distance to point $(1, 1, 1)$, whose definition is $|x - 1| + |y - 1| + |z - 1|$.

Solution. The problem is to minimize $|x - 1| + |y - 1| + |z - 1|$ such that $(x, y, z) \in \Omega$. We introduce three auxiliary variables u_1, u_2, and u_3. The model is then equivalent to minimizing $u_1 + u_2 + u_3$ such that $u_1 \geq x - 1$, $u_1 \geq 1 - x$, $u_2 \geq y - 1$, $u_2 \geq 1 - y$, $u_3 \geq z - 1$, $u_3 \geq 1 - z$, and $(x, y, z) \in \Omega$.

Some nonlinear optimization models can be approximated by linear optimization models. For example, if we minimize $x^2 + y$, and we know that $1 \leq x \leq 3$, then we might use the maximum of the three tangent lines at the points $(1, 1)$, $(2, 4)$, and $(3, 9)$ in Figure 16.13 to approximate x^2. The objective function will be approximated by $\max\{2x - 1, 4x - 4, 9x - 18\} + y$. It should be noted that we will slightly underestimate the objective function, and the optimal solution may also be slightly changed due to the change in the objective function. Nevertheless, if we use a large number of tangent lines, the error caused by approximation will be very small**.

If a constraint is nonlinear, then we might use secant lines to approximate the curve. For example, if we have a constraint $x^2 + y \leq 3$, and we know that $1 \leq x \leq 3$, then we might use the maximum of the two secant lines connecting $(1, 1)$ and $(2, 4)$, and $(2, 4)$ and $(3, 9)$ in Figure 16.14 to approximate x^2. The constraint will be approximated by $\max\{3x - 2, 5x - 6\} + y \leq 3$. It should be noted that some feasible solutions will be infeasible due to the approximation. If we had used tangent lines, then some infeasible solutions would be feasible and it would be possible that the optimal solution to the approximation model is infeasible to the original nonlinear model.

It may also be necessary to use tangent planes. For example, we minimize $x \ln x + y \ln y - (x + y) \ln(x + y)$ with $x > 0, y > 0$. The term $x \ln x$ is convex in x. If we let $z = x + y$ and replace the objective function by $x \ln x + y \ln y - z \ln z$, then this

---

¶The notation "max" followed by a set means the element in the set with the largest value. Some people use $\max(x, y)$ instead of $\max\{x, y\}$.
**An application can be found in Reference 3.

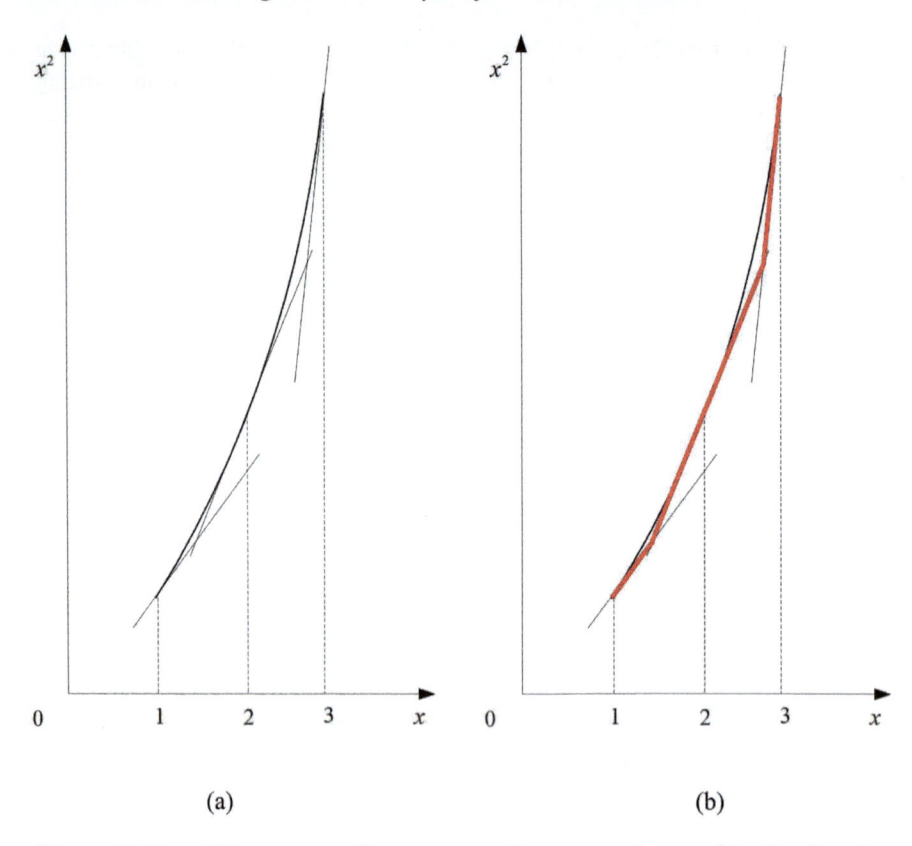

Figure 16.13   Using tangent lines to approximate a nonlinear objective function

function is no longer convex. However, we can easily check that the Hessian of $x \ln x + y \ln y - (x+y) \ln(x+y)$ is positive semi-definite. Hence, it is convex in x and y and can be linearized using tangent planes.

Finally, some nonlinear constraints cannot be approximated by linear constraints with high accuracy, e.g., $x_1 \geq \sin x_2$ and $x_1 \leq x_2^2$, because they are non-convex.

16.4 Practice

Example 16.21: Consider the following service with the port rotation:

$Shanghai(S, 1) \rightarrow HongKong(H, 2) \rightarrow Singapore(P, 3)$
$\rightarrow Rotterdam(R, 4) \rightarrow Shanghai(S, 1)$

Ships with a capacity of 6 000 (TEUs) are deployed to provide a weekly frequency. The container shipment demand (TEUs/week) is as follows: Shanghai to Rotterdam $q^{SR} = 3500$, Hong Kong to Rotterdam $q^{HR} = 2500$, Singapore to Rotterdam $q^{PR} = 1500$, and Rotterdam to Shanghai $q^{RS} = 2500$. The profit of transporting one

Advanced linear optimization 251

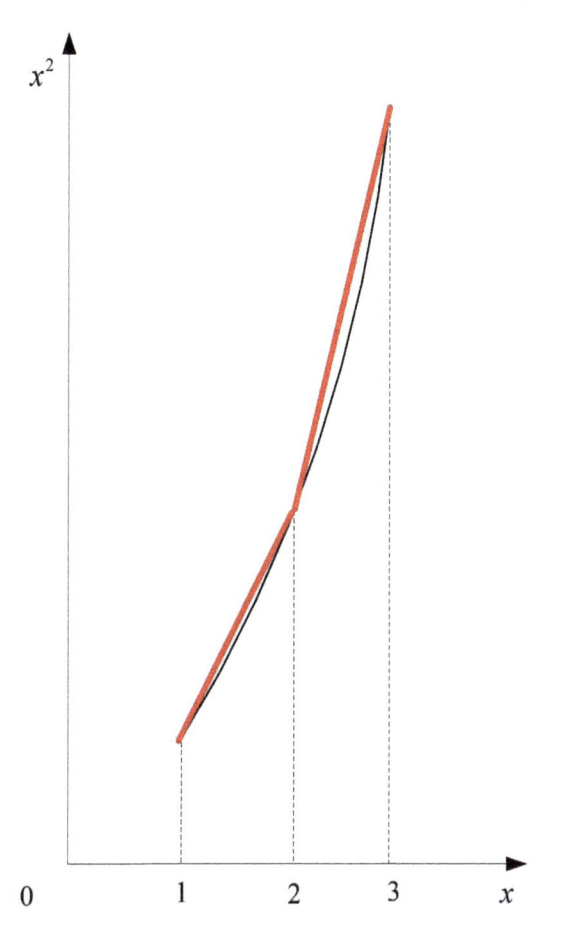

Figure 16.14   Using secant lines to approximate a nonlinear constraint

TEU ($/TEU) from Shanghai to Rotterdam $g^{SR} = 2500$, Hong Kong to Rotterdam $g^{HR} = 2300$, Singapore to Rotterdam $g^{PR} = 2000$, and Rotterdam to Shanghai $g^{RS} = 1500$. Develop an optimization model to evaluate the maximum profit ($/week) the company can make.

Solution. Let y^{SR}, y^{HR}, y^{PR}, and y^{RS} be the decision variables representing the volumes of containers transported from S to R, H to R, P to R, and R to S, respectively. The model is as follows:

$$\max 2500 y^{SR} + 2300 y^{HR} + 2000 y^{PR} + 1500 y^{RS}$$

subject to

$$y^{SR} \leq 6000 \text{ (ship capacity constraint on the leg from S to H)}$$
$$y^{SR} + y^{HR} \leq 6000 \text{ (ship capacity constraint on the leg from H to P)}$$

$$y^{SR} + y^{HR} + y^{PR} \leq 6000 \text{ (ship capacity constraint on the leg from P to R)}$$
$$y^{RS} \leq 6000 \text{ (ship capacity constraint on the leg from R to S)}$$
$$0 \leq y^{SR} \leq 3500$$
$$0 \leq y^{HR} \leq 2500$$
$$0 \leq y^{PR} \leq 1500$$
$$0 \leq y^{RS} \leq 2500.$$

Note: It is also correct if you only have the following constraints because the other three constraints are implied by them:

$$y^{SR} + y^{HR} + y^{PR} \leq 6000 \text{ (ship capacity constraint on the leg from P to R)}$$
$$0 \leq y^{SR} \leq 3500$$
$$0 \leq y^{HR} \leq 2500$$
$$0 \leq y^{PR} \leq 1500$$
$$0 \leq y^{RS} \leq 2500.$$

Example 16.22: Consider the following service with the port rotation:

Shanghai (S, 1) → Hong Kong (H, 2) → Singapore (P, 3)
→ Rotterdam (R, 4) → Shanghai (S, 1)

Ships with a capacity of 6000 (TEUs) are deployed to provide a weekly frequency. The container shipment demand (TEUs/week) is as follows: Shanghai to Rotterdam $q^{SR} = 3500$, Hong Kong to Rotterdam $q^{HR} = 2500$, Singapore to Rotterdam $q^{PR} = 1500$, Rotterdam to Shanghai $q^{RS} = 2500$, Rotterdam to Hong Kong $q^{RH} = 1500$, Rotterdam to Singapore $q^{RP} = 500$, and Shanghai to Singapore $q^{SP} = 1400$. The profit of transporting one TEU ($/TEU) from Shanghai to Rotterdam $g^{SR} = 2500$, Hong Kong to Rotterdam $g^{HR} = 2300$, Singapore to Rotterdam $g^{PR} = 2000$, Rotterdam to Shanghai $g^{RS} = 1500$, Rotterdam to Hong Kong $g^{RH} = 1500$, Rotterdam to Singapore $g^{RP} = 1500$, and Shanghai to Singapore $g^{SP} = 1000$. Develop an optimization model to evaluate the maximum profit ($/week) the company can make.

Solution. Let y^{SR}, y^{HR}, y^{PR}, y^{RS}, y^{RH}, y^{RP}, and y^{SP} be the decision variables representing the volumes of containers transported from S to R, H to R, P to R, R to S, R to H, R to P, and S to P, respectively. The model is as follows:

$$\max 2500 y^{SR} + 2300 y^{HR} + 2000 y^{PR} + 1500 y^{RS} + 1500 y^{RH} + 1500 y^{RP} + 1000 y^{SP}$$

subject to

$$y^{SR} + y^{RH} + y^{RP} + y^{SP} \leq 6000 \text{ (ship capacity constraint on the leg from S to H)}$$
$$y^{SR} + y^{HR} + y^{RP} + y^{SP} \leq 6000 \text{ (ship capacity constraint on the leg from H to P)}$$
$$y^{SR} + y^{HR} + y^{PR} \leq 6000 \text{ (ship capacity constraint on the leg from P to R)}$$

Advanced linear optimization 253

$$y^{RS} + y^{RH} + y^{RP} \leq 6000 \text{ (ship capacity constraint on the leg from R to S)}$$
$$0 \leq y^{SR} \leq 3500$$
$$0 \leq y^{HR} \leq 2500$$
$$0 \leq y^{PR} \leq 1500$$
$$0 \leq y^{RS} \leq 2500$$
$$0 \leq y^{RH} \leq 1500$$
$$0 \leq y^{RP} \leq 500$$
$$0 \leq y^{SP} \leq 1400.$$

Example 16.23: Consider the following service with the port rotation:

$$Shanghai(S, 1) \rightarrow HongKong(H, 2) \rightarrow Singapore(P, 3)$$
$$\rightarrow Rotterdam(R, 4) \rightarrow HongKong(H, 5) \rightarrow Shanghai(S, 1)$$

Ships with a capacity of 6000 (TEUs) are deployed to provide a weekly frequency. The container shipment demand (TEUs/week) is as follows: Shanghai to Rotterdam $q^{SR} = 3500$, Hong Kong to Rotterdam $q^{HR} = 2500$, Singapore to Rotterdam $q^{PR} = 1500$, Rotterdam to Shanghai $q^{RS} = 2500$, Rotterdam to Hong Kong $q^{RH} = 1500$, Rotterdam to Singapore $q^{RP} = 500$, and Shanghai to Singapore $q^{SP} = 1400$. The profit of transporting one TEU ($/TEU) from Shanghai to Rotterdam $g^{SR} = 2500$, Hong Kong to Rotterdam $g^{HR} = 2300$, Singapore to Rotterdam $g^{PR} = 2000$, Rotterdam to Shanghai $g^{RS} = 1500$, Rotterdam to Hong Kong $g^{RH} = 1500$, Rotterdam to Singapore $g^{RP} = 1500$, and Shanghai to Singapore $g^{SP} = 1000$.

(i) Develop an optimization model to evaluate the maximum profit ($/week) the company can make.

(ii) Write down two distinct feasible solutions, and calculate their objective function values.

(iii) Is this model infeasible? Why? Is this model unbounded? Why? Does this model have an optimal solution? Why?

Solution. (i) Let y^{SR}, y^{HR}, y^{PR}, y^{RS}, y^{RH}, y^{RP}, and y^{SP} be the decision variables representing the volumes of containers transported from S to R, H to R, P to R, R to S, R to H, R to P, and S to P, respectively. The model is as follows:

$$\max 2500y^{SR} + 2300y^{HR} + 2000y^{PR} + 1500y^{RS} + 1500y^{RH} + 1500y^{RP} + 1000y^{SP}$$

subject to

$$y^{SR} + y^{RP} + y^{SP} \leq 6000 \text{ (ship capacity constraint on the leg from S to H)}$$
$$y^{SR} + y^{HR} + y^{RP} + y^{SP} \leq 6000 \text{ (ship capacity constraint on the leg from H to P)}$$
$$y^{SR} + y^{HR} + y^{PR} \leq 6000 \text{ (ship capacity constraint on the leg from P to R)}$$
$$y^{RS} + y^{RH} + y^{RP} \leq 6000 \text{ (ship capacity constraint on the leg from R to H)}$$
$$y^{RS} + y^{RP} \leq 6000 \text{ (ship capacity constraint on the leg from H to S)}$$

$$0 \le y^{SR} \le 3500$$
$$0 \le y^{HR} \le 2500$$
$$0 \le y^{PR} \le 1500$$
$$0 \le y^{RS} \le 2500$$
$$0 \le y^{RH} \le 1500$$
$$0 \le y^{RP} \le 500$$
$$0 \le y^{SP} \le 1400.$$

(ii) One feasible solution is, and its objective function value is 0. Another feasible solution is, and its objective function value is 2 500. (iii) The model is feasible, because we have found two feasible solutions in (ii). The feasible set of the model is bounded, because letting, any feasible solution satisfies the following. Hence, the model is not unbounded. Since this is a linear optimization model that is feasible and not unbounded, it has an optimal solution.

Example 16.24: Reconsider the CC1 service with the port rotation:

Shanghai (S, 1) → Kwangyang (K, 2) → Pusan (P, 3) → Los Angeles (L, 4)
→ Oakland(O, 5) → Pusan (P, 6) → Kwangyang(K, 7) → Shanghai (S, 1)

Ships with a capacity of (TEUs) are deployed to provide a weekly frequency. The container shipment demand (TEUs/week) is as follows: Shanghai to Los Angeles q^{SL}, Pusan to Los Angeles q^{PL}, Oakland to Shanghai q^{OS}, and Shanghai to Oakland q^{SO}. The profit of transporting one TEU ($/TEU) from Shanghai to Los Angeles is g^{SL}, Pusan to Los Angeles g^{PL}, Oakland to Shanghai g^{OS}, and Shanghai to Oakland g^{SO}. The values of E, q's, and g's are all given.

(i) Develop an optimization model to evaluate the maximum profit ($/week) the company can make.

(ii) If q^{SL} is increased, how will the feasible set of the model in (i) change, and how will the optimal objective function value change?

(iii) If E is increased, how will the feasible set of the model in (i) change, and how will the optimal objective function value change?

(iv) If the values of g^{SL}, g^{PL}, g^{OS}, g^{SO} and are all doubled simultaneously, how will the feasible set of the model in (i) change, and how will the optimal objective function value change?

Solution. (i) Define the set of OD pairs $W = \{(S, L), (P, L), (O, S), (S, O)\}$. Let y^{od} be the decision variables representing the volumes of containers transported between OD pairs $(o, d) \in W$. The model is as follows:

$$\max \sum_{(o,d) \in W} g^{od} y^{od}$$

subject to

$$y^{SL} + y^{SO} \le E \text{ (ship capacity constraint on the leg from S to K)}$$
$$y^{SL} + y^{SO} \le E \text{ (ship capacity constraint on the leg from K to P)}$$

$$y^{SL} + y^{PL} + y^{SU} \leq E \text{ (ship capacity constraint on the leg from P to L)}$$
$$y^{SO} \leq E \text{ (ship capacity constraint on the leg from L to O)}$$
$$y^{OS} \leq E \text{ (ship capacity constraint on the leg from O to P)}$$
$$y^{OS} \leq E \text{ (ship capacity constraint on the leg from P to K)}$$
$$y^{OS} \leq E \text{ (ship capacity constraint on the leg from K to S)}$$
$$0 \leq y^{od} \leq q^{od}, (o, d) \in W.$$

Note: It is also correct if you only have the following constraints because the other five constraints are implied by them:

$$y^{SL} + y^{PL} + y^{SO} \leq E \text{ (ship capacity constraint on the leg from P to L)}$$
$$y^{OS} \leq E \text{ (ship capacity constraint on the leg from K to S)}$$
$$0 \leq y^{od} \leq q^{od}, (o, d) \in W.$$

(ii) The feasible set will be larger or unchanged; the optimal objective function value will be larger or unchanged.

(iii) The feasible set will be larger or unchanged; the optimal objective function value will be larger or unchanged.

(iv) The feasible set will be unchanged; the optimal objective function value will be doubled.

Example 16.25: (i) Write down an example of a linear optimization model with two variables and three constraints that is infeasible.

(ii) Write down an example of a linear optimization model with two variables and three constraints that is unbounded.

(iii) Write down an example of a linear optimization model with two variables and three constraints that has an infinite number of optimal solutions.

(iv) Write down an example of a linear optimization model with two variables and three constraints whose optimal objective function value is -123.

Solution. Note: The answers are not unique.
(i) Minimize subject to $x + y$ subject to $x \leq 0$, $x \geq 2$, $y \geq o$.
(ii) Minimize subject to $x + y$ subject to $x \leq 0$, $x \leq -2$, $y \geq 0$.
(iii) Minimize subject to $x + y$ subject to $x + y \geq 1$, $x \geq 0$, $y \geq 0$.
(iv) Minimize subject to $x + y$ subject to $x \geq -123$, $x \geq -200$, $y \geq 0$.

Example 16.26: We have the following linear optimization model:

$$\min x - y$$

subject to

$$x + y \geq 3$$
$$y \leq 2$$
$$x - 2y = 1.$$

(i) Transform the model to the canonical form that maximizes $c^T x$ subject to $Ax \leq b$ and $x \geq 0$.

(ii) Transform the model to the standard form that maximizes $c^T x$ subject to $Ax = b$ and $x \geq 0$.

Note: The answers are not unique.

Solution. Letting $x = u - v, u \geq 0, v \geq 0$ and $w = 2 - y, w \geq 0$, the model is transformed to

$$\min (u-v) - (2-w)$$

subject to

$$(u-v) + (2-w) \geq 3$$
$$(u-v) - 2(2-w) = 1$$
$$u \geq 0$$
$$v \geq 0$$
$$w \geq 0.$$

(i) The model is equivalent to

$$\max -(u-v) + (2-w)$$

subject to

$$-(u-v) - (2-w) \leq -3$$
$$(u-v) - 2(2-w) \leq 1$$
$$-(u-v) + 2(2-w) \leq -1$$
$$u \geq 0$$
$$v \geq 0$$
$$w \geq 0$$

which is equivalent to

$$\max -u + v - w$$

subject to

$$-u + v + w \leq -1$$
$$u - v + 2w \leq 5$$
$$-u + v - 2w \leq -5$$
$$u \geq 0$$
$$v \geq 0$$
$$w \geq 0.$$

Hence, in the canonical form, $x = \begin{bmatrix} u & v & w \end{bmatrix}^T$, $A = \begin{bmatrix} -1 & 1 & 1 \\ 1 & -1 & 2 \\ -1 & 1 & -2 \end{bmatrix}$, $b = \begin{bmatrix} -1 & 5 & -5 \end{bmatrix}^T$, and $c = \begin{bmatrix} -1 & 1 & -1 \end{bmatrix}^T$, with $x = u - v$ and $y = 2 - w$.

(ii) The model is equivalent to

$$\max -(u-v)+(2-w)$$

subject to

$$(u-v)+(2-w)-t=3$$
$$(u-v)-2(2-w)=1$$
$$u \geq 0$$
$$v \geq 0$$
$$w \geq 0$$
$$t \geq 0$$

which is equivalent to

$$\max -u+v-w$$

subject to

$$u-v-w-t=1$$
$$u-v+2w=5$$
$$u \geq 0$$
$$v \geq 0$$
$$w \geq 0$$
$$t \geq 0.$$

Hence, in the standard form, $\mathbf{x} = \begin{bmatrix} u & v & w & t \end{bmatrix}^T$, $\mathbf{A} = \begin{bmatrix} 1 & -1 & -1 & -1 \\ 1 & -1 & 2 & 0 \end{bmatrix}$, $\mathbf{b} = \begin{bmatrix} 1 & 5 \end{bmatrix}^T$, and $\mathbf{c} = \begin{bmatrix} -1 & 1 & -1 & 0 \end{bmatrix}^T$, with $x = u - v$ and $y = 2 - w$.

Example 16.27: Consider the model below:

$$\max x + 2y$$

subject to

$$x + y \geq 1$$
$$x + y \leq 2$$
$$x \geq 0$$
$$y \geq 0.$$

(i) Use the graphical method to find the optimal solution.

(ii) Propose another objective function that will make the model have an infinite number of optimal solutions. You are not allowed to propose a function that is a constant, e.g., minimize 1.

(iii) Write down a linear optimization model satisfying: first, it has two variables; second, its objective function is not a constant; third, its feasible set is a pentagon; and fourth, it has exactly one optimal solution. Use the graphical method to find the optimal solution.

258 *Machine learning and data analytics for maritime studies*

(iv) Write down a linear optimization model satisfying: first, it has two variables; second, its objective function is not a constant; third, its feasible set is a pentagon; and fourth, it has an infinite number of optimal solutions. Use the graphical method to find an optimal solution.

Solution. (i) See Figure 16.15. The optimal solution corresponds to the intersection of the lines $x = 0$ and $x + y = 2$. Therefore, the optimal solution is $x^* = 0, y^* = 2$. The optimal objective function value is 4.

(ii) Any of the following solutions are correct:
$\min x$, $\min y$, $\min x + y$, or $\max x + y$.

(iii) An example is

$\min x + y$
subject to

$$x \leq 2$$
$$y \leq 2$$
$$x + y \leq 3$$
$$x \geq 0$$
$$y \geq 0.$$

The graphical method is not provided here. The optimal solution is $x^* = 0, y^* = 0$, and the optimal objective function value is 0.

(iv) We can change the objective function in the model in (iii) to $\min x$. The graphical method is not provided here. An optimal solution is $x^* = 0, y^* = 0$, and the optimal objective function value is 0.

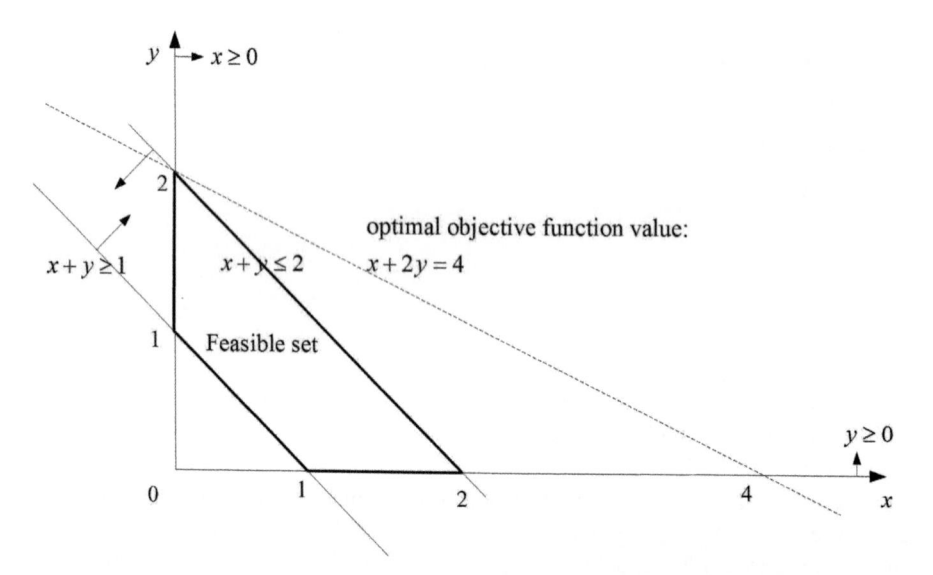

Figure 16.15 Graphical method

Advanced linear optimization 259

Example 16.28: Use the graphical method to find the optimal solution to the model below:

$$\max x + 2y$$

subject to

$$x + y = 1$$
$$x + y \leq 2$$
$$x \geq 0$$
$$y \geq 0.$$

Solution. See Figure 16.16. The optimal solution corresponds to the intersection of the lines $x = 0$ and $x + y = 1$. Therefore, the optimal solution is $x^* = 0, y^* = 1$. The optimal objective function value is 2.

Example 16.29: Use Excel or any other software to solve the following linear optimization model. You only need to write down the final answer:

$$\min x + 2y + 3z$$

subject to

$$4x + 5y + 6z \geq 10000$$
$$7x + 8y + 9z \geq 100$$
$$10x + 11y + 12z \geq 200$$

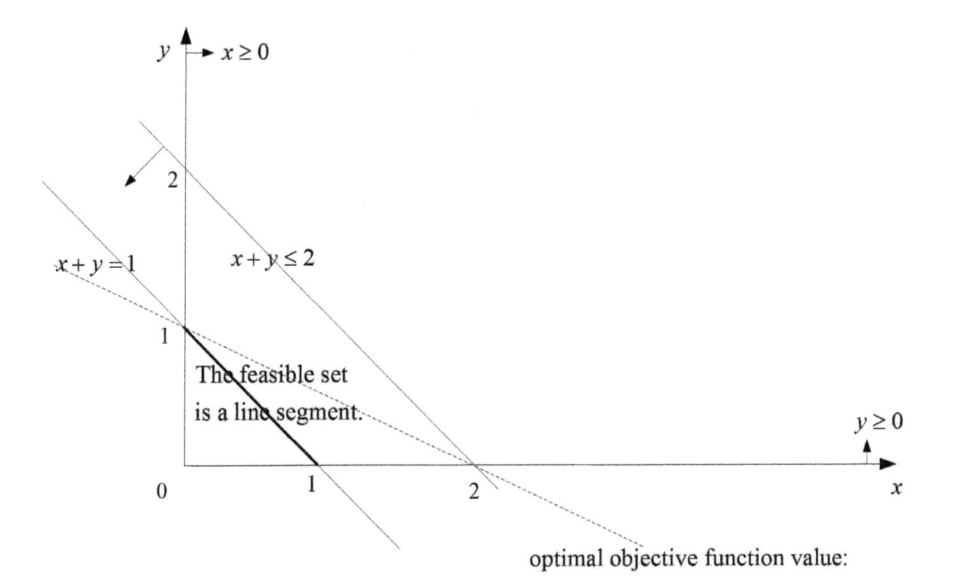

Figure 16.16   Graphical method

260 Machine learning and data analytics for maritime studies

$$13x + 14y + 15z \geq 300$$
$$x \geq 0$$
$$y \geq 0$$
$$z \geq 0.$$

Solution. The optimal solution is $x^* = k/4, y^* = 0, z^* = 0$. The optimal objective function value is $k/4$.

Example 16.30: We have a transportation network shown in Figure 16.17, where the set of nodes $\mathcal{N}$ and the set of arcs $\mathcal{A}$ are shown. Each arc $(i,j) \in \mathcal{A}$ has a known capacity denoted by $s_{ij} \geq 0$. Develop a link-flow linear optimization model to find the maximum flow from A to F. You must write down the details of the objective function and constraints without using "$\sum$" or "$\forall$" except in the nonnegativity constraints.

Solution. Let f_{ij} be the decision variables representing the flow on arcs $(i,j) \in \mathcal{A}$. The model is as follows:

$$\max f_{AB} + f_{AC} + f_{AF} \text{ (maximize the total outflow from node A)}$$

subject to

$$f_{AB} + f_{DB} = f_{BC} + f_{BE}$$
(flow conservation at node B)
$$f_{AC} + f_{BC} = f_{CD} + f_{CE}$$
(flow conservation at node C)
$$f_{CD} + f_{ED} = f_{DB} + f_{DF}$$
(flow conservation at node D)
$$f_{BE} + f_{CE} = f_{ED} + f_{EF}$$
(flow conservation at node E)
$$f_{AB} \leq s_{AB}, f_{AC} \leq s_{AC}, f_{AF} \leq s_{AF}, f_{BC} \leq s_{BC}, f_{BE} \leq s_{BE}$$
$$f_{CD} \leq s_{CD}, f_{CE} \leq s_{CE}, f_{DB} \leq s_{DB}, f_{DF} \leq s_{DF}, f_{ED} \leq s_{ED}, f_{EF} \leq s_{EF}$$
$$f_{ij} \geq 0, \forall (i,j) \in \mathcal{A}.$$

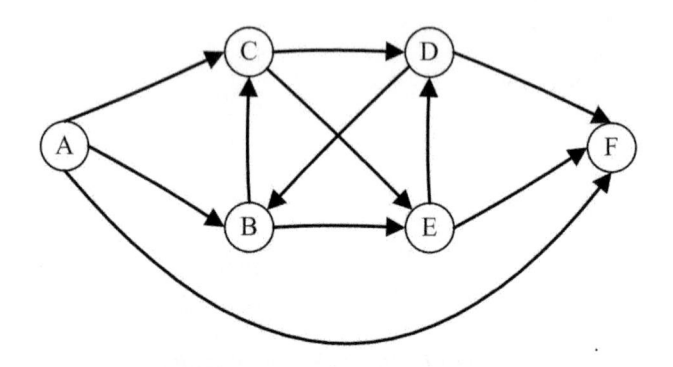

Figure 16.17   A transportation network

Advanced linear optimization 261

Example 16.31: We have a transportation network shown in Figure 16.17, where the set of nodes $\mathcal{N}$ and the set of arcs $\mathcal{A}$ are shown. Each arc $(i,j) \in \mathcal{A}$ has a known unit cost denoted by $c_{ij} \geq 0$ and a known capacity denoted by $s_{ij} \geq 0$. We need to send 10 units of cargo from A to E (note it is from A to E rather than to F). Develop a link-flow linear optimization model to find the minimum total cost. You must write down the details of the objective function and constraints without using "$\sum$" or "$\forall$" except in the nonnegativity constraints.

Solution. Let f_{ij} be the decision variables representing the flow on arcs $(i,j) \in \mathcal{A}$. The model is as follows:

$$\min c_{AB}f_{AB} + c_{AC}f_{AC} + c_{AF}f_{AF} + c_{BC}f_{BC} + c_{BE}f_{BE}$$
$$+ c_{CD}f_{CD} + c_{CE}f_{CE} + c_{DB}f_{DB} + c_{DF}f_{DF} + c_{ED}f_{ED} + c_{EF}f_{EF}$$

subject to

$$f_{AB} + f_{AC} + f_{AF} = 10$$
$$f_{AB} + f_{DB} = f_{BC} + f_{BE}$$
$$f_{AC} + f_{BC} = f_{CD} + f_{CE}$$
$$f_{CD} + f_{ED} = f_{DB} + f_{DF}$$
$$f_{BE} + f_{CE} = f_{ED} + f_{EF} + 10$$
$$f_{AF} + f_{DF} + f_{EF} = 0$$
$$f_{AB} \leq s_{AB}, f_{AC} \leq s_{AC}, f_{AF} \leq s_{AF}, f_{BC} \leq s_{BC}, f_{BE} \leq s_{BE}$$
$$f_{CD} \leq s_{CD}, f_{CE} \leq s_{CE}, f_{DB} \leq s_{DB}, f_{DF} \leq s_{DF}, f_{ED} \leq s_{ED}, f_{EF} \leq s_{EF}$$
$$f_{ij} \geq 0, \forall (i,j) \in \mathcal{A}.$$

It is correct to say that there is no flow on (E,D), (A,F), (D,F), and (E,F). Therefore, the following model is also correct:

$$\min c_{AB}f_{AB} + c_{AC}f_{AC} + c_{BC}f_{BC} + c_{BE}f_{BE} +$$
$$c_{CD}f_{CD} + c_{CE}f_{CE} + c_{DB}f_{DB}$$

subject to

$$f_{AB} + f_{AC} = 10$$
$$f_{AB} + f_{DB} = f_{BC} + f_{BE}$$
$$f_{AC} + f_{BC} = f_{CD} + f_{CE}$$
$$f_{CD} = f_{DB}$$
$$f_{BE} + f_{CE} = 10$$
$$f_{AB} \leq s_{AB}, f_{AC} \leq s_{AC}, f_{BC} \leq s_{BC}, f_{BE} \leq s_{BE}$$
$$f_{CD} \leq s_{CD}, f_{CE} \leq s_{CE}, f_{DB} \leq s_{DB}$$
$$f_{ij} \geq 0, \forall (i,j) \in \mathcal{A} \setminus \{(E,D),(A,F),(D,F),(E,F)\}.$$

Example 16.32: Figure 16.18 shows a liner container shipping network. The circles represent ports. There are four routes $r_1, r_2, r_3,$ and r_4 that provide weekly shipping services.

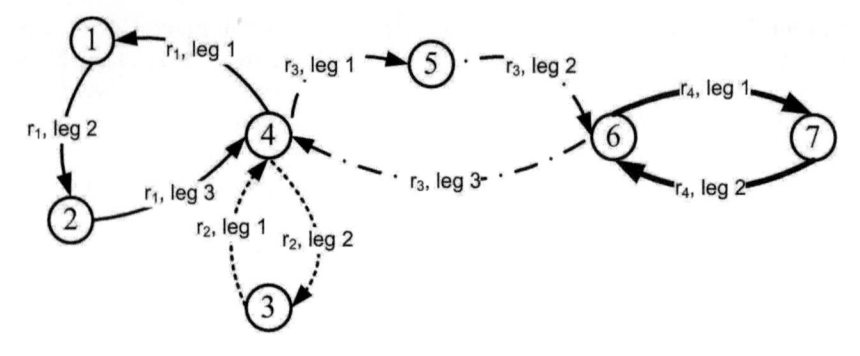

262 Machine learning and data analytics for maritime studies

Positive demand: $q^{15}, q^{17}, q^{25}, q^{36}, q^{46}$

Figure 16.18   A complex hub and spoke liner shipping network

The capacity of a ship deployed on route r_1 is E_1 (TEUs). E_2, E_3, and E_4 have similar meanings. The volumes of containers between different OD pairs are shown in the figure. For example, q^{15} is the demand (TEUs/week) from port 1 to port 5. The profit for transporting one TEU from port 1 to port 5 is g^{15} ($/TEU). g^{17}, g^{25}, g^{36}, and g^{46} have similar meanings. Assume that container handling costs are 0. Develop a path-flow linear optimization model to find the maximum profit that can be gained from transporting containers. You must write down the details of the objective function and constraints without using "$\sum$" or "$\forall$" except in the nonnegativity constraints.

Solution. The set of OD pairs is $\mathcal{W} = \{(1,5), (1,7), (2,5), (3,6), (4,6)\}$. Let $\mathcal{H}^{od}$ be the set of itineraries (paths) for OD pair $(o, d) \in \mathcal{W}$. To simplify the notation, we use $<r, i>$ to represent leg i of route r. Then $\mathcal{H}^{15}$ consists of the following:

$h_1 : <1, 2> \to <1, 3> \to <3, 1>$

$\mathcal{H}^{17}$ consists of the following:

$h_2 : <1, 2> \to <1, 3> \to <3, 1> \to <3, 2> \to <4, 1>$

$\mathcal{H}^{25}$ consists of the following:

$h_3 : <1, 3> \to <3, 1>$

$\mathcal{H}^{36}$ consists of the following:

$h_4 : <2, 1> \to <3, 1> \to <3, 2>$

$\mathcal{H}^{46}$ consists of the following:

$h_5 : <3, 1> \to <3, 2>$.

Define $\mathcal{H} := \{1, 2, 3, 4, 5\}$. Let y_h be the decision variables representing the flow on itinerary $h \in \mathcal{H}$ (TEUs/week). The model is as follows:

max $g^{15} y_1 + g^{17}(y_2) + g^{25}(y_3) + g^{36}(y_4) + g^{46}(y_5)$ (maximize the total profit)

subject to

$$y_1 \leq q^{15}$$
$$y_2 \leq q^{17}$$
$$y_3 \leq q^{25}$$
$$y_4 \leq q^{36}$$
$$y_5 \leq q^{46}$$
$$y_1 + y_2 \leq E_1 \text{(capacity on leg } <1,2>)$$
$$y_1 + y_2 + y_3 \leq E_1 \text{(capacity on leg } <1,3>)$$
$$y_4 \leq E_2 \text{(capacity on leg } <2,1>)$$
$$y_1 + y_2 + y_3 + y_4 + y_5 \leq E_3 \text{(capacity on leg } <3,1>)$$
$$y_2 + y_4 + y_5 \leq E_3 \text{(capacity on leg } <3,2>)$$
$$y_2 \leq E_4 \text{(capacity on leg } <4,1>)$$
$$y_h \leq 0, h \in \mathcal{H}.$$

Example 16.33: Propose a linear optimization model that satisfies all of the following requirements or answer "such a model does not exist": (i) it is infeasible, (ii) after removing the first constraint, it has exactly one optimal solution; (iii) after removing the first two constraints, it has an infinite number of optimal solutions; and (iv) after removing the first three constraints, it is unbounded.

Solution. An example can be

$$\max x$$

subject to

$$x \leq -1$$
$$x + y \leq 1$$
$$x \leq 1$$
$$x \geq 0$$
$$y \geq 0.$$

Example 16.34: We have a transportation network shown in Figure 16.19, where the set of nodes $\mathcal{N}$ and the set of arcs $\mathcal{A}$ are shown. Each arc $(i,j) \in \mathcal{A}$ has a capacity denoted by $s_{ij} \geq 0$. We do not know s_{ij}, but we know the maximum flow from A to D is 11, the maximum flow from D to G is 12, and the maximum flow from G to J is 13. What is the maximum flow from A to J? You do not need to show the calculation.

Solution. The answer is 11.

Example 16.35: Consider a problem with three decision variables x, y, z and a set of linear constraints that defines a non-empty and bounded feasible set. Suppose that there are three potential objective functions to maximize: $x + y + z$, $2x - y$, and $-y - z$. How to use a linear optimization solver to check whether there is a feasible solution that could maximize the three objectives at the same time?

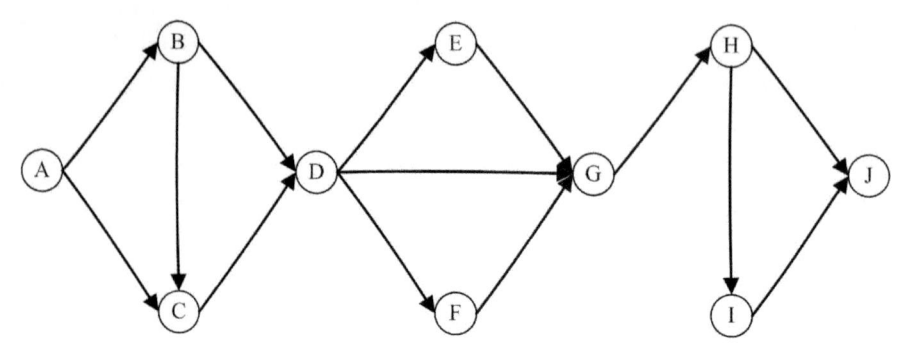

Figure 16.19   A transportation network

Solution. Solve the maximization problem with the objective function $x+y+z$ and the given constraints, and let (x_1^*, y_1^*, z_1^*) be the optimal solution. Solve the maximization problem with the objective function $2x - y$ and the given constraints, and let (x_2^*, y_2^*, z_2^*) be the optimal solution. Solve the maximization problem with the objective function $-y - z$ and the given constraints, and let (x_3^*, y_3^*, z_3^*) be the optimal solution. Solve the following model: maximizing 0 subject to the given constraints and $x + y + z = x_1^* + y_1^* + z_1^*, 2x - y = 2x_2^* - y_2^*, -y - z = -y_3^* - z_3^*$. If this model has a feasible solution, then the answer is yes; otherwise, the answer is no.

Another approach is as follows. Let C_1 be the optimal objective function value of the maximization model with the objective function $x+y+z$ and the given constraints, C_2 be the one of the models with the objective function $2x - y$ and the given constraints, C_3 be the one of the models with the objective function $-y - z$ and the given constraints, and C_4 be the one of the models with the objective function $(x + y + z) + (2x - y) + (-y - z)$ and the given constraints. If $C_4 = C_1 + C_2 + C_3$, then the answer is yes; otherwise, the answer is no.

Example 16.36: We have an optimization model with decision variables x, y, z, linear constraints and the objective function that minimizes

$$\max\{x, y, z\} - \min\{x, y, z\}.$$

The feasible set is non-empty. Transform the model to a linear optimization model.

Solution. Define two auxiliary decision variables u and v. The model is equivalent to

$$\min u - v$$

subject to

$$u \geq x, u \geq y, u \geq z$$
$$v \leq x, v \leq y, v \leq z$$

and the given constraints.

Example 16.37: A linear optimization model has the following decision variables x_1, x_2 and constraints:

$$a_{11}x_1 + a_{12}x_2 \geq b_1$$
$$a_{21}x_1 + a_{22}x_2 \geq b_2$$
$$a_{31}x_1 + a_{32}x_2 \geq b_3$$
$$\vdots$$
$$a_{m1}x_1 + a_{m2}x_2 \geq b_m$$
$$0 \leq x_1 \leq 1$$
$$0 \leq x_2 \leq 1.$$

Suppose that we know all points in the feasible set form a square. Given a linear optimization solver, how to calculate the area of the square?

Example 16.38: We have a linear optimization model with an infinite number of optimal solutions. How to find two optimal solutions with the largest L_1 distance?

Example 16.39: Given a linear optimization model, how to use a linear optimization solver to check whether all points in the feasible set form a line segment?

Example 16.40: Walmart has three warehouses (W1, W2, and W3) that store the same type of product and five supermarkets (S1–S5) that need the products in a city. The number of products available at each warehouse, the number of products needed at each supermarket, and the transportation cost per unit product from each warehouse to each supermarket are shown below. Develop a linear optimization model to help Walmart make the decision of how to transport the products. You must write down the details of the objective function and constraints without using "$\sum$" or "$\forall$" except in the nonnegativity constraints.

Warehouse	Number of products available
W1	100
W2	200
W3	50

Supermarket	Number of products needed
S1	80
S2	90
S3	70
S4	60
S5	50

Unit cost ($)	S1	S2	S3	S4	S5
W1	1	2	4	3	6
W2	5	2	4	4	4
W3	5	1	1	3	2

Solution. Let f_{ij} be the decision variables representing the number of products transported from warehouse $i = 1, 2, 3$ to supermarket $j = 1, 2, 3, 4, 5$. The model is as follows:

$$\min f_{11} + 2f_{12} + 4f_{13} + 3f_{14} + 6f_{15} + 5f_{21} + 2f_{22} +$$
$$4f_{23} + 4f_{24} + 4f_{25} + 5f_{31} + f_{32} + f_{33} + 3f_{34} + 2f_{35}$$

subject to

$$f_{11} + f_{12} + f_{13} + f_{14} + f_{15} \leq 100$$
$$f_{21} + f_{22} + f_{23} + f_{24} + f_{25} \leq 200$$
$$f_{31} + f_{32} + f_{33} + f_{34} + f_{35} \leq 50$$
$$f_{11} + f_{21} + f_{31} = 80$$
$$f_{12} + f_{22} + f_{32} = 90$$
$$f_{13} + f_{23} + f_{33} = 70$$
$$f_{14} + f_{24} + f_{34} = 60$$
$$f_{15} + f_{25} + f_{35} = 50$$
$$f_{ij} \geq 0, i = 1, 2, 3, j = 1, 2, 3, 4, 5.$$

Note: The following constraints are incorrect because evidently all f_{ij} will be 0 in the optimal solution:

$$f_{11} + f_{12} + f_{13} + f_{14} + f_{15} \leq 100$$
$$f_{21} + f_{22} + f_{23} + f_{24} + f_{25} \leq 200$$
$$f_{31} + f_{32} + f_{33} + f_{34} + f_{35} \leq 50$$
$$f_{11} + f_{21} + f_{31} \leq 80$$
$$f_{12} + f_{22} + f_{32} \leq 90$$
$$f_{13} + f_{23} + f_{33} \leq 70$$
$$f_{14} + f_{24} + f_{34} \leq 60$$
$$f_{15} + f_{25} + f_{35} \leq 50$$
$$f_{ij} \geq 0, i = 1, 2, 3, j = 1, 2, 3, 4, 5.$$

Example 16.41: We have the following linear optimization model:

$$\min x - y$$

subject to

$$x + y \geq 3$$
$$y \leq 2$$
$$x - 2y = 1.$$

Transform the model to the standard form that maximizes $c^T x$ subject to $Ax = b$ and $x \geq 0$.

Solution. The model is equivalent to

$$\max -(u - v) + (2 - w)$$

subject to
$$(u-v)+(2-w)-t=3$$
$$(u-v)-2(2-w)=1$$
$$u \geq 0$$
$$v \geq 0$$
$$w \geq 0$$
$$t \geq 0$$

which is equivalent to
$$\max -u+v-w$$
subject to
$$u-v-w-t=1$$
$$u-v+2w=5$$
$$u \geq 0$$
$$v \geq 0$$
$$w \geq 0$$
$$t \geq 0.$$

Hence, in the standard form, $x = \begin{bmatrix} u & v & w & t \end{bmatrix}^T$, $A = \begin{bmatrix} 1 & -1 & -1 & -1 \\ 1 & -1 & 2 & 0 \end{bmatrix}$, $b = \begin{bmatrix} 1 & 5 \end{bmatrix}^T$, and $c = \begin{bmatrix} -1 & 1 & -1 & 0 \end{bmatrix}^T$, with $x = u - v$ and $y = 2 - w$.

Note 1: The solution to this question is not unique.

Note 2: The following interesting solutions are also correct: Let $u = x+y-3 \geq 0$ and $v = 2-y \geq 0$. Solving this equation system, we have $x = u+v+1, y = 2-v$. Therefore, the model is equivalent to min $u + 2v - 1$ subject to $u + 3v - 3 = 1, u \geq 0, v \geq 0$, etc.

Note 3: It is incorrect to say $y \leq 2$ is equivalent to $w = -y, -w \leq 2, w \geq 0$.

Example 16.42: Use the graphical method to find the optimal solution to the model below:
$$\max 2x + y$$
subject to
$$x+y \geq 1$$
$$x \leq 2$$
$$y \leq 1$$
$$x \geq 0$$
$$y \geq 0.$$

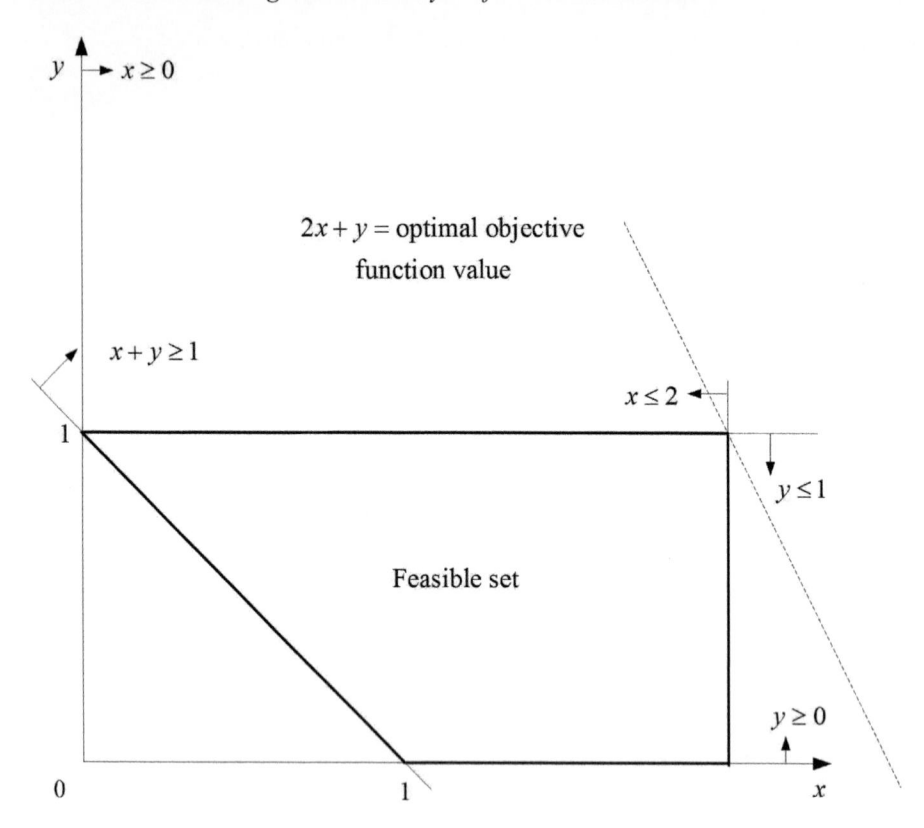

Figure 16.20   Graphical method

Solution. See Figure 16.20. The optimal solution corresponds to the intersection of the lines $x = 2$ and $y = 1$. Therefore, the optimal solution is $x^* = 2, y^* = 1$. The optimal objective function value is 5.

Note: The slope of the line $2x + y = k$ is -2 rather than 2.

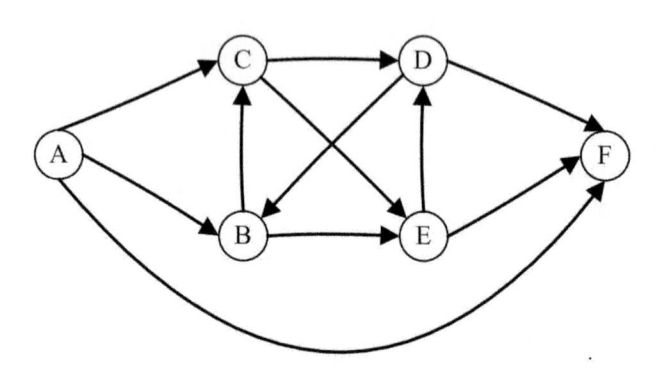

Figure 16.21   A transportation network

Advanced linear optimization 269

Example 16.43: We have a transportation network shown in Figure 16.21, where the set of nodes $\mathcal{N}$ and the set of arcs $\mathcal{A}$ are shown. Each arc $(i,j) \in \mathcal{A}$ has a known unit transportation cost denoted by $c_{ij} \geq 0$ and a known capacity denoted by $s_{ij} \geq 0$. We need to send 10 units of cargo from A to F. Develop a link-flow linear optimization model to find the minimum total cost. You must write down the details of the objective function and constraints without using "$\sum$" or "$\forall$" except in the nonnegativity constraints.

Solution. Let f_{ij} be the decision variables representing the flow on arcs $(i,j) \in \mathcal{A}$. The model is as follows:

$$\min c_{AB}f_{AB} + c_{AC}f_{AC} + c_{AF}f_{AF} + c_{BC}f_{BC} + c_{BE}f_{BE} +$$
$$c_{CD}f_{CD} + c_{CE}f_{CE} + c_{DB}f_{DB} + c_{DF}f_{DF} + c_{ED}f_{ED} + c_{EF}f_{EF}$$

subject to

$$f_{AB} + f_{AC} + f_{AF} = 10$$
$$f_{AB} + f_{DB} = f_{BC} + f_{BE}$$
$$f_{AC} + f_{BC} = f_{CD} + f_{CE}$$
$$f_{CD} + f_{ED} = f_{DB} + f_{DF}$$
$$f_{BE} + f_{CE} = f_{ED} + f_{EF}$$
$$f_{AF} + f_{DF} + f_{EF} = 10$$
$$f_{AB} \leq s_{AB}, f_{AC} \leq s_{AC}, f_{AF} \leq s_{AF}, f_{BC} \leq s_{BC}, f_{BE} \leq s_{BE}$$
$$f_{CD} \leq s_{CD}, f_{CE} \leq s_{CE}, f_{DB} \leq s_{DB}, f_{DF} \leq s_{DF}, f_{ED} \leq s_{ED}, f_{EF} \leq s_{EF}$$
$$f_{ij} \geq 0, \forall (i,j) \in \mathcal{A}.$$

Example 16.44: Figure 16.22 shows a liner container shipping network. The circles represent ports. There are five routes r_1, r_2, r_3, r_4, and r_5 that provide weekly shipping services. The capacity of a ship deployed on route r_1 is E_1 (TEUs). E_2, E_3, E_4, and E_5 have similar meanings. The volumes of containers between different OD pairs are shown in the figure. For example, q^{12} is the demand (TEUs/week) from port 1 to port 2. The profit for transporting one TEU from port 1 to port 2 is g^{12} ($/TEU). g^{14} and g^{15} have similar meanings. Assume that container handling costs are 0. Develop a path-flow linear optimization model to find the maximum profit that can be gained from transporting containers. You must write down the details of the objective function and constraints without using "$\sum$" or "$\forall$" except in the nonnegativity constraints.

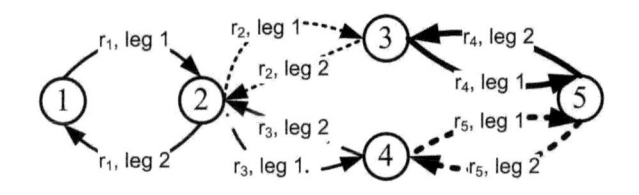

Positive demand: q^{12}, q^{14}, q^{15}

Figure 16.22 A complex transshipment liner shipping network

Solution. The set of OD pairs is $\mathcal{W} = \{(1,2),(1,4),(1,5)\}$. Let $\mathcal{H}^{od}$ be the set of itineraries (paths) for OD pair $(o,d) \in \mathcal{W}$. To simplify the notation, we use $<r, i>$ to represent leg i of route r. Then $\mathcal{H}^{12}$ consists of the following:

$h_1 :< 1, 1 >$

$\mathcal{H}^{14}$ consists of the following:

$$h_2 :< 1, 1 > \to < 3, 1 >$$
$$h_3 :< 1, 1 > \to < 2, 1 > \to < 4, 1 > \to < 5, 2 >$$

$\mathcal{H}^{15}$ consists of the following:

$$h_4 :< 1, 1 > \to < 2, 1 > \to < 4, 1 >$$
$$h_5 :< 1, 1 > \to < 3, 1 > \to < 5, 1 >.$$

Define $\mathcal{H} := \{1, 2, 3, 4\}$. Let y_h be the decision variables representing the flow on itinerary (TEUs/week). The model is as follows:

$$\max g^{12} y_1 + g^{14}(y_2 + y_3) + g^{15}(y_4 + y_5) \text{ (maximize the total profit)}$$

subject to

$$y_1 \leq q^{12}$$
$$y_2 + y_3 \leq q^{14}$$
$$y_4 + y_5 \leq q^{15}$$
$$y_1 + y_2 + y_3 + y_4 + y_5 \leq E_1 \text{ (capacity on leg} < 1, 1 >)$$
$$y_3 + y_4 \leq E_2 \text{ (capacity on leg} < 2, 1 >)$$
$$y_2 + y_5 \leq E_3 \text{ (capacity on leg} < 3, 1 >)$$
$$y_3 + y_4 \leq E_4 \text{ (capacity on leg} < 4, 1 >)$$
$$y_5 \leq E_5 \text{ (capacity on leg} < 5, 1 >)$$
$$y_3 \leq E_5 \text{ (capacity on leg} < 5, 2 >)$$
$$y_h \geq 0, h \in \mathcal{H}.$$

Note: It is wrong to say $y_4 \leq q^{15}, y_5 \leq q^{15}$.

Example 16.45: Propose a linear optimization model that satisfies all of the following requirements or answer "such a model does not exist": (i) it is infeasible; (ii) after removing the first constraint, it has an infinite number of optimal solutions; (iii) after removing the first two constraints, it has exactly one optimal solution; and (iv) after removing the first three constraints, it is unbounded.

Solution. Note: The answer is not unique. An example can be

$$\max x$$

subject to

$$x \leq -1$$
$$x \leq 1$$
$$x + y \leq 2$$
$$x \geq 0$$
$$y \geq 0.$$

Example 16.46: We have a transportation network shown in Figure 16.23, where the set of nodes $\mathcal{N}$ and the set of arcs $\mathcal{A}$ are shown. Each arc $(i,j) \in \mathcal{A}$ has a capacity denoted by $s_{ij} \geq 0$. We do not know s_{ij}, but we know the maximum flow from A to D is 11, the maximum flow from D to G is 12, the maximum flow from G to J is $\frac{234,554}{1,000,000}$, the maximum flow from A to M is 14, and the maximum flow from M to G is 15. What is the maximum flow from A to J? You do not need to show the calculation.

Solution.

$$\frac{234,554}{1,000,000}$$

Example 16.47: A linear optimization model has decision variables $x_1, x_2, \cdots, x_n$ and constraints:

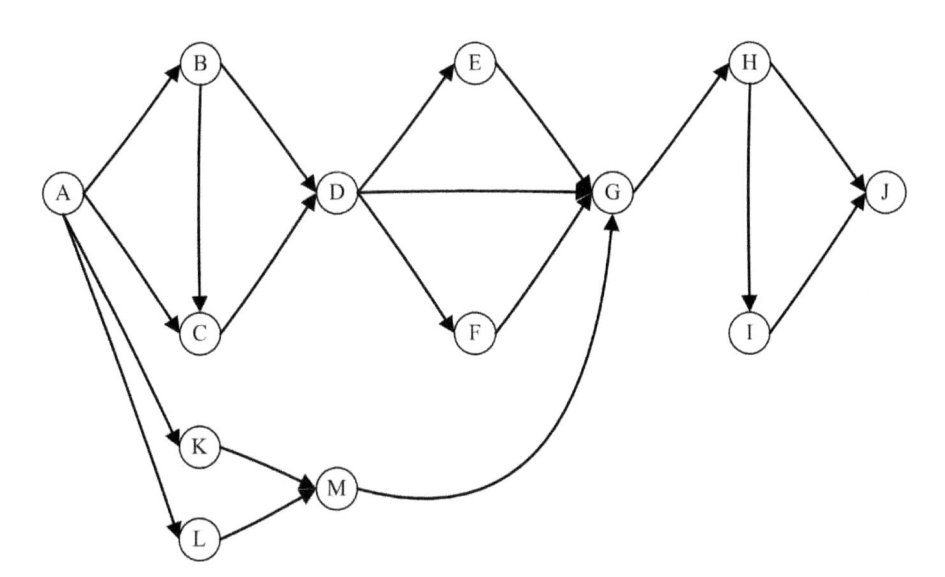

Figure 16.23   A transportation network

$$a_{11}x_1 + a_{12}x_2 + \cdots + a_{1n}x_n = b_1$$
$$a_{21}x_1 + a_{22}x_2 + \cdots + a_{2n}x_n \geq b_2$$
$$a_{31}x_1 + a_{32}x_2 + \cdots + a_{3n}x_n \geq b_3$$
$$\vdots$$
$$a_{m1}x_1 + a_{m2}x_2 + \cdots + a_{mn}x_n \geq b_m$$
$$0 \leq x_1 \leq 1$$
$$0 \leq x_2 \leq 1$$
$$\vdots$$
$$0 \leq x_n \leq 1$$

where a_{ij}, b_i are all constants, $i = 1, 2, \cdots, m$, $j = 1, 2, \cdots, n$. We do not know its objective function yet. How to use a linear optimization solver to check whether the first constraint $a_{11}x_1 + a_{12}x_2 + \cdots + a_{1n}x_n = b_1$ is redundant?

Solution. First, we minimize $a_{11}x_1 + a_{12}x_2 + \cdots + a_{1n}x_n$ subject to the second to the last constraints. Second, we maximize $a_{11}x_1 + a_{12}x_2 + \cdots + a_{1n}x_n$ subject to the second to the last constraints. If both models have the same optimal objective function value b_1, then the first constraint is redundant. Otherwise, it is not redundant.

Example 16.48: We have an optimization model with decision variables x, y, z, linear constraints, and the objective function that maximizes

$$\max\{x, y, z\} - \min\{x, y, z\}.$$

The feasible set is non-empty. How to use a linear optimization solver to address the problem? Hint: you can use the linear optimization solver several times.

Solution. There are six possible cases: $x \geq y \geq z$, $x \geq z \geq y$, $y \geq x \geq z$, $y \geq z \geq x$, $z \geq x \geq y$, and $z \geq y \geq x$. Taking the first case as an example, we maximize $x - z$ subject to all the given constraints and $x \geq y, y \geq z$. Taking the last case as an example, we maximize $z - x$ subject to all the given constraints and $z \geq y, y \geq x$. Hence, we solve six linear optimization models. The best one of the six optimal solutions is what we want.

Note 1: The following answer is also correct: we maximize $x - y$ subject to all the given constraints; we maximize $x - z$ subject to all the given constraints; and we do so with the objective functions $y - x$, $y - z$, $z - x$, $z - y$. Hence, we solve six linear optimization models. The best one of the six optimal solutions is what we want.

Note 2: The following answer is incorrect: solve $\max(u - v)$ subject to $u \geq x, u \geq y, u \geq z, v \leq x, v \leq y, v \leq z$ and all of the other constraints. It is wrong because the problem will always be unbounded: u can be infinity and v can be negative infinity.

Note 3: The following answer is incorrect: u^* is the best objective function value of the following three models: $\max x$ subject to all of the constraints, $\max y$ subject to all of the constraints, and $\max z$ subject to all of the constraints. v^* is the best objective function value of the following three models: $\min x$ subject to all of the constraints, $\min y$ subject to all of the constraints, and

Advanced linear optimization 273

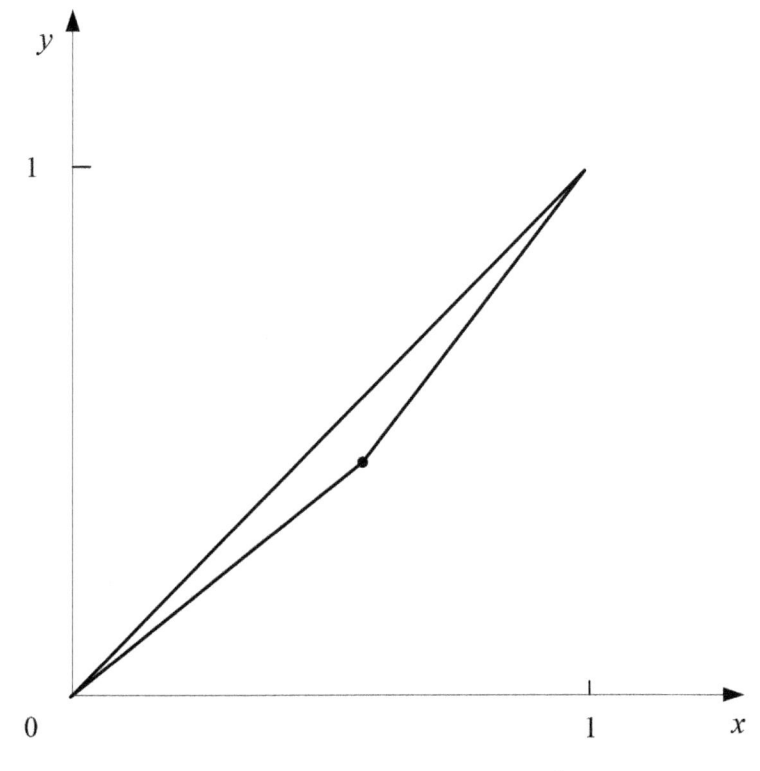

Figure 16.24   A counter-example

min z subject to all of the constraints. Then we solve the model min 0 subject to $u \leq u^*, u \geq x, u \geq y, u \geq z, v \geq v^*, v \leq x, v \leq y, v \leq z$ and all of the other constraints, whose optimal solution is also optimal to the original model.

It is wrong because there may not exist a feasible solution $(\bar{x}, \bar{y}, \bar{z})$ such that $\max(\bar{x}, \bar{y}, \bar{z}) = u^*, \min(\bar{x}, \bar{y}, \bar{z}) = v^*$. To appreciate this point, consider a model with only two decision variables x and y, the feasible set is the triangle in Figure 16.24, and we want to maximize $\max(x, y) - \min(x, y)$. In plain words, we want to find a point that is as far away from the line $y = x$ as possible. We can see that the optimal solution is the dot in the figure, and the optimal objective function value is very small (somewhere between 0.1 and 0.2). If we use the aforementioned wrong approach, we have $u^* = 1, v^* = 0$ and the optimal objective function value is 1.

A detailed explanation of the solution to the above example:

There are six possible cases: $x \geq y \geq z$, $x \geq z \geq y$, $y \geq x \geq z$, $y \geq z \geq x$, $z \geq x \geq y$, and $z \geq y \geq x$.

We solve Model 1: maximize $x - z$ subject to all the given constraints and $x \geq y, y \geq z$, and its optimal solution is $(x^{(1)}, y^{(1)}, z^{(1)})$.

We solve Model 2: maximize $x - y$ subject to all the given constraints and $x \geq z, z \geq y$, and its optimal solution is $(x^{(2)}, y^{(2)}, z^{(2)})$.

We solve Model 3: maximize $y - z$ subject to all the given constraints and $y \geq x, x \geq z$, and its optimal solution is $(x^{(3)}, y^{(3)}, z^{(3)})$.

We solve Model 4: maximize $y - x$ subject to all the given constraints and $y \geq z, z \geq x$, and its optimal solution is $(x^{(4)}, y^{(4)}, z^{(4)})$.

We solve Model 5: maximize $z - y$ subject to all the given constraints and $z \geq x, x \geq y$, and its optimal solution is $(x^{(5)}, y^{(5)}, z^{(5)})$.

We solve Model 6: maximize $z - x$ subject to all the given constraints and $z \geq y, y \geq x$, and its optimal solution is $(x^{(6)}, y^{(6)}, z^{(6)})$.

The optimal solution is $(x^{(i^*)}, y^{(i^*)}, z^{(i^*)})$, in which $i^* \in \{1, 2, 3, 4, 5, 6\}$ and model i^* has the largest optimal objective function value among these six models.

References

[1] Wang S., Meng Q., Sun Z. 'Container routing in liner shipping'. *Transportation Research Part E*. 2013;**49**(1):1–7.

[2] Wang S. 'A novel hybrid-link-based container routing model'. *Transportation Research Part E*. 2014;**61**:165–75.

[3] Wang S., Meng Q. 'Sailing speed optimization for container ships in a liner shipping network'. *Transportation Research Part E*. 2012;**48**(3):701–14.

Chapter 17
Integer optimization

In linear optimization, we assume that all of the decision variables are continuous. However, in reality, some are not. For example, the number of crude oil tankers used to transport crude oil from the Middle East to the US is an integer, and rounding down 6.5 ships to 6 ships can lead to considerable errors. Therefore, a natural extension to linear optimization models is integer linear optimization models, which are the same as linear optimization models except that the decision variables can only take integer values. We also have mixed-integer linear optimization models, in which some decision variables can only take integer values, and the others are continuous. We often use "integer optimization models" to refer to integer linear optimization models or both integer linear optimization models and mixed-integer linear optimization models. However, one should keep in mind that integer linear optimization models are not linear; in other words, integer linear optimization models belong to the category of nonlinear optimization models.

We use $\mathbb{Z}_+$ to represent the set of non-negative integers[*]. Hence, $x \in \mathbb{Z}_+$ means that x is non-negative and can only take integer values.

17.1 Formulation I: natural integer decision variables

Example 17.1: Suppose that there is 1 million tons of coal to transport from Indonesia to Japan. The following bulk carriers are available:

Ship type	Capacity (1000 tons)	Number of ships	Cost per trip (million $)
Capesize	100	8	1
Panamax	60	8	0.8
Handymax	40	7	0.6
Handysize	10	5	0.3

---

[*]One often uses $\mathbb{Z}_{++}$ to represent the set of positive integers.

Each bulk carrier can only complete one trip from Indonesia to Japan due to the deadline for delivering the coal. The company needs to determine how many ships in each type to use. Formulate an integer optimization model that minimizes the total cost of transporting the coal.

Solution: Let x, y, z, and w be the numbers of Capesize, Panamax, Handymax, and Handysize ships to use, respectively. The model is:

$$\min 1x + 0.8y + 0.6z + 0.3w$$

subject to

$$100x + 60y + 40z + 10w \geq 1000$$
$$x \leq 8$$
$$y \leq 8$$
$$z \leq 7$$
$$w \leq 5$$
$$x \in \mathbb{Z}_+$$
$$y \in \mathbb{Z}_+$$
$$z \in \mathbb{Z}_+$$
$$w \in \mathbb{Z}_+.$$

Example 17.2: Suppose that there is 2 million tons of crude oil to transport from Saudi Arabia to the US. The following crude oil tankers are available (capacity in terms of 1000 tons and cost per trip in terms of million $):

Ship type	Capacity	Number	Cost per trip
Very large crude carrier (VLCC)	200	8	2
Suezmax crude tanker	120	8	1.5
Aframax crude tanker	80	7	1.2
Panamax crude tanker	60	5	1

Each ship can only complete one trip from Saudi Arabia to the US due to the deadline for delivering the crude oil. The company needs to determine how many ships in each type to use. Formulate an integer optimization model that minimizes the total cost of transporting the crude oil.

Solution: Let x, y, z, and w be the numbers of VLCC, Suezmax, Aframax, and Panamax ships to use, respectively. The model is

$$\min 2x + 1.5y + 1.2z + 1w$$

subject to

$$200x + 120y + 80z + 60w \geq 2000$$
$$x \leq 8$$
$$y \leq 8$$
$$z \leq 7$$
$$w \leq 5$$
$$x \in Z_+$$
$$y \in Z_+$$
$$z \in Z_+$$
$$w \in Z_+.$$

Example 17.3: Reconsider Example 17.2. Suppose that there is a further requirement that at most 10 ships can be used. Formulate an integer optimization model that minimizes the total cost of transporting the crude oil.

Solution: We only need to add the constraint $x + y + z + w \leq 10$ to the above model.

Example 17.4: This is a ship repositioning problem. Suppose that there are three liner services ($r_1, r_2,$ and r_3) on which container ships are to be deployed. The service r_1 needs four ships, r_2 needs five ships, and r_3 needs six ships. The available ships are at five locations: location i has $i + 1$ ships, $i = 1, 2, 3, 4, 5$. The cost ($1000/ship) of repositioning a ship from its current location to a service is shown below:

Location	r1	r2	r3
1	11	12	13
2	25	28	37
3	0	50	40
4	20	20	0
5	31	14	2

Formulate an integer optimization model to determine how to reposition the ships to minimize the total repositioning cost.

Solution: Let $x_{ij}, i = 1, 2, 3, 4, 5, j = 1, 2, 3$, be the number of ships to reposition from location i to service j. The model is

$$\min 11x_{11} + 12x_{12} + 13x_{13} + 25x_{21} + 28x_{22} + 27x_{23} + 50x_{32}$$
$$+ 40x_{33} + 20x_{41} + 20x_{42} + 31x_{51} + 14x_{52} + 2x_{53}$$

subject to

$$x_{11} + x_{12} + x_{13} \leq 2$$
$$x_{21} + x_{22} + x_{23} \leq 3$$
$$x_{31} + x_{32} + x_{33} \leq 4$$
$$x_{41} + x_{42} + x_{43} \leq 5$$
$$x_{51} + x_{52} + x_{53} \leq 6$$

$$x_{11} + x_{21} + x_{31} + x_{41} + x_{51} = 4$$
$$x_{12} + x_{22} + x_{32} + x_{42} + x_{52} = 5$$
$$x_{13} + x_{23} + x_{33} + x_{43} + x_{53} = 6$$
$$x_{ij} \in \mathbb{Z}_+, i = 1, 2, 3, 4, 5, j = 1, 2, 3.$$

17.2 Formulation II: 0–1 decision variables

Some integer optimization models have only binary decision variables, i.e., decision variables which can only take the value 0 or 1. Such models are called 0–1 integer optimization models or binary integer optimization models. 0–1 integer optimization models have wide applications for expressing logical constraints.

Example 17.5: Consider a teaching venue allocation problem. Suppose that there are six courses (C1–C6) to be taught in three classrooms (R1–R3). Each classroom can only be used to teach two courses due to reservations, e.g., meetings. Not all classrooms are suitable for all courses because of available teaching equipment, as shown in the table below where "Y" means the classroom is suitable for the course and "N" means not suitable. Formulate an integer optimization model to find a feasible classroom allocation plan for all the courses.

	R1	R2	R3
C1	Y	Y	N
C2	Y	N	N
C3	N	N	Y
C4	Y	Y	Y
C5	N	N	Y
C6	Y	Y	N

Solution: Let x_{ij}, $i = 1, 2, 3, 4, 5, 6, j = 1, 2, 3$, be a binary decision variable which equals 1 if course i is taught in classroom j, and 0 otherwise. The model is

$$\min 0$$

subject to

$$x_{11} + x_{12} + x_{13} = 1$$
$$x_{21} + x_{22} + x_{23} = 1$$
$$x_{31} + x_{32} + x_{33} = 1$$
$$x_{41} + x_{42} + x_{43} = 1$$
$$x_{51} + x_{52} + x_{53} = 1$$
$$x_{61} + x_{62} + x_{63} = 1$$

$$x_{11} + x_{21} + x_{31} + x_{41} + x_{51} + x_{61} \le 2$$
$$x_{12} + x_{22} + x_{32} + x_{42} + x_{52} + x_{62} \le 2$$
$$x_{13} + x_{23} + x_{33} + x_{43} + x_{53} + x_{63} \le 2$$
$$x_{13} = x_{22} = x_{23} = x_{31} = x_{32} = x_{51} = x_{52} = x_{63} = 0$$
$$x_{ij} \in \{0, 1\}, i = 1, 2, 3, 4, 5, 6, j = 1, 2, 3.$$

Example 17.6: Consider a service with the port rotation below:

$$Shanghai(S, 1) \to Pusan(P, 2) \to LosAngeles(L, 3) \to Shanghai(S, 1)$$

The container shipment demand is: S to P $q^{SP} = 800$ and S to L $q^{SL} = 6000$. The revenue of transporting one TEU (twenty-foot equivalent unit) from S to P is $200 and from S to L is $1 300. Suppose that the company needs to determine which of the following three types of container ship to use:

Ship type	Capacity (TEUs)	Cost (thousand $/week)
Post-Panamax	8 000	500
Panamax type 2	5 100	350
Panamax type 1	4 800	330

Exactly one type of ship will be used. The company aims to maximize its profit. Formulate a mixed-integer linear optimization model to help the company make the decision of which type of ship to use.

Solution: Let x_1, x_2, x_3 be binary decision variables that equal 1 if and only post-Panamax, Panamax type 2, and Panamax type 1 ships are used, respectively, and 0 otherwise. Let y^{SP} and y^{SL} be the decision variables representing volumes of containers transported from S to P and S to L, respectively. The model is

$$\max 200 y^{SP} + 1300 y^{SL} - (500,000 x_1 + 350,000 x_2 + 330,000 x_3)$$

subject to

$$y^{SP} + y^{SL} \le 8000 x_1 + 5100 x_2 + 4800 x_3$$
$$y^{SL} \le 8000 x_1 + 5100 x_2 + 4800 x_3$$
$$y^{SP} \le 800$$
$$y^{SL} \le 6000$$
$$x_1 + x_2 + x_3 = 1$$
$$x_1 \in \{0, 1\}, x_2 \in \{0, 1\}, x_3 \in \{0, 1\}$$
$$y^{SP} \ge 0, y^{SL} \ge 0.$$

The above example explains why we assume continuous numbers of containers in some problems: our purpose is to decide the type of ship to use rather than to transport containers according to the calculated numbers. After the ships are deployed and the service is operated, we will know the exact demand for each week (which will fluctuate with time and be different from the forecast value in our example) and

280 Machine learning and data analytics for maritime studies

then plan the delivery of containers by formulating integer numbers of containers (moreover, we should also consider the type of containers, the storage of containers on the ship, etc.).

Example 17.7: Figure 17.1 shows a liner container shipping network. The circles represent ports. There are three candidate routes r_1, r_2, and r_3. The capacity of a ship deployed on route r_1 is E_1 (TEUs). E_2 and E_3 have similar meanings. The volumes of containers between different OD (origin-destination) pairs are shown in the figure. For example, q^{12} is the demand (TEUs/week) from port 1 to port 2. The revenue for transporting one TEU from port 1 to port 2 is g^{12} (\$/TEU). g^{13} and g^{32} have similar meanings. Assume that the container handling costs are 0. Suppose that the company needs to determine which ones of the three routes r_1, r_2, and r_3 should be operated. The costs for operating r_1, r_2, and r_3 are known and represented by c_1, c_2, and c_3 (\$/week), respectively. If some routes are not profitable, then they will not be operated. Formulate a mixed-integer linear optimization model to help the company make the decision of which routes to operate.

Solution: The set of OD pairs are $\mathcal{W} = \{(1,2),(1,3),(3,2)\}$. Let $\mathcal{H}^{od}$ be the set of itineraries (paths) for OD pair $(o,d) \in \mathcal{W}$. To simplify the notation, we use, $i >$ to represent leg i of route r. Then $\mathcal{H}^{12}$ consists of:

$$h_1 :< 1,1 >$$
$$h_2 :< 3,1 > \rightarrow < 2,2 >$$

$\mathcal{H}^{13}$ consists of :

$$h_3 :< 3,1 >$$
$$h_4 :< 1,1 > \rightarrow < 2,1 >$$

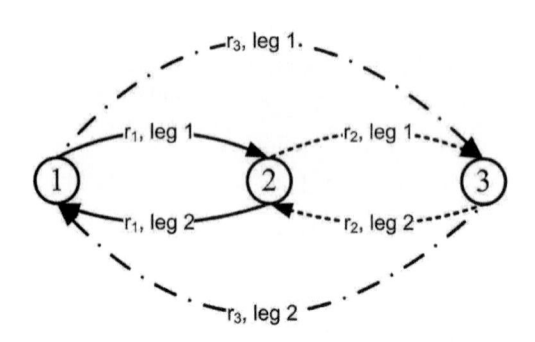

Positive demand: q^{12}, q^{13}, q^{32}

Figure 17.1 A container liner shipping network with direct delivery and transshipment

$\mathcal{H}^{32}$ consists of :

$$h_5 : <2,2>$$
$$h_6 : <3,2> \to <1,1>.$$

Define $\mathcal{H} := \cup_{(o,d)\in\mathcal{W}} \mathcal{H}^{od}$. Let x_r be a binary decision variable that equals 1 if and only if route r is operated, $r = 1, 2, 3$. Let y_h be the decision variable representing the flow on itinerary $h \in \mathcal{H}$ (TEUs/week). The model is

$$\max \sum_{(o,d)\in\mathcal{W}} g^{od} \sum_{h\in\mathcal{H}^{od}} y_h - (c_1 x_1 + c_2 x_2 + c_3 x_3)$$

subject to

$$\sum_{h\in\mathcal{H}^{od}} y_h \leq q^{od}, (o,d) \in \mathcal{W}$$
$$y_1 + y_4 + y_6 \leq E_1 x_1 \text{ (capacity on leg } <1,1>)$$
$$y_4 \leq E_2 x_2 \text{ (capacity on leg } <2,1>)$$
$$y_2 + y_5 \leq E_2 x_2 \text{ (capacity on leg } <2,2>)$$
$$y_2 + y_3 \leq E_3 x_3 \text{ (capacity on leg } <3,1>)$$
$$y_6 \leq E_3 x_3 \text{ (capacity on leg } <3,2>)$$
$$x_1 \in \{0,1\}, x_2 \in \{0,1\}, x_3 \in \{0,1\}$$
$$y_h \geq 0, h \in \mathcal{H}.$$

17.3 Formulation III: complex logical constraints

In an optimization model, the relation between the constraints is "AND," which means all of them must be satisfied. Binary variables can be used to formulate more complex logic as linear constraints.

Suppose that we have three candidate constraints $x + y \leq 2$, $2x - y \leq 3$, $-3x + 4y \leq 4$ that do not necessarily hold. If we require that at least one of them must hold, then we can define M as a large positive number, e.g., $M = 10^{10}$ (note that M is a constant rather than a decision variable), and define three binary decision variables z_1, z_2, z_3, and the requirement can be formulated as

$$x + y - 2 \leq M z_1$$
$$2x - y - 3 \leq M z_2$$
$$-3x + 4y - 4 \leq M z_3$$
$$z_1 + z_2 + z_3 \leq 2$$
$$z_1 \in \{0,1\}, z_2 \in \{0,1\}, z_3 \in \{0,1\}.$$

Because of M, we often say we use "the big-M method" to formulate the constraints.

It might be attempting to set M at a very large value to be on the safe side. However, a larger M will make the model more time-consuming to solve. In reality, we should try to find the smallest possible M. For example, if the above model further has the following constraints: $0 \leq x \leq 2$ and $-1 \leq y \leq 3$, then we know that $\max\{x + y - 2, 2x - y - 3, -3x + 4y - 4\} = 8$, and we can hence set $M = 8$.

Example 17.8: We have a model. One of its decision variables x can either be greater than or equal to 2, or less than or equal to 1. How to use the big-M method to formulate this requirement as linear constraints with binary variables?

Solution: The requirement is at least one of the following two constraints must hold: $2 - x \leq 0$ and $x - 1 \leq 0$. Hence, it can be formulated as

$$2 - x \leq Mz_1$$
$$x - 1 \leq Mz_2$$
$$z_1 + z_2 \leq 1$$
$$z_1 \in \{0, 1\}, z_2 \in \{0, 1\}.$$

Example 17.9: We have a model. One of its decision variables x can either be greater than or equal to 2, or between 0 and 1 (inclusive of 0 and 1). How to use the big-M method to formulate this requirement as linear constraints with binary variables?

Solution: The requirement is: $x \geq 0$ and at least one of the following two constraints must hold: $2 - x \leq 0$ and $x - 1 \leq 0$. Hence, it can be formulated as

$$2 - x \leq Mz_1$$
$$x - 1 \leq Mz_2$$
$$z_1 + z_2 \leq 1$$
$$z_1 \in \{0, 1\}, z_2 \in \{0, 1\}$$
$$x \geq 0.$$

17.4 Solving mixed-integer optimization models

The model obtained by removing the integrality constraints of the variables in an integer optimization model is called the linear programming relaxation of the integer optimization model, which is a linear optimization model.

Example 17.10: Write down the linear programming relaxation of the model in Example 17.1.

Solution:
$$\min 1x + 0.8y + 0.6z + 0.3w$$

subject to
$$100x + 60y + 40z + 10w \geq 1000$$
$$x \leq 8$$
$$y \leq 8$$
$$z \leq 7$$

$$w \leq 5$$
$$x \geq 0$$
$$y \geq 0$$
$$z \geq 0$$
$$w \geq 0.$$

Note that the last four non-negativity constraints are the corresponding linear programming relaxation of the constraints on the decision variables that they must take non-negative integer values.

It is easy to understand that the optimal objective function value of the linear programming relaxation is at least as good as that of the corresponding integer or mixed-integer optimization model. It is even possible that the linear programming relaxation is unbounded, while the corresponding integer optimization model is infeasible. For example, the following mixed-integer optimization model has no feasible solution, while its linear programming relaxation is unbounded: maximizing $x + y$ subject to $x \geq 0$, $0.5 \leq y \leq 0.9$ and $y \in \mathbb{Z}_+$.

Most algorithms for solving integer optimization models are based on repeatedly solving linear programming relaxations that are related to the integer optimization models, such as the cutting-plane method, the branch-and-bound method, and the branch-and-cut method. Most linear optimization solvers, such as Excel, lpsolve, MATLAB®, and CPLEX, can also solve integer optimization models. We must bear in mind that while linear optimization models are generally easy to solve, integer optimization models are generally difficult to solve. It is not uncommon that a linear optimization model can be solved in 1 s, and after imposing that all the decision variables are integers, the model cannot be solved in 1 week. As a rule of thumb, one might expect a model with 20 integer variables and 1 000 constraints to be solved in 1 min and expect that a model with 100 integer variables and 1 000 constraints cannot be solved in 1 day.

17.5 Formulation IV: challenging problems

In contrast to the fact that linear optimization models are generally easy to formulate, (i) it may be difficult to formulate an integer optimization model, (ii) one may often formulate an incorrect integer optimization model, and (iii) it is possible to formulate a bad integer optimization model when there are good formulations that can be solved efficiently.

Example 17.11: There is a model with the following constraints

$$x_1 = x_2$$
$$x_1 = x_3$$
$$x_1 = x_4$$

$x_1 \in \{0, 1\}, x_2 \in \{0, 1\}, x_3 \in \{0, 1\}, x_4 \in \{0, 1\}.$

Formulate one constraint to replace the first three constraints.

Solution: We can use $3x_1 = x_2 + x_3 + x_4$.

Note that the above example simply aims to show that there may be different integer optimization formulations for the same problem. It does not mean that fewer constraints are better.

Example 17.12: Reconsider Example 17.11. Which constraints can be dropped from the set of constraints?

Solution: We can drop $x_2 \in \{0,1\}, x_3 \in \{0,1\}, x_4 \in \{0,1\}$.

Example 17.13: This is the traveling salesman problem. Suppose that a delivery man is at node 1 (a KFC store), needs to visit $n-1$ customers, denoted by nodes $2, 3, \cdots, n$, and returns to node 1. Define the set of nodes $N := \{1, 2, \cdots, n\}$ and the set of arcs $A := \{(i,j), i \in N, j \in N, i \neq j\}$. The distance of arc (i,j) is d_{ij}. Develop an integer optimization model to determine the shortest tour to visit all customers.

Beginners might work as follows: let x_{ij} be a binary variable which equals 1 if and only if arc (i,j) is traversed, and the model is

$$\min \sum_{(i,j) \in A} d_{ij} x_{ij}$$

subject to

$$\sum_{j \in N \setminus \{i\}} x_{ij} = 1, i \in N$$

$$\sum_{i \in N \setminus \{j\}} x_{ij} = 1, j \in N$$

$$x_{ij} \in \{0,1\}, (i,j) \in A.$$

The above formulation is incorrect because its solution may contain subtours, see the example in Figure 17.2.

Solution: Let x_{ij} be a binary variable which equals 1 if and only if arc (i,j) is traversed, and let u_i represent the number of customers that have been served after visiting i. The purpose of the decision variables u_i is to eliminate subtours. The model is

$$\min \sum_{(i,j) \in A} d_{ij} x_{ij}$$

$n = 4, x_{12}^* = x_{21}^* = x_{34}^* = x_{43}^* = 1$, **two subtours in the solution**

Figure 17.2 Subtours in TSP

subject to
$$\sum_{j\in M\setminus\{i\}} x_{ij} = 1, i \in N$$
$$\sum_{i\in M\setminus\{j\}} x_{ij} = 1, j \in N$$
$$x_{ij} \in \{0,1\}, (i,j) \in A$$
$$u_j \geq u_i + 1 - (n-1)(1-x_{ij}), i \in N, j \in N\setminus\{1\}, j \neq i$$
$$u_i \geq 1, i \in N\setminus\{1\}$$
$$u_1 = 0.$$

Example 17.14: In the traveling salesman problem, write down the model for the case of $n = 4$. You are not allowed to use "$\sum$" or "$\forall$" except in the non-negativity/binary constraints.

Solution: Let x_{ij} be a binary variable which equals 1 if and only if arc (i,j) is traversed, and let u_i represent the number of customers that have been served after visiting i. The model is

$$\min d_{12}x_{12} + d_{13}x_{13} + d_{14}x_{14} + d_{21}x_{21} + d_{23}x_{23} + d_{24}x_{24} +$$
$$d_{31}x_{31} + d_{32}x_{32} + d_{34}x_{34} + d_{41}x_{41} + d_{42}x_{42} + d_{43}x_{43}$$

subject to
$$x_{12} + x_{13} + x_{14} = 1$$
$$x_{21} + x_{23} + x_{24} = 1$$
$$x_{31} + x_{32} + x_{34} = 1$$
$$x_{41} + x_{42} + x_{43} = 1$$
$$x_{21} + x_{31} + x_{41} = 1$$
$$x_{12} + x_{32} + x_{42} = 1$$
$$x_{13} + x_{23} + x_{43} = 1$$
$$x_{14} + x_{24} + x_{34} = 1$$
$$x_{ij} \in \{0,1\}, (i,j) \in A$$
$$u_2 \geq u_1 + 1 - (n-1)(1-x_{12})$$
$$u_2 \geq u_3 + 1 - (n-1)(1-x_{32})$$
$$u_2 \geq u_4 + 1 - (n-1)(1-x_{42})$$
$$u_3 \geq u_1 + 1 - (n-1)(1-x_{13})$$
$$u_3 \geq u_2 + 1 - (n-1)(1-x_{23})$$
$$u_3 \geq u_4 + 1 - (n-1)(1-x_{43})$$
$$u_4 \geq u_1 + 1 - (n-1)(1-x_{14})$$
$$u_4 \geq u_2 + 1 - (n-1)(1-x_{24})$$
$$u_4 \geq u_3 + 1 - (n-1)(1-x_{34})$$

$$u_2 \geq 1$$
$$u_3 \geq 1$$
$$u_4 \geq 1$$
$$u_1 = 0.$$

Example 17.15: Revisit the shortest path problem. A transportation network has a set of nodes $N := \{1, 2, \cdots, n\}$ and a set of arcs $A := \{(i,j), i \in N, j \in N, i \neq j\}$. The distance of arc (i,j) is d_{ij}. Develop an integer optimization model to determine the shortest path from node 1 to node n. Note the differences between the shortest path problem and the TSP (travelling salesman problem); in the shortest path problem, not all nodes have to be visited, and the person does not return to node 1.

Solution: Let x_{ij} be a binary variable which equals 1 if and only if arc (i,j) is traversed. The model is

$$\min \sum_{(i,j) \in A} d_{ij} x_{ij}$$

subject to

$$\sum_{j \in M\{1\}} x_{1j} = 1$$

$$\sum_{i \in M\{n\}} x_{in} = 1$$

$$\sum_{j \in M\{i\}} x_{ij} = \sum_{j \in M\{i\}} x_{ji}, i \in N \setminus \{1, n\}$$

$$x_{ij} \in \{0, 1\}, (i,j) \in A.$$

Note that the second constraint can be removed as it is implied by other constraints. Note further that you can add the following constraints to the model if you like: $x_{i1} = 0, i \in N \setminus \{1\}$ and $x_{nj} = 0, j \in N \setminus \{n\}$.

Example 17.16: In the shortest path problem, write down the model for the case of $n = 4$. You are not allowed to use "$\sum$" or "$\forall$" except in the non-negativity/binary constraints.

Solution: The set of arcs are $\mathcal{A} := \{(1,2), (1,3), (1,4), (2,3), (2,4), (3,2), (3,4)\}$. Let x_{ij} be a binary variable which equals 1 if and only if arc $(i,j) \in \mathcal{A}$ is traversed. The model is

$$\min d_{12}x_{12} + d_{13}x_{13} + d_{14}x_{14} + d_{23}x_{23} + d_{24}x_{24} + d_{32}x_{32} + d_{34}x_{34}$$

subject to

$$x_{12} + x_{13} + x_{14} = 1$$
$$x_{14} + x_{24} + x_{34} = 1$$
$$x_{23} + x_{24} = x_{12} + x_{32}$$
$$x_{32} + x_{34} = x_{13} + x_{23}$$
$$x_{ij} \in \{0, 1\}, (i,j) \in \mathcal{A}.$$

Example 17.17: Ships need to be deployed on three services r_1, r_2, and r_3 to maintain a weekly frequency. The operating cost of a service given the number of ships deployed is shown below.

Cost (million $/week)	Five ships	Six ships	Seven ships
r_1	1.5	1.3	1.1
r_2	2.0	1.9	1.6
r_3	2.3	2.2	2.0

Suppose that the company has ten ships and can charter in additional ships at the cost of 0.1 million $/week per ship. Develop a model to determine how to deploy ships on the three services at minimum cost.

Solution: Let x_r and y_r be the numbers of owned ships and chartered ships to deploy on route $r = 1, 2, 3$, respectively. Let z_{rs} be a binary variable that equals 1 if and only if $s = 5, 6, 7$ ships are deployed on route $r = 1, 2, 3$. The model is

min

$$00.1(y_1 + y_2 + y_3) +$$
$$1.5z_{15} + 1.3z_{16} + 1.1z_{17} + 2z_{25} + 1.9z_{26} + 1.6z_{27} + 2.3z_{35} + 2.2z_{36} + 2z_{37}$$

subject to

$$x_r + y_r = \sum_{s=5}^{7} s z_{rs}, r = 1, 2, 3$$

$$\sum_{s=5}^{7} z_{rs} = 1, r = 1, 2, 3$$

$$x_r \in \mathbb{Z}_+, r = 1, 2, 3$$

$$\sum_{r=1}^{3} x_r \leq 10$$

$$y_r \in \mathbb{Z}_+, r = 1, 2, 3$$

$$z_{rs} \in \{0, 1\}, r = 1, 2, 3, s = 5, 6, 7.$$

The problem with the above formulation is that it has a large number of symmetrical optimal solutions. For example, if $(x_1^*, y_1^*, x_2^*, y_2^*, x_3^*, y_3^*) = (4, 1, 4, 2, 2, 5)$ is an optimal solution, then $(x_1^*, y_1^*, x_2^*, y_2^*, x_3^*, y_3^*) = (3, 2, 4, 2, 3, 4)$ is also an optimal solution. In fact, there will be 35 optimal solutions. Generally speaking, having so many symmetrical feasible and optimal solutions will make the model very difficult to solve (the reason is beyond the scope of the subject) Readers can refer to Reference [1] to see a comparison of efficiency of the two models. A better formulation is as follows.

Solution: Let m_r be the number of ships (including owned and chartered ships) to deploy on route $r = 1, 2, 3$. Let x be the total number of owned ships deployed on the three routes and y be the total number of chartered ships deployed on the three routes. Let z_{rs} be a binary variable that equals 1 if and only if $s = 5, 6, 7$ ships are deployed on route $r = 1, 2, 3$. The model is

min

$$00.1y+$$
$$1.5z_{15} + 1.3z_{16} + 1.1z_{17} + 2z_{25} + 1.9z_{26} + 1.6z_{27} + 2.3z_{35} + 2.2z_{36} + 2z_{37}$$

subject to

$$m_r = \sum_{s=5}^{7} sz_{rs}, r = 1, 2, 3$$
$$\sum_{s=5}^{7} z_{rs} = 1, r = 1, 2, 3$$
$$\sum_{r=1}^{3} m_r = x + y$$
$$x \leq 10$$
$$m_r \in \mathbb{Z}_+, r = 1, 2, 3$$
$$x \in \mathbb{Z}_+$$
$$y \in \mathbb{Z}_+$$
$$z_{rs} \in \{0, 1\}, r = 1, 2, 3, s = 5, 6, 7.$$

We can further prove that the integrality constraints on x and y can be removed.

Moreover, for some problem, natural integer optimization formulations have an exponential number of variables, while more clever formulations have only a polynomial number of variables.

17.6 Formulation V: linearizing binary variables multiplied by another variable

Given $z_1 \in \{0, 1\}, z_2 \in \{0, 1\}$, the constraint $z \geq z_1 z_2$ is equivalent to $z \geq z_1 + z_2 - 1, z \geq 0$; the constraint $z \leq z_1 z_2$ is equivalent to $z \leq z_1, z \leq z_2$; and the constraint $z = z_1 z_2$ is equivalent to $z \leq z_1, z \leq z_2, z \geq z_1 + z_2 - 1, z \geq 0$.

Given $z \in \{0, 1\}, 0 \leq x \leq M$, the constraint $y \geq zx$ is equivalent to $y \geq 0, y \geq x - M(1 - z)$; the constraint $y \leq zx$ is equivalent to $y \leq x, y \leq Mz$; and the constraint $y = zx$ is equivalent to $y \geq 0, y \geq x - M(1 - z), y \leq x, y \leq Mz$.

Example 17.18: A model has decision variables $z_1 \in \{0, 1\}, z_2 \in \{0, 1\}, 0 \leq x \leq M$, an objective function that minimizes $3z_1 z_2 - 4z_1 x$, and some linear constraints. How do you linearize the objective function?

Solution: We introduce two new variables u_1 and u_2. The model is

min $3u_1 - 4u_2$

subject to
$$u_1 = z_1 z_2,$$
$$u_2 = z_1 x,$$

and other relevant constraints. This model is equivalent to:
$$\min 3u_1 - 4u_2$$
subject to
$$u_1 \geq z_1 z_2,$$
$$u_2 \leq z_1 x,$$

and other relevant constraints. This model is equivalent to:
$$\min 3u_1 - 4u_2$$
subject to
$$u_1 \geq z_1 + z_2 - 1,$$
$$u_1 \geq 0,$$
$$u_2 \leq x,$$
$$u_2 \leq M z_1,$$

and other relevant constraints.

17.7 Practice

Example 17.19: A student has at most 20 days for preparing the final exams of four subjects. Each subject is worth six credit points. The student needs to determine how many days to spend on each subject. It is required that the number of days allocated to each subject must be an integer.

Suppose that the final score of a subject is proportional to the number of days spent on it. Let c_i be the scores the student can get if he spends 1 day on subject $i = 1, 2, 3, 4$. For instance, if $c_1 = 11$, then the student can get 99 marks if he spends 9 days on subject 1; as the student cannot get more than 100 marks, the student will not spend more than 9 days on the subject. We therefore let T_i be the maximum days the student spends on subject $i = 1, 2, 3, 4$ is the largest integer that does not exceed $\frac{100}{c_i}$. The student knows the values of c_i and T_i, $i = 1, 2, 3, 4$.

The student wants to maximize his average mark. How should he allocate time to the four subjects?

Solution: Let $x_i, i = 1, 2, 3, 4$, be the number of days spent on subject i. The model is
$$\max \frac{c_1 x_1 + c_2 x_2 + c_3 x_3 + c_4 x_4}{4}$$

subject to

$$x_1 + x_2 + x_3 + x_4 \leq 20$$
$$x_i \leq T_i, i = 1, 2, 3, 4$$
$$x_i \in \mathbb{Z}_+, i = 1, 2, 3, 4.$$

Example 17.20: In the context of Example 17.19, the student must make sure that he passes all the subjects (at least 50 marks). How should he allocate time to the four subjects in order to maximize his average mark?

Solution: Let $x_i, i = 1, 2, 3, 4$, be the number of days spent on subject i. The model is

$$\max \frac{c_1 x_1 + c_2 x_2 + c_3 x_3 + c_4 x_4}{4}$$

subject to

$$x_1 + x_2 + x_3 + x_4 \leq 20$$
$$c_i x_i \geq 50, i = 1, 2, 3, 4$$
$$x_i \leq T_i, i = 1, 2, 3, 4$$
$$x_i \in \mathbb{Z}_+, i = 1, 2, 3, 4.$$

Example 17.21: In the context of Example 17.19, the student must make sure that he passes all the subjects. How should he allocate time to the four subjects in order to get the most number of "HD" (at least 85 marks)? (Hint: Let $x_i, i = 1, 2, 3, 4$, be the number of days spent on subject i. Let $z_i, i = 1, 2, 3, 4$, be a binary variable that equals 1 if and only if he gets HD for subject i. Then we maximize $z_1 + z_2 + z_3 + z_4$. We may have the constraint $c_i x_i \geq 85 z_i$, so that z_i can be 1 only if $c_i x_i \geq 85$.)

Solution: Let $x_i, i = 1, 2, 3, 4$, be the number of days spent on subject i. Let $z_i, i = 1, 2, 3, 4$, be a binary variable that equals 1 if and only if he gets HD for subject i. The model is

$$\max z_1 + z_2 + z_3 + z_4$$

subject to

$$x_1 + x_2 + x_3 + x_4 \leq 20$$
$$c_i x_i \geq 50, i = 1, 2, 3, 4$$
$$c_i x_i \geq 85 z_i, i = 1, 2, 3, 4$$
$$x_i \leq T_i, i = 1, 2, 3, 4$$
$$x_i \in \mathbb{Z}_+, i = 1, 2, 3, 4$$
$$z_i \in \{0, 1\}, i = 1, 2, 3, 4.$$

Example 17.22: In the context of Example 17.19, develop a model to help the student check whether he can get "D" (at least 75 marks) for all the subjects.

Solution: Let $x_i, i = 1, 2, 3, 4$, be the number of days spent on subject i. The model is

$$\min 0$$

subject to

$$x_1 + x_2 + x_3 + x_4 \leq 20$$
$$c_i x_i \geq 75, i = 1, 2, 3, 4$$
$$x_i \leq T_i, i = 1, 2, 3, 4$$
$$x_i \in \mathbb{Z}_+, i = 1, 2, 3, 4.$$

If the above model has a feasible solution, he can get "D" for all the subjects, otherwise he cannot.

Example 17.23: In the context of Example 17.19, develop a model to help the student check whether he can get at least two "D" and two "HD" (i.e., two "D" and two "HD," one "D" and three "HD," or four "HD").

Solution: Similar to Example 17.23, this question requires that the score for each subject should at least be 75 and checks whether the maximum number of "HD" exceeds two. Let $x_i, i = 1, 2, 3, 4$, be the number of days spent on subject i. Let $z_i, i = 1, 2, 3, 4$, be a binary variable that equals 1 if and only if he gets HD for subject i. The model is

$$\max z_1 + z_2 + z_3 + z_4$$

subject to

$$x_1 + x_2 + x_3 + x_4 \leq 20$$
$$c_i x_i \geq 75, i = 1, 2, 3, 4$$
$$c_i x_i \geq 85 z_i, i = 1, 2, 3, 4$$
$$x_i \leq T_i, i = 1, 2, 3, 4$$
$$x_i \in \mathbb{Z}_+, i = 1, 2, 3, 4$$
$$z_i \in \{0, 1\}, i = 1, 2, 3, 4.$$

If the optimal objective function value is greater than or equal to 2, the student can get at least two "D" and two "HD," otherwise he cannot.

Example 17.24: In the context of Example 17.19, it may not be reasonable to assume that the score of a subject is proportional to the number of days spent on it. Therefore, we let c_{ij} be the total score of subject $i = 1, 2, 3, 4$ the student can get if $j = 0, 1, \cdots, 20$ days is spent on it. Of course, $0 \leq c_{i0} \leq c_{i1} \leq c_{i2} \leq \cdots \leq c_{i,20} \leq 100$. The student knows the values of c_{ij}. How should he allocate time to the four subjects in order to maximize his average mark? (Hint: Let $y_{ij}, i = 1, 2, 3, 4$ and $j = 0, 1, \cdots, 20$, be a binary variable that equals 1 if and only if j days is spent on subject i.)

Solution: Let $y_{ij}, i = 1, 2, 3, 4$ and $j = 0, 1, \cdots, 20$, be a binary variable that equals 1 if and only if j days is spent on subject i. The model is:

$$\max \frac{\sum_{i=1}^{4} \sum_{j=0}^{20} c_{ij} y_{ij}}{4}$$

subject to

$$\sum_{i=1}^{4}\sum_{j=0}^{20} jy_{ij} \leq 20$$

$$\sum_{j=0}^{20} y_{ij} = 1, i = 1, 2, 3, 4$$

$$y_{ij} \in \{0, 1\}, i = 1, 2, 3, 4, j = 0, 1, \cdots, 20.$$

Example 17.25: M is a given number. We have the constraints $y \geq x_1 z + x_2 z^2 + x_3 z^3$, $z \in \{0, 1\}, 0 \leq x_1 \leq M, 0 \leq x_2 \leq M, 0 \leq x_3 \leq M$. How to linearize the first one?

Solution: Since z is binary, $z = z^2 = z^3$. Therefore, the constraint is $y \geq x_1 z + x_2 z + x_3 z = (x_1 + x_2 + x_3) z$, which is equivalent to $y \geq x_1 + x_2 + x_3 - 3M(1-z)$, $y \geq 0$.

Example 17.26: M is a given positive number. We have the constraints $y = xu, u \in \mathbb{Z}_+, 0 \leq x \leq M, u \leq M$. How to linearize the first one?

Solution: Define binary variables $z_i \in \{0, 1\}, i = 0, 1 \cdots M$. The constraints can be written as $y = x \sum_{i=0}^{M} i z_i$, $\sum_{i=0}^{M} z_i = 1, 0 \leq x \leq M$.

Define $u_i = x z_i, i = 0, 1 \cdots M$, which is equivalent to $u_i \geq 0, u_i \geq x - M(1 - z_i)$, $u_i \leq x, u_i \leq M z_i, i = 0, 1 \cdots M$. Then $y = \sum_{i=0}^{M} i u_i$.

In sum, the linearized constraints are

$$y = \sum_{i=0}^{M} i u_i$$

$$u_i \geq 0, i = 0, 1 \cdots M$$

$$u_i \geq x - M(1 - z_i), i = 0, 1 \cdots M$$

$$u_i \leq x, i = 0, 1 \cdots M$$

$$u_i \leq M z_i, i = 0, 1 \cdots M$$

$$\sum_{i=0}^{M} z_i = 1$$

$$z_i \in \{0, 1\}, i = 0, 1 \cdots M$$

$$0 \leq x \leq M.$$

Example 17.27: Suppose that there is a mixed-integer linear optimization solver that can address mixed-integer linear optimization models with at most one integer variable in each constraint. For example, if $x \in \mathbb{R}, y \in \mathbb{R}, z_1 \in \mathbb{Z}, z_2 \in \mathbb{Z}$, then the constraint $x + 2y - 3z_1 \geq 4$ has only one integer variable and therefore can be addressed; the constraint $x + z_2 = 4$ can also be addressed. How do you use this solver to address a mixed-integer linear optimization model with at most one integer variable in each constraint except the following one $x + 2y - 3z_1 + 5z_2 \geq 4$?

Solution: Define a continuous decision variable $u \in \mathbb{R}$. The constraint is equivalent to

$$x + 2y - 3z_1 + 5u \geq 4$$

$$u = z_2.$$

Integer optimization 293

Example 17.28: Consider a different version of the shortest path problem. A transportation network has a set of nodes $N := \{1, 2, \cdots, n\}$ and a set of arcs $A := \{(i,j), i \in N, j \in N, i \neq j\}$, $n \geq 10$. The distance of arc (i,j) is d_{ij}. We need to find the shortest path from node 1 to node n subject to the requirement that at least three nodes between 1 and n must be visited. For instance, the tour $1 \to 5 \to 2 \to n$ is not valid. Develop an integer optimization model to solve the problem.

Solution: The model below is incorrect because the resulting solution may contain subtours. Let x_{ij} be a binary variable which equals 1 if and only if arc (i,j) is traversed. Let z_i be a binary variable which equals 1 if and only if node $i \in N \setminus \{1, n\}$ is visited. The model is $\min \sum_{(i,j) \in A} d_{ij} x_{ij}$

subject to

$$\sum_{j \in N \setminus \{1\}} x_{1j} = 1$$

$$\sum_{i \in N \setminus \{n\}} x_{in} = 1$$

$$\sum_{j \in N \setminus \{i\}} x_{ij} = z_i, i \in N \setminus \{1, n\}$$

$$\sum_{j \in N \setminus \{i\}} x_{ji} = z_i, i \in N \setminus \{1, n\}$$

$$\sum_{i \in N \setminus \{1, n\}} z_i \geq 3$$

$$x_{ij} \in \{0, 1\}, (i, j) \in A$$

$$z_i \in \{0, 1\}, i \in N \setminus \{1, n\}.$$

The solution below is correct.

Let z_i be a binary variable which equals 1 if and only if node i is visited, $i \in N \setminus \{1, n\}$, let x_{ij} be a binary variable which equals 1 if and only if arc (i,j) is traversed, and let u_i be auxiliary variables for eliminating subtours, $i \in N$. The model is

$$\min \sum_{(i,j) \in A} d_{ij} x_{ij}$$

subject to

$$\sum_{j \in N \setminus \{1\}} x_{1j} = 1$$

$$\sum_{i \in N \setminus \{n\}} x_{in} = 1$$

$$\sum_{j \in N \setminus \{i\}} x_{ij} = z_i, i \in N \setminus \{1, n\}$$

$$\sum_{i \in N \setminus \{j\}} x_{ij} = z_j, j \in N \setminus \{1, n\}$$

$$\sum_{i \in N \setminus \{1, n\}} z_i \geq 3$$

$$x_{ij} \in \{0, 1\}, (i, j) \in A$$

$$u_j \geq u_i + 1 - (n-1)(1 - x_{ij}), i \in N, j \in N \setminus \{1\}, j \neq i$$

$$z_i \in \{0, 1\}, i \in N \setminus \{1, n\}$$

$$u_i \geq 0, i \in N.$$

Example 17.29: A student has at most 20 days for preparing the final exams of four subjects. Each subject is worth six credit points. The student needs to determine how many days to spend on each subject. It is required that the number of days allocated to each subject must be an integer.

Suppose that the final mark of a subject is proportional to the number of days spent on it. Let c_i be the mark the student can get if he spends 1 day on subject $i = 1, 2, 3, 4$. For instance, if $c_1 = 11$, then the student can get 99 marks if he spends 9 days on subject 1; as the student cannot get more than 100 marks, the student will not spend more than 9 days on the subject. We therefore let T_i be the maximum number of days the student spends on subject $i = 1, 2, 3, 4$ is the largest integer that does not exceed $\frac{100}{c_i}$. The student knows the values of c_i and T_i, $i = 1, 2, 3, 4$. All c_i, $i = 1, 2, 3, 4$, are integers. Therefore, the final mark of a subject must also be an integer.

The student wants to maximize his average mark while (i) not failing any subject (i.e., the final mark of each subject is at least 50) and (ii) not getting a "credit" in any subject (i.e., the final mark of each subject is either less than or equal to 64, or greater than or equal to 75). In sum, the final mark of each subject is either between 50 (inclusive) and 64 (inclusive), or at least 75 (inclusive). Formulate an integer linear optimization model to help the student allocate time to the four subjects to achieve his goal.

Solution: Let x_i, $i = 1, 2, 3, 4$, be the number of days spent on subject i. The model is

$$\max \frac{c_1 x_1 + c_2 x_2 + c_3 x_3 + c_4 x_4}{4}$$

subject to

$$x_1 + x_2 + x_3 + x_4 \leq 20$$
$$c_i x_i \geq 50, i = 1, 2, 3, 4$$
$$c_i x_i \leq 64 \text{ or } c_i x_i \geq 75, i = 1, 2, 3, 4$$
$$x_i \leq T_i, i = 1, 2, 3, 4$$
$$x_i \in \mathbb{Z}_+, i = 1, 2, 3, 4.$$

To linearize the "or" constraints, we introduce new binary variables y_i and z_i, $i = 1, 2, 3, 4$. The integer linear optimization model is

$$\max \frac{c_1 x_1 + c_2 x_2 + c_3 x_3 + c_4 x_4}{4}$$

subject to

$$x_1 + x_2 + x_3 + x_4 \leq 20$$
$$c_i x_i \geq 50, i = 1, 2, 3, 4$$
$$c_i x_i - 64 \leq 100 y_i, i = 1, 2, 3, 4$$
$$75 - c_i x_i \leq 100 z_i, i = 1, 2, 3, 4$$
$$y_i + z_i \leq 1, i = 1, 2, 3, 4$$
$$x_i \leq T_i, i = 1, 2, 3, 4$$
$$x_i \in \mathbb{Z}_+, i = 1, 2, 3, 4$$

$$y_i \in \{0,1\}, i = 1,2,3,4$$
$$z_i \in \{0,1\}, i = 1,2,3,4.$$

Example 17.30: A salesman is at node 1. There are $n-1$ customers, denoted by nodes $2,3,\cdots,n$, $n \geq 4$. The salesman only needs to visit $n-3$ customers (the remaining two customers can be visited by other salesmen) and then return to node 1. Define the set of nodes $N := \{1,2,\cdots,n\}$ and the set of arcs $A := \{(i,j), i \in N, j \in N, i \neq j\}$. The distance of arc $(i,j) \in$ is d_{ij}. Develop an integer optimization model to help the salesman determine which $n-3$ customers to visit and in what sequence to visit them so that the total distance of visiting them and then returning to node 1 is the shortest.

Solution: Let z_i be a binary variable which equals 1 if and only if customer i is visited, $i = 2,3\cdots,n$, let x_{ij} be a binary variable which equals 1 if and only if arc (i,j) is traversed, and let u_i represent the number of customers that have been served after visiting i. The model is

$$\min \sum_{(i,j) \in A} d_{ij} x_{ij}$$

subject to

$$\sum_{j \in N \setminus \{1\}} x_{1j} = 1$$
$$\sum_{i \in N \setminus \{1\}} x_{i1} = 1$$
$$\sum_{j \in N \setminus \{i\}} x_{ij} = z_i, i \in N \setminus \{1\}$$
$$\sum_{i \in N \setminus \{j\}} x_{ij} = z_j, j \in N \setminus \{1\}$$
$$x_{ij} \in \{0,1\}, (i,j) \in A$$
$$u_j \geq u_i + 1 - (n-1)(1 - x_{ij}), i \in N, j \in N \setminus \{1\}, j \neq i$$
$$u_i \geq 1, i \in N \setminus \{1\}$$
$$u_1 = 0$$
$$z_i \in \{0,1\}, i \in N \setminus \{1\}$$
$$\sum_{i=2}^{n} z_i = n - 3.$$

Example 17.31: (i) An optimization model has decision variables $z_1 \in \{0,1\}, z_2 \in \{0,1\}, z_3 \in \{0,1\}$ and $z \in \{0,1\}$. One of its constraints is $z \geq z_1 z_2 z_3$. How do you linearize this constraint?

(ii) An optimization model has decision variables $z_1 \in \{0,1\}, z_2 \in \{0,1\}, z_3 \in \{0,1\}$ and $z \in \{0,1\}$. One of its constraints is $z = z_1 z_2 z_3$. How do you linearize this constraint?

(iii) An optimization model has decision variables $z_1 \in \{0,1\}, z_2 \in \{0,1\}, z_3 \in \{0,1\}$ and $z \in \{0,1\}$. One of its constraints is $z \leq z_1 z_2 z_3$. How do you linearize this constraint?

Solution: (i) $z \geq z_1 + z_2 + z_3 - 2$.
(ii) $z \geq z_1 + z_2 + z_3 - 2, z \leq z_1, z \leq z_2, z \leq z_3$.
(iii) $z \leq z_1, z \leq z_2, z \leq z_3$.

Reference

[1] Wang S., Wang T., Meng Q. 'A note on liner ship fleet deployment'. *Flexible Services and Manufacturing Journal*. 2011;**23**(4):422–30.

Chapter 18
Conclusion

18.1 Summary of this book

Machine Learning and Data Analytics in Maritime Studies: Models, Algorithms, and Applications aims to explore the fundamental principle of analyzing practical problems in maritime transportation using data-driven models, especially using machine learning (ML) models and operations research models. The book first introduces the state-of-the-art data-enabled methodologies, technologies, and applications in maritime transportation in a way that is easy to understand by maritime researchers and practitioners. To achieve this, plain words are used to present the algorithms and models, and real examples of ship inspection by port state control and operations planning of container ships are accompanied. By doing so, readers are expected to learn how to solve practical problems in maritime transportation by data-driven models while taking shipping domain knowledge and black box model explanations into consideration.

A total of 5 parts comprising 17 chapters are contained in this book. In Part I, the status quo of maritime transportation, including key players in the shipping industry, in addition to an overview of container liner shipping, is introduced in Chapter 1. In Chapter 2, the comprehensive issues in ship inspection by port state control are covered after discussing key issues in maritime transport, and the dataset used in this book is also introduced.

Part II aims to give an overview of data-driven models. In Chapter 3, typical predictive problems in maritime transport are first discussed, and popular data-driven approaches to solving them are also covered. In Chapter 4, detailed procedure of developing ML models to address maritime transport problems is introduced, where comprehensive explanations and illustrations to each step are also given.

Part III introduces popular ML models that are used to tackle practical problems in maritime transport. Especially, linear regression models (Chapter 5), Bayesian networks (Chapter 6), support vector machine (Chapter 7), artificial neural network (Chapter 8), tree-based models (Chapter 9), association rule learning (Chapter 10), and cluster analysis (Chapter 11) are introduced in details. Applications and result explanations of all the above models to predict ship risk in port state control are given as examples, and the core code for model construction is also given.

298 Machine learning and data analytics for maritime studies

Part IV explores data-driven models to address practical issues in maritime transportation more deeply. Chapter 12 first summarizes topics in maritime transportation research and popular research methods to solve practical problems. Issues of adopting data-driven models to address these practical problems are also discussed from the perspectives of data, model, user, and target. Chapter 13 further discusses the issue of incorporating shipping domain knowledge into data-driven model developing using examples in port state control and ship energy efficiency prediction. Chapter 14 further illustrates issues regarding explanations of ML models within the context of maritime transportation, where examples of decision tree explanation and using SHAP to explain ML models in black box nature are given as examples.

Part V introduces another important data analytics approach based on operations research, which is also widely used to address practical problems in maritime transportation. Chapter 15 introduces linear optimization techniques as the foundation of operations research. Chapter 16 presents more advanced linear optimization techniques. Chapter 17 covers integer optimization.

18.2 Future research agenda

Several future research directions are discussed in this section. It should be noted that these topics are just the tip of the iceberg. There is still a long way to go to realize digitization transformation in the traditional maritime industry.

1. Data play the key role and act as the foundation of constructing data-driven models to solve practical problems in maritime transport. However, there are digital inefficiencies in the maritime transport chains, where there are rare public data sources while data provided by commercial companies can be very expensive. Efforts should be put to promote digital collaborations between ship to port, port to port, and ship to hinterland for data sharing. Data-driven models developed to address practical problems in maritime transport should also comply with shipping domain knowledge and are expected to take such knowledge explicitly into the procedure of data construction.
2. Data-driven models are usually of black box nature, which means that they are opaque to model users or even model developers regarding how the prediction model works and why a certain prediction is made. This issue should also be taken into account when developing and explaining data-driven prediction models to improve model transparency and acceptance.
3. In many practical problems, accurate prediction is far from enough. Instead, it is expected that predictions given by data-driven models should be input to the following optimization problems to prescribe better decisions. Unfortunately, there is still a big gap between making a good prediction and making a good decision, where future research efforts should be put into.
4. Up to now, as covered in Chapter 1, there are only a few maritime transportation problems that have been dealt with by ML approaches, and most of these

problems are prediction-based. Even if there are such prediction models, realizing accurate and real-time prediction is still rare and difficult. There are several important issues from both shipping and port sides that have not yet been addressed, such as vessel routing and scheduling, fleet planning and development, shipping network design, port congestion prediction and management, port resource allocation management, and port safety and security management. In future research, these issues are expected to be addressed.

Index

Abuja MoU 11, 12
Accumulated error backpropagation 100
Accuracy 44
Adaboost 122–125
Agglomerative algorithm 157–162
Apriori algorithm 133–141
Area under ROC curve (AUC) 46–47
Artificial neural network (ANN)
  concepts 93–97
  model
    hyperparameters 100–101, 103
    training 97–100
  structure 93–97
Asia Pacific Computerized Information System (APCIS) 171
Association rule learning 131
  Apriori algorithm 133–141
  definition 132–133
  FP-growth algorithm 141–142
Automatic identification system (AIS) 33, 167–169

Backpropagation algorithm 98
Backward sequential elimination and joining (BSEJ) 69
Backward sequential elimination (BSE) 69
Baltic dry index (BDI) 168
Base learners 121
Bayesian network (BN) classifiers 73–77
  BN classifiers 73–77
  Naive Bayes classifier 63–68
  semi-naive Bayes classifiers 68–73
Binary variables 290–291

Black-box ML models
  advantages 195–196
  explanation 193–198
  elements 196–197
  forms 198–200
  intrinsic explanation model using DT 200–202
  SHAP method 202–208
Black Sea MoU 11, 12
Boosting tree 125–130
Bulk carrier 1

C4.5 algorithm 109–110
Central China 1 (CC1) service 6
Central China 2 (CC2) service 5, 7
Centroid-linkage clustering 149
Chi-squared test (c^2 test) 41–42
Classic clustering algorithm 151, 152
Classification and regression tree (CART) 110–113
Classification decision tree
  C4.5, 109–110
  classification and regression tree (CART) 110–113
  iterative dichotomizer 3, 106–109
Classification society 4
Cluster analysis 145
  clustering algorithm performance evaluation metrics for 149–151
  agglomerative algorithm 157–162
  classic algorithm 151, 152
  density-based spatial clustering of applications with noise (DBSCAN) 154–157
  K-means algorithm 151, 153

distance measure
    of clusters 148–149
    of examples 145–148
Complete-linkage clustering 149
Container liner shipping 5–7
Container ship 2–3
Convex integration 186–190
Counterfactual explanation 199
Cross validation 47
Cubic law 22

Data cleaning 35, 36
Data-driven models
    adoption issues 169–170
        data 170–171
        model 171–172
        target 174–175
        user 172–174
    ML modeling (see ML modeling)
    statistical modeling 23–25
    theory-based modeling vs., 23
Davies–Bouldin index (DBI) 149–150
Decision tree (DT) 105–106
    advantages 115
    black-box ML models 200–202
    boosting and bagging based on 128
    construction algorithms 117
    disadvantages 115–116
Decision variables 211, 212, 215, 280–283
Deep learning models 29, 93, 102, 103
Deep neural networks (DNN) 29, 100, 103
Deficiency number 18, 52–54, 61, 62, 64, 67, 184–186, 200–207
Density-based spatial clustering of applications with noise (DBSCAN) algorithm 154–157
Dropout 100
Dunn index (DI) 150

Elastic net regression 23
Euclidean distance 146–147
Excel 227–229

eXtreme Gradient Boosting (XGBoost) 182, 184–186

Feature engineering 34
    Chi-squared test (c^2 test) 41–42
    data cleaning 35, 36
    encoding 38, 40
    extraction 35
    mutual information (MI) 42
    Pearson's correlation coefficient (Pearson's r) 41
    pre-processing 37–39
    recursive feature elimination (RFE) 42
    scaling 38, 40
    selection 40–41
        based on regularization 42
        on importance 42
    Spearman's rank coefficient (Spearman's r) 41
Feature extraction 35
Flag state 3–4
Forward sequential elimination (FSE) 69
Fowlkes and Mallows index (FMI) 150–151
Freight forwarder 4
Frequent pattern (FP)-growth algorithm 141–142

Gaussian kernel 89, 91
Gini index 110–111
Gradient boosting decision tree (GBDT) 125–130
Grid search 49

Handymax carriers 38
Hard margin support vector machine 79–83

Integer optimization 277–298
Intrinsic explanation 199
Iterative dichotomizer 3, 106–109

Kernel principal components analysis 35
Kernel trick 86–89
k-fold cross validation 47
K-means algorithm 151, 153, 158, 160–162

Lagrange multipliers 85
Laplacian kernel 89
Lasso regression 23, 25
Leaky ReLU activation function 98
Least absolute shrinkage and selection operator (LASSO) regression 61–62
Least squares method 51–52
Leave-one-out cross validation 47
Linear discriminant analysis (LDA) 35
Linear kernel 86, 88, 89
Linear optimization
  advanced 252–276
  basics 211–214
  classification 215–217
  dummy nodes and links 247–250
  equivalence between different formulations 217–220
  Excel 227–229
  network flow optimization 233–247
  for nonlinear problems 250–252
  solver 227–229, 230–232
  with two variables using graphs 220–227
Linear regression 23, 25, 55–59
  and least squares 51–52
  multiple linear regression 53–55
    logistic regression 56–59
    polynomial regression 55–56
  shrinkage 59
    LASSO regression 61–62
    ridge regression 60–61

Machine learning (ML) modeling 23, 26–28
  to address maritime transport problems 29, 31, 32
  assessment 50

data collection 32–34
feasibility assessment 32
feature engineering (*see* Feature engineering)
interpretation 50
model construction 43–48
model refinement 48–50
problem specification 32
Manhattan distance 146
Maritime Labor Convention 2006 (MLC2006) 10
Maritime transport
  key issues
    marine pollution control 10
    maritime safety management 9–10
    seafarers' management 10
  methods and applications 165–169
  overview 1, 2
  predictive problem 21–24
  research topics 163–164
Mean absolute error (MAE) 43
Mean absolute percentage error (MAPE) 43
Mean normalization 40
Mean-squared error (MSE) 43, 51–52, 201
Mediterranean MoU 11, 12
Memorandum of Understanding (MoU) 11, 12, 15, 165
Min-max scaling 40, 59–61
Mixed-integer optimization models 284–285
Monotonic constraint
  integration
    into artifical neural network (ANN) 186–190
    into XGBoost 184–186
MSE. *See* Mean-squared error (MSE)
Multidimensional scaling (MDS) 35
Multiple linear regression 53–55
  logistic regression 56–59
  polynomial regression 55–56
Mutual information (MI) 42

Naive Bayes classifier 63–68
Naïve-Bayes Decision Tree (NBTree) 69

Natural integer decision variables 277–280
Network flow optimization 233–247
Neuron 93–94
New inspection regime (NIR) 12, 13, 182

Objective function 211, 212, 215
Ocean freight market condition prediction 168–169
Oil tanker 2
One-hot encoding 40, 54, 85–86, 101
Orient Overseas Container Line (OOCL) 5–6
Origin-destination (OD) pair 7
Out-of-bag error 48

Paris MoU 11, 12, 14, 15
Pearson's correlation coefficient (Pearson's r) 41
Polynomial kernel 88, 89, 91
Polynomial regression 23, 25, 55–56
Port condition prediction 169
Port state control (PSC)
   memorandum of understanding (MoU) on 11
   ship inspection 10
      background 11
      data set 15–18
      development 11
      onboard procedure 12
      results 14–15
      ship selection 11–14
Port state control officer (PSCO) 11
Port statistics 34
Posthoc explanation 199
Precision 44–45
Predictive accuracy (PDR) 196
Predictive problem 21–23
Principal components analysis (PCA) 35, 40
PyDotPlus API 200

Rand index (RI) 150–151
Random forest (RF) 105, 118–119, 186, 203, 207
Receiver operating characteristic (ROC) curve 45–46
Rectified linear units (ReLU) activation function 97
Recursive feature elimination (RFE) 42
Remote follow-up inspection 12
Ridge regression 23, 25, 60–61
Root-mean-squared error (RMSE) 43

Safety management 9–10
Scikit-learn API 52, 54, 55, 59–61, 85, 88, 91, 101, 112, 114, 119, 126, 158, 200
Semi-naive Bayes classifiers 68–73
Shapley additive explanations (SHAP) 200, 202–208, 300
Ship accident data 34
Ship broker 5
Ship destination port and arrival time at port prediction 169
Ship energy efficiency prediction 168
Ship management company 3
Ship operator 3
Ship owner 3
Shipping industry
   key roles
      charterer 4
      classification society 4
      flag state 3–4
      freight forwarder 4
      ship broker 5
      ship management company 3
      ship operator 3
      ship owner 3
Ship risk prediction
   feature monotonicity 181
   oveview 182–184
   safety management 167–168
Ship risk profile (SRP) 12, 165

Ship sailing records 33–34
Ship trajectory prediction 167
Shrinkage linear regression models 59
    LASSO regression 61–62
    ridge regression 60–61
Sigmoid activation function 95
Sigmoid kernel 89
Simple linear regression 51–52
Single-linkage clustering 148–149
Soft margin support vector machine 83–86
Spearman's rank coefficient (Spearman's r) 41
Standardization 40
Statistical modeling 23
Stepwise regression 23, 25
Support vector machine (SVM) 79
    hard margin 79–83
    Kernel trick 86–89
    soft margin 83–86
    support vector regression 90–92
Support vector regression (SVR) 79, 90–92
Synthetic minority oversampling technique (SMOTE) 59

Tanh activation function 96
Target encoding 40
Tokyo MoU 11–15, 18, 101, 139, 140, 171, 172, 174, 182–184, 204–206
Tree augment naive Bayes (TAN) classifier 68, 70, 71, 73
Tree-based models
    decision tree 105–106
    ensemble learning 116
        Adaboost 122–125
        bagging 116–221
        boosting 121–130
    node splitting
        in classification decision tree (*see* Classification decision tree)
        in regression trees 113–117

Unweighted average-linkage clustering 149

World fleet structure
    bulk carrier 1
    container ship 2–3
    oil tanker 2